PENGUIN BOOKS

MAKERS OF MATHEMATICS

Stuart Hollingdale was born in Ealing in 1910. He was educated at Latymer Upper School, Hammersmith, and Christ's College, Cambridge. He then moved to Imperial College, London, where he obtained a PhD for research in fluid dynamics. In 1936 he joined the Aerodynamics Department of the Royal Aerospace Establishment at Farnborough. During and immediately after the war he was engaged in operational research and personnel administration. In 1948 he returned to Farnborough to become head of the newly formed Mathematics Department, a post he occupied until 1967 when he moved to Birmingham to become Director of the University Computer Centre. He retired from full-time employment in 1974 and was, for some years, a part-time consultant and tutor for the Open University's history of mathematics course.

His publications include *High Speed Computing: Methods and Applications* (1959) and, with a colleague, Geoff Tootill, *Electronic Computers* (1965, revised 1970 and 1975), which was published in Pelican.

Dr Hollingdale is a widower and has two daughters and three grandchildren.

Makers of
Mathematics

STUART HOLLINGDALE

Penguin Books

PENGUIN BOOKS

Published by the Penguin Group
Penguin Books Ltd, 27 Wrights Lane, London W8 5TZ, England
Penguin Books USA Inc., 375 Hudson Street, New York, New York 10014, USA
Penguin Books Australia Ltd, Ringwood, Victoria, Australia
Penguin Books Canada Ltd, 10 Alcorn Avenue, Toronto, Ontario, Canada M4V 3B2
Penguin Books (NZ) Ltd, 182–190 Wairau Road, Auckland 10, New Zealand

Penguin Books Ltd, Registered Offices: Harmondsworth, Middlesex, England

First published in Pelican Books 1989
Reprinted in Penguin Books with a new Preface, two new Appendices and minor corrections 1991
Reprinted with a new Appendix and minor corrections 1994
10 9 8 7 6 5 4 3 2

Filmset in Times New Roman (Monotype Lasercomp)
Printed in England by Clays Ltd, St Ives plc

Contents

Preface to the 1989 Edition

The origins of this book are to be found in a number of essays (many of them anniversary tributes) and book reviews which have appeared in the *Bulletin of the Institute of Mathematics and its Applications* over a period of some ten years from 1976. The articles were written primarily for a specialist readership – the Fellows and members of the Institute, the great majority of whom are graduates in mathematics. The material has been edited and reworked to meet the needs of the 'general reader', which I take to be a broadly educated and lively-minded person whose mathematical knowledge may not extend much beyond that required for the General Certificate of Education at Ordinary Level (or a broadly equivalent standard in the new GCSE). More advanced ideas and procedures are carefully explained and illustrated as they are introduced.

Most of the IMA articles are set in the seventeenth, eighteenth or early nineteenth centuries, so some gaps in the story needed to be filled. Seven chapters (1, 2, 4, 5, 7, 15 and 16) are completely new. Even so, the book is in no sense a balanced history of mathematics, but rather an informal, personal – even idiosyncratic – attempt to present the main features of a long story through the lives and achievements of some of its great men, from Pythagoras to Einstein. We have recently been celebrating the tercentenary of the publication of Newton's *Principia*, so it seemed appropriate to accord this unique masterpiece a chapter to itself. Some more specialized and technically demanding material has been placed in five short appendices. For the first of these I am indebted to J.H. Cadwell, a close colleague for more than twenty years who died in 1984.

While writing this book, two objectives have been kept in mind. The first is to convey something of the fascination of mathematics – of its austere beauty, intellectual power and infinite variety. The second is to draw attention to the eventful lives and colourful personalities of so many of its leading practitioners. The common view of the mathematician as a desiccated, passionless, cloistered creature of narrow interests, out of touch with the real world, is indeed very far from the truth. In selecting the main characters of the story, I have not attempted to hide my personal preferences. For me, Newton stands alone at the summit, closely flanked by Archimedes, Gauss and Einstein; Eudoxus, Fermat, Euler and Cantor stand not far below. Nor have I sought to disguise my affection for such lesser-known figures as Archytas, Tartaglia, Cotes and d'Alembert.

The intense specialization of contemporary mathematics is a fairly recent phenomenon, albeit an inevitable one. Few of our subjects made much, if any, distinction between what Thomas Jefferson, when founding the University of Virginia, called 'Mathematics, pure' and 'Physico-mathematics'. Indeed, the subject of the final chapter, Albert Einstein, was not primarily a mathematician at all, but he is too important a figure to be excluded. In this book the whole of mathematics, together with its many applications, is to be perceived as a seamless robe.

Although there is much biographical information, I have not hesitated to include too a substantial measure of 'solid' mathematics, in particular demonstrations of some of the more historically significant, elegant or unexpected theorems and procedures. Indeed, some readers may find parts of the book heavy going. I hope they will not give up too easily. To follow a chain of reasoning of moderate length and complexity undoubtedly demands persistence and determination, but it can be argued that the exercise of these faculties is one of the pleasures to be obtained from the study of mathematics – and hence, it is to be hoped, from reading a book of this kind.

I have not thought it necessary (or indeed desirable) to include scholarly footnotes; the number of individual references to the works quoted has been kept to a minimum. The list of references given at the end has been limited, with a few exceptions, to sources from my own library which I consulted while writing this book. I am much indebted to the authors listed there, and also to my former colleagues at the Open University who were responsible for providing the course material for the first OU history of mathematics course. A wide-ranging reader (Reference 2) has recently been published for the second such course; it is a mine of information and critical comment. I am also indebted to David Nelson for his valuable comments on my draft text. Finally, I would like to express my thanks to Catherine Richards, the Secretary and Registrar of the IMA, and her staff for encouragement, for typing services and for general administrative assistance – and, not least, for sending me so many good books to review.

June 1989

Preface to the 1991 Edition

A new edition provides a welcome opportunity to correct misprints (how *do* they get through?), clarify or amplify a few paragraphs and add two new Appendices (6 and 7). Many of the changes derive from comments made by correspondents and I would like to thank them collectively for their valuable help. Three of them must, however, be mentioned individually.

Soon after the book was published in September 1989, I received a letter from Dr C. Kenneth Thornhill, an old colleague in the Scientific Civil Service, expressing his surprise at my statement (now deleted) on page 104 that the method used by Leonardo of Pisa to solve the court 'squares' problem was not known. He writes: 'I know how to obtain his solution and I should think it fairly certain that my way is the same method that Leonardo used.' In further correspondence Dr Thornhill communicated several of his methods for solving the generalized 'squares' problem, of which Leonardo's 'court' problem is a special case. He then consulted a polymath friend of his, Mr A. Roger Thatcher, who retired a few years ago as Registrar-General after a distinguished career in the Civil Service. He pointed out that Leonardo's treatment of the general problem is discussed in detail in Chapter XVI, Vol. 2, of L. K. Dickson's monumental three-volume treatise, *History of the Theory of Numbers* (1919–23). Another old colleague, Professor Paul Samet of University College London, was kind enough to provide me with the text of the relevant chapter. Mr Thatcher then sent me some further information and pointed out that 'there is a large literature on the "squares" problem, much of it written since Dickson's history'.

Selected material from all these sources – and also a recent book on Leonardo – provides the subject-matter of Appendix 6.

Appendix 7, which can be savoured by the most non-mathematical reader, tells a remarkable story of mistaken identity in high places: it may be allowed to speak for itself.

Several correspondents from the teaching profession have queried my remark in the 1989 Preface that readers need to have little mathematical knowledge beyond that required for the GCE at ordinary level. To some, indeed, my claim was wildly optimistic. I must plead guilty: I was, I regret to say, showing my age! I had not realized to what extent the 'classical' mathematical content had been reduced since I'd taken the General Schools Certificate (and London Matric) examinations in the 1920s. In those days, Elementary Mathematics, Mathematics (more advanced) and Mechanics could be offered as three separate subjects. The process of emasculation (if that is not too strong a word) appears to have started a few years after the war with the introduction of the GCE and to have been carried further by the recent change to the GCSE. I hope that younger readers will not allow a limited school background to deter them from tackling the book. Some readers will, no doubt, be able to rectify any gaps in their knowledge without too much difficulty as they work through the book. For others, perhaps, some judicious skipping would be appropriate.

Stuart Hollingdale
Lower Froyle, Alton, Hampshire

January 1991

List of illustrations

1   The beginnings

'Counting is a symbolic process employed only by man, the sole symbol-creating animal.' R. L. Wilder

1. Prehistory

Palaeolithic man undoubtedly possessed a rudimentary sense of number, as indeed do some mammals, birds and insects. He could recognize that something had changed when a pebble was added to or removed from a small heap of pebbles. His next step, a major intellectual achievement, was to create the abstract concept of *number*: to realize that three goats, three fingers and three days shared a common property of 'threeness'. This gave rise to the idea of matching the members of one collection against those of another. If there is perfect matching, the two collections are the same size: they have the same number of members. There is a *one-to-one correspondence* between the members of one collection and those of the other. True *counting* entails a further step: that of setting up a correspondence between the objects to be counted and an ordered sequence of *symbols*. Such symbols may take a variety of forms – spoken number-words, notches on a tally-stick, knots on a cord, the position of the fingers (as in finger-counting) or written characters known as *numerals*. To count efficiently we need a system of numeration to specify rules of succession so that large numbers can be expressed using only a small number (ten in our system) of different numerals. We see here the genesis of our dual concept of number. We make a distinction between *cardinal numbers*, based on the one-to-one correspondence principle, and *ordinal numbers*, based on the principle of ordered succession. Although both principles are at the heart of mathematical thinking, it is the creation of the ordinal number concept that may be taken to mark the beginning of mathematics as we understand it today.

It seems that what we loosely call 'civilization' first came into being along the river valleys of Egypt, Mesopotamia, India and China. The high fertility of the soil enabled the communities to produce more than was required for bare subsistence, so allowing some people to be released from primary production to perform a variety of specialized functions: as craftsmen, administrators, priests, scribes, surveyors – and, eventually, mathematicians. There are no reliable records of the early oriental civilizations, so we shall start our story on the banks of the Nile, and then move to the land of the two great rivers – the Tigris and the Euphrates.

Before doing so, however, we should point out that some authorities would date the beginnings of mathematics far back in prehistoric times. In the 1950s in the village of Isango in Zaïre a bone was dug up which has a large number of notches carved on it. To some scholars – but by no means all – the patterning of the notches suggests the use of a decimal counting system, and even a knowledge of a few prime numbers. On the geometrical side, the situation is broadly similar. Statistical analysis of recent surveys of megalithic sites in Western Europe has led some enthusiasts to claim a very high level of geometrical sophistication for the Stone Age builders of the fourth and third millennia BC. Once again, there is much scholarly disagreement over the interpretation of the evidence. An excellent survey of the main facts and contending views may be found in Reference 2.

2. Egypt

A distinctive Egyptian civilization began to emerge in the fifth millennium BC, and lasted until the Roman conquests of the first century BC. During most of that time Egyptian society was largely self-contained, resistant to change, practical rather than speculative, and intensely religious. Why, we may ask, did the ancient Egyptians concern themselves with mathematical matters at all?

Herodotus, 'the Father of History', believed that Egyptian mathematics arose from the practical need to resurvey the land every year after the annual flooding of the Nile Valley. Aristotle, however, put the emphasis on the existence of a priestly leisured class with intellectual interests. The question remains open.

Most of our knowledge of Egyptian mathematics comes from two primary sources: the Rhind Papyrus and the Moscow Papyrus. The first of these is a practical handbook written in about 1700 BC by a scribe called Ahmes. It contains 84 worked problems which include a variety of arithmetical calculations with whole numbers and fractions, the solution of linear equations, and the mensuration of simple areas and volumes.

The Egyptians' system of numeration was based on the scale of ten, but they had no concept of *place value*. There was simple repetition within each decade, with separate symbols for 10, 100, etc., up to a million. Thus, if we denote the symbols for 1, 10 and 100 by I, T and H, then 234 might be written as HHTTTIIII. No symbol for zero was needed, the numerals could be written in any order, and special arrangements had to be made to represent very large numbers.

Egyptian arithmetic was based on the operations of adding, doubling and halving; multiplication was achieved by successive doubling. So, to multiply 29 by 13, Ahmes would proceed thus:

1	29 ←	
2	58	Since $13 = 8 + 4 + 1$, the result
4	116 ←	is obtained by adding the marked
8	232 ←	numbers in the second column.
13	377	

This method, know as *duplation*, has a long history: a hundred years ago it was known as the Russian peasant's method of multiplication. More recently it has been given a new lease of life with the arrival of the electronic computer, in which internal operations are carried out in the binary system of numeration.

Division entails operating with fractions, and here we meet a distinctive feature of Egyptian arithmetic. All fractions were reduced to sums of *unit fractions*, i.e. fractions of the form $1/n$. The only exception was 2/3, which was accorded a special symbol. In Problem 24 Ahmes wishes to divide 19 by 8; in Problem 25 to divide 16 by 3. The calculations are carried out as follows:

	19 : 8		16 : 3
1	8	1	3 ←
2	16 ←	2	6
1/2	4	4	12 ←
1/4	2 ←	2/3	2
1/8	1 ←	1/3	1 ←
Quotient:	2 + 1/4 + 1/8		5 + 1/3

The objective is, starting with 8 (or 3), to produce a set of numbers that add to 19 (or 16). In Problem 24 Ahmes uses the sequence 1/2, 1/4, 1/8, . . . ; in Problem 25 the sequence 2/3, 1/3, 1/6, ... These are, indeed, the two basic 'halving sequences' of Egyptian arithmetic.

When carrying out such calculations it is often necessary to double a unit fraction (i.e. to compute $2/n$ as a sum of unit fractions). If n is even, the operation is trivial; if n is odd, it may be quite difficult. (Egyptian scribes were not allowed, for some reason, simply to write 2/7, for example, as 1/7 + 1/7.) No less than a third of the complete Rhind Papyrus consists of a table for expressing fractions of the form $2/(2n + 1)$, for values of n from 2 to 50, as sums of unit fractions, each with a different denominator. The method of calculation, as well as the final result, is given for each value of n.

Thus to evaluate 2/7, Ahmes proceeds as follows:

	1 : 7
1/2	3 + 1/2
1/4	1 + 1/2 + 1/4 ←
1/7	1
1/14	1/2
1/28	1/4 ←
2/7	= 1/4 + 1/28

With the aid of the table, any division by 7 becomes a fairly simple matter. Thus, for example, if the problem is to divide 16 loaves of bread equally among 7 men, we have

$$16/7 = (2 \underset{\text{3 doublings}}{\times 2 \times} 2)(1/4 + 1/28) = 2 + 2/7 = 2 + 1/4 + 1/28$$

as the share of each man. Here are three more examples:

$$2/19 = 1/12 + 1/76 + 1/114$$
$$2/59 = 1/36 + 1/236 + 1/531$$
$$2/97 = 1/56 + 1/679 + 1/776$$

Multiplication of sums of unit fractions was also covered. Problem 13, for example, asks for the product of $1/16 + 1/112$ and $1 + 1/2 + 1/4$ and obtains the correct result of $1/8$. Some of the Rhind problems require considerable skill and ingenuity for their solution: for example, the division of 7 by 29 to yield

$$7/29 = 1/6 + 1/24 + 1/58 + 1/87 + 1/232$$

It is worth noting that $7/29$ may be expressed more simply as $1/5 + 1/29 + 1/145$, but the use of the Ahmes $2/(2n + 1)$ table leads to the former result.

The algebraic problems in the papyrus are mainly concerned with the solution of linear equations. Problem 23 calls, in present-day notation, for the solution of the equation $x + x/7 = 19$. Ahmes uses the *method of false position*: he guesses a value of x which he knows to be incorrect, and then scales the result appropriately. Here he assumes that $x = 7$, giving a value of 8 for the left side of the equation. He needs, therefore, to multiply his guessed value by $19/8$, i.e. by $2 + 1/4 + 1/8$. The answer is found to be $16 + 1/2 + 1/8$, which Ahmes checks by augmenting his answer by $1/7$ to obtain 19. The Papyrus also contains a few geometrical problems, but there is no mention of any numerical instance of Pythagoras' theorem. However, the 'rope stretchers', as the Egyptian surveyors were called, almost certainly used knotted ropes with lengths in the ratio of $3:4:5$ to lay out a right angle. In Problem 50, Ahmes equates the area of a square field of side 8 units with that of a circle of diameter

9 units. This leads to a value of π (to use the modern symbol) of $4 \times (8/9)^2$, or very nearly 3.16. More information on the Rhind Papyrus is given in References 2, 4 and 5.

The Moscow Papyrus, which was written about 1900 BC, contains 25 problems. One of these is of exceptional interest in that it contains – albeit in somewhat disguised form – the correct formula for the volume of a truncated pyramid, namely $V = \frac{1}{3}h(a^2 + ab + b^2)$, where a and b are the sides of the two parallel square faces, and h is the distance between them. We would, of course, expect the Egyptians to know a lot about pyramids, and we shall return to the subject in Chapter 3.

3. Mesopotamia

In marked contrast to Ancient Egypt, Mesopotamia was the scene of incessant conflict: invasions, pillage, forced migrations and dynastic upheavals. Even so, the cultural pre-eminence of the region lasted for some 4000 years, during which time the indigenous Sumerians were succeeded by Akkadians, Hittites, Medes, Persians and Chaldeans, among others. The convenient term 'Babylonian' is commonly applied to both the region and its culture during the period of greatest achievement – roughly from 2200 to 500 BC. The Babylonian Empire actually came to an end in 538 BC when Babylon fell to Cyrus of Persia, but a distinctive Babylonian culture survived for another five centuries.

In our study of Babylonian mathematics we are fortunate in having an abundance of primary source material in the form of inscribed clay tablets. The writing is known as *cuneiform* ('wedge shaped') because it was produced by pressing a stylus into a block of soft clay which was then baked or allowed to harden in the sun. Most of the surviving tablets containing mathematical material come from one of two widely separated periods: either the first half of the second millennium BC (the 'Old Babylonian' age) or the last third of the first millennium BC (the 'Seleucid' period). However, the oldest tablets go back to about 2100 BC, when the region was ruled by a Sumerian dynasty centred on Ur, the city of Abraham. Babylonian mathematics seems to have reached its zenith very early, around 1800 BC, about the time of the reign of Hammurabi, the great lawgiver.

The Babylonian number system is entirely different from the Egyptian. It is sexagesimal and positional, and uses only two numerals, made by pressing the triangular end of a stylus into the soft clay in two different ways. Let us denote these numerals by V and X; they basically stand for 1 and 10. The positional (or place value) feature means that V can also represent 1×60^n, and X can represent 10×60^n, where n can take any integral value – either positive or zero for whole numbers, or negative for fractions. The correct value of n in a particular case must be deduced (or guessed) from the context. Thus, for example, the ordered sequence of numerals VXXVVV could represent $60 + 2(10) + 3 = 83$, or $3600 + 2(10 \times 60) + 3/60 = 4800\frac{1}{20}$, or $1/60 + 20/60^2 + 3/60^3$. It will be convenient to transcribe Babylonian numbers into a more readable form, so let us write the numbers quoted above as 1,23 and 1,20,0;3 and 0;1,20,3. The cause of the ambiguity is the lack of a symbol for *zero*, a symbol which in our number language enables us to distinguish between, say, 32, 302 and 320. The omission was eventually rectified, but only in part. During the Seleucid period a separator symbol was introduced to indicate an empty space inside a number, but it was not used in a terminal position. In spite of its deficiencies, the Babylonian system offered so many advantages, especially for dealing with large numbers and fractions, that it was used by Greek and Arab astronomers for many centuries. This is why we still retain a sexagesimal system to measure time and angles.

Many of the mathematical tablets contain tables for multiplication, and tables of reciprocals, squares, square roots, etc. Here are the first few entries in a table of reciprocals:

2	30	8	7,30	The reciprocals of the 'irregular'
3	20	9	6,40	numbers 7 and 11 are omitted
4	15	10	6	because they are non-terminating
5	12	12	5	in a sexagesimal system – as is 1/3
6	10			in our decimal system.

By 1900 BC the Babylonians had a well-established algebra. They could solve quadratic equations (positive coefficients and positive solutions only) and some types of higher-degree equation as well. Thus one problem asks for the side of a square whose area exceeds the side by 14,30. This leads to the equation $x^2 - x = 870$. The solution is set out thus:

Take half of 1 and multiply it by itself to give 0;15. Add the result to 14,30 to give 14,30;15. This is the square of 29;30. To this add 0;30 to give 30, which is the desired result.

The procedure is exactly equivalent to our method of solving quadratic equations by 'completing the square'.

Some Babylonian mathematics reached a high level of sophistication. For example, the tablet known as Plimpton 322 contains a table of 15 rows of which the first 5 and the last 2 are (with a couple of obvious clerical errors corrected):

1,59,0,15	1,59	2,49	1
1,56,56,58,14,50,6,15	56,7	1,20,25	2
1,55,7,41,15,33,45	1,16,41	1,50,49	3
1,53,10,29,32,52,16	3,31,49	5,9,1	4
1,48,54,1,40	1,5	1,37	5
⋮	⋮	⋮	⋮
1,25,48,51,35,6,40	29,31	53,49	14
1,23,13,46,40	56	1,46	15

What, we may ask, does this table mean? Let us consider a right-angled triangle ABC with sides x, y and z (Figure 1.1). If the numbers in the second and third columns are taken to be x and z, then the first column turns out to be z^2/y^2 – or $\sec^2 A$, in modern notation. (A more logical notation would be $(\sec A)^2$, but we shall adhere to the established convention for denoting powers of the trigonometric functions. We define $\sec A$ as the reciprocal of $\cos A$.) Clearly, not only were the Babylonians familiar with Pythagoras' theorem, but they also knew how to construct what are known as *Pythagorean number triples*. These are solutions (x, y, z) in positive integers of the equation $x^2 + y^2 = z^2$. To form such a triple we take any two positive integers p and q, where $p > q$, such that p and q have no common factor and are of opposite parity, i.e. one is odd

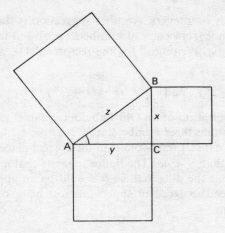

Figure 1.1.

and the other even. (The symbols > and < denote 'greater than' and 'less than' respectively.) Then the expressions

$$x = p^2 - q^2, \quad y = 2pq \quad \text{and} \quad z = p^2 + q^2$$

will produce all primitive number triples with no repetitions. By 'primitive' we mean that the numbers comprising the triple have no common factor. Thus $p = 2$, $q = 1$ gives the well-known (3, 4, 5), while $p = 3$, $q = 2$ gives (5, 12, 13). The triple corresponding to the first row of our table is (119, 120, 169), derived from $p = 12$, $q = 5$; that corresponding to the fifth row is (65, 72, 970), from $p = 9$, $q = 4$. We don't know how or for what purpose the table was constructed, but the numbers are so large that they must have been determined by some rule. The numbers in the first column decrease steadily, with the first nearly equal to $\sec^2 45°$ ($= 2$) and the last to $\sec^2 31°$. We have a table of $\sec^2 A$ where the angle A in Figure 1.1 decreases from 45° to 31° in 1° steps: an impressive example of what could be achieved nearly 4000 years ago. However, not all authorities accept this interpretation of the significance of the Plimpton tablet. One rival theory is that it is a 'teacher's aid' for setting and solving problems involving right-angled triangles which would 'come out' nicely in integers. Another suggestion is that the tablet was computed, not from pairs of numbers (p, q), but from a single parameter $s = p/q$. Writing s^R for the reciprocal $1/s$, we see that

$$x/y = \tfrac{1}{2}(s - s^R) \quad \text{and} \quad z/y = \tfrac{1}{2}(s + s^R)$$

Figure 1.2 is a transcription of an Old Babylonian tablet now at Yale University. It contains three numbers: $a = 30$, $b = 1,24,51,10$ and $c = 42,25,35$. If we insert semi-colons in the correct places and take a to be $\tfrac{1}{2}$, we see that $c = ab$. The figure suggests that a is the side of a square and c is the diagonal, so b should be an approximation to $\sqrt{2}$. It is indeed an excellent approximation, being correct to one part in a million.

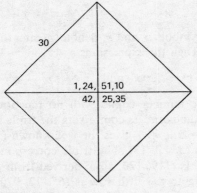

Figure 1.2.

How did the Babylonians achieve such accuracy? It seems likely that they used an iterative method, along the following lines. To approximate to $\sqrt{a}$ we guess a first approximation, a_1, and compute $a/a_1 = b_1$. Then the mean, $a_2 = \tfrac{1}{2}(a_1 + b_1)$, is a better approximation. The procedure can be continued by computing $b_2 = a/a_2$, then $a_3 = \tfrac{1}{2}(a_2 + b_2)$, and so on. The approximation to $\sqrt{2}$ shown in Figure 1.2 is in fact the value of a_3 when a_1 is taken to be $1;30$.

It used to be thought that the Babylonians had contributed little to geometry, but discoveries made this century, particularly those in 1936 at Susa, have changed the picture. One Susa tablet compares the areas of some regular polygons with the squares on their sides. For the pentagon the ratio is given as 1;40, for the hexagon as 2;37,30 and for the heptagon as 3;41. The figure of 3;7,30 (or $3\frac{1}{8}$) is used as an approximation for π.

The mathematics of the Egyptians and the Babylonians, while impressive by virtue of its antiquity, suffers from several deficiencies. The extant material treats specific problems only; there is little generalization; the concept of proof is conspicuous by its absence; and there is no clear distinction between exact and approximate results. The extent to which pre-Hellenic mathematics was entirely practical, or was, in part, cultivated for its own sake, is still a matter of scholarly dispute, just as it was in the days of Herodotus and Aristotle.

2   Early Greek mathematics

'The Greeks were the first mathematicians who are still "real" to us today. Oriental mathematics may be an interesting curiosity, but Greek mathematics is the real thing. The Greeks first spoke a language which modern mathematicians can understand.'
G. H. Hardy

1. Introduction

The civilization of Classical Greece is one of the glories of the human story. To account for such an outburst of creative energy presents historians with a major and largely unsolved problem. In the history of mathematics – as in that of philosophy, literature and the plastic arts – 'the Greeks are the supreme event'. The first records of the Olympic Games are usually said to date from 776 BC, by which time Greek literature was highly developed; it was another 200 years before Greek mathematics made its entrance on the stage of history. It is convenient to distinguish two periods of Greek mathematics: the Classical period, from 600 to 300 BC, and the Hellenistic (or Alexandrian) period, from 300 BC to about AD 600. During the whole of this time the cultural influence of Greece extended far beyond the Greek mainland. By the sixth century BC Greek-speaking peoples were settled all around the Eastern Mediterranean: in Sicily, southern Italy, Asia Minor and North Africa, and on the shores of the Black Sea and the Aegean Islands. Many Greeks lived in independent city-states such as Athens, Corinth, Syracuse and Miletus. Such, however, was the cultural unity of the Greek world that we can speak of 'Greek' mathematics without bothering about exact geographical locations.

2. The Greek contribution

Of what, we may ask, does the unique Greek contribution to mathematics consist? We can identify three elements. The first is an insistence that all mathematical results must be established by *deductive* reasoning. The argument must proceed by logical steps from an agreed starting point, thereby guaranteeing acceptance of the conclusion. This procedure is in sharp contrast to what usually happens in ordinary life. Here we proceed *inductively*, from the particular to the general; we learn from experience by trial and error; we reason by analogy. The second element, closely related to the first, is that the Greeks made mathematics *abstract*. The search for abstract perfection is a dominant theme of Greek thought and is exemplified above all by Plato. He conceived a spiritual world of abstract ideas and ideals – timeless, changeless and indestructible – over and above the imperfect and transitory world of matter as perceived by the senses. Mathematics belongs, *par excellence*, to the ideal world and so should be studied by all who seek to lead society or to influence their fellow men. Hence the well-known motto commonly held to grace the entrance to Plato's Academy at Athens: 'Let no one ignorant of geometry enter here.'

This brings us to the third notable feature of Greek mathematics: the emphasis on geometry and the use of geometrical methods for solving problems. The other side of the coin is the Greek failure to develop a symbolic notation of the kind needed to make real progress in algebra and the calculus. Even so, the overriding questions must be: Why did the Greeks create so much first-class mathematics? What were the sources of their inspiration? What were their objectives?

The answers to these questions are to be found in the Greeks' attitude to the world about them, in their desire to understand the nature of reality. They valued and cultivated mathematics because they believed that the world was designed (or, at any rate, operated) in accordance with mathematical laws. The study of mathematics was, therefore, the key to the comprehension of the natural world. This attitude to nature – secular, critical and rational – was embraced by many Greek intellectuals as early as the sixth century BC, although not, of course, by the common people. They remained deeply attached to their mythologies, mysteries and magic. Gods and demons controlled events and nature was capricious, chaotic and terrifying.

Although there were considerable differences between the philosophers, from the 'idealism' of Plato to the 'atomism' of Democritus, most Greek thinkers (Socrates was perhaps an exception) placed mathematics squarely at the centre of the cultural scene. Most of the leading Greek mathematicians were also astronomers, and many applied their talents to the study of music, optics, mechanics or geography. Mathematics was twice blessed: it was valued both for its own sake and as the key to unlock the secrets of nature.

3. The textual sources

When evaluating the Greek mathematical achievement, there is one thing we must always bear in mind: the vast majority of Greek mathematical works have been lost. Indeed, from the Classical period only a few scattered fragments have survived, and we have to rely on references by commentators and editors, often writing centuries later. For the Hellenistic period we are rather better served, but even here we have no primary sources, whereas we do for Ancient Egypt and Babylonia. The oldest surviving Greek texts are 'copies of copies, many times removed'. The third century was the Golden Age of Greek mathematics, but most of our knowledge of it derives from the extant works (far from complete) of three masters, Euclid, Apollonius and Archimedes, who will be the subjects of the next two chapters.

The problems of establishing a firm documentary basis for the study of Greek mathematics may be illustrated by the textual history of the most influential mathematical work of all time, the *Elements* of Euclid. This great treatise was written (or more likely, put together and edited) in about 300 BC, but the earliest surviving texts are Byzantine and date from the tenth century AD. Most of them derive from an edition written by Theon of Alexandria in the fourth century AD. By the end of the eighth century the *Elements* had been translated into Arabic; during the twelfth century some Arabic texts were translated into Latin and introduced into Europe. The first printed edition of the *Elements* was published in 1484; the first Latin translation direct from the Greek appeared in 1505; the Greek text itself was published ten years later. A definitive edition, based on a close study of all available texts, did not emerge until the end of the nineteenth century. The great Danish scholar J. L. Heiberg published his critical edition of the *Elements* between 1883 and 1888. It has formed the basis of all subsequent editions, including Sir Thomas Heath's English translation of 1926. This tortuous sequence of events is typical of the textual history of the relatively few Greek works that have come down to us. Indeed, sometimes – as with some of the writings of Apollonius and Archimedes, for example – no Greek text survives and scholars have to make do with an Arabic translation.

4. Thales and Pythagoras

The earliest Greek philosopher and mathematician known to us by name is Thales of Miletus (*c*. 634–548 BC), who is credited – probably incorrectly – with predicting the solar eclipse of 585 BC. He is said to have proved a number of geometrical theorems, such as the equality of the base angles of an isosceles triangle. The next great figure is Pythagoras of Samos (*c*. 580–500 BC), saluted by Bertrand Russell as 'intellectually one of the most important men that ever lived'. In about 540 BC Pythagoras founded a school of mathematics and philosophy at Crotona in southern Italy. The explicit recognition that mathematics is primarily concerned with abstract ideas is usually attributed to the Pythagoreans. Indeed, Aristotle says of them: 'They supposed the elements of number to be the elements of all things, and the whole heavens to be a musical scale and a number.' One is reminded of the well-known epigram of the nineteenth-century mathematician Leopold Kronecker: 'God made the integers; all the rest is the work of man.'

It is not surprising, therefore, that the Pythagoreans made a close study of the properties of numbers. Their work is a curious mixture of good mathematics and 'number mysticism' – where, for example,

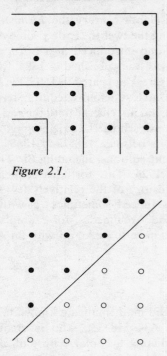

Figure 2.1.

Figure 2.2.

odd numbers are regarded as male and even numbers as female, so 5 is the number of marriage. Two examples of their 'good' mathematics must suffice. The Pythagoreans were in the habit of representing numbers by pebbles on the sand. Figure 2.1 illustrates how successive *square numbers* can be built up by adding *gnomons* (to use the Greek term) of odd numbers. The general result is that the sum of the first *n* odd numbers is equal to the *n*th square number, i.e.

$$1 + 3 + 5 + \ldots + (2n - 1) = n^2$$

Figure 2.2 shows how successive triangular numbers (1, 3, 6, 10, ...) can be built up, and how a second, equal triangular number can be added to form an oblong, thus demonstrating that the value of the *n*th triangular number is

$$1 + 2 + 3 + \ldots + n = \tfrac{1}{2}n(n + 1)$$

5. The classification of numbers

It is convenient at this point to digress somewhat in order to define and explain the various kinds of number that we recognize today: we shall then know what we are talking about in later chapters. It is intuitively acceptable to assume that a continuously varying quantity, such as distance or time, may be represented along a straight line; that the length OP in Figure 2.3, measured from an origin O, specifies a unique point P to which we can assign a unique number, called a *real number*. The converse is also true: any real number corresponds to a unique point on the line. Real numbers can be subdivided into numbers of various kinds. Starting with the *positive integers*, the *natural* (or *counting*) *numbers* 1, 2, 3, . . . , we add first the number 0 corresponding to the origin, and then the *negative integers*, as indicated in Figure 2.3. An obvious next step yields the *rational fractions*, denoted by p/q, where p and q are integers and $q \neq 0$ or 1. However, we have not finished yet. We shall see in the next section that some real numbers are neither integers nor rational fractions. Such numbers are called *irrational numbers* (an unfortunate term that we are stuck with for historical reasons); $\sqrt{2}$, $\sqrt[3]{5}/\sqrt{2}$ and π are examples. Irrational numbers may be further divided into those that provide solutions of *polynomial equations* of the form

$$a_0 x^n + a_1 x^{n-1} + \ldots + a_{n-1}x + a_n = 0$$

where the a's are integers, and those that do not. The former are called *algebraic irrational numbers*; the latter *transcendental numbers* (they 'transcend' the ordinary algebraic processes). Of the three

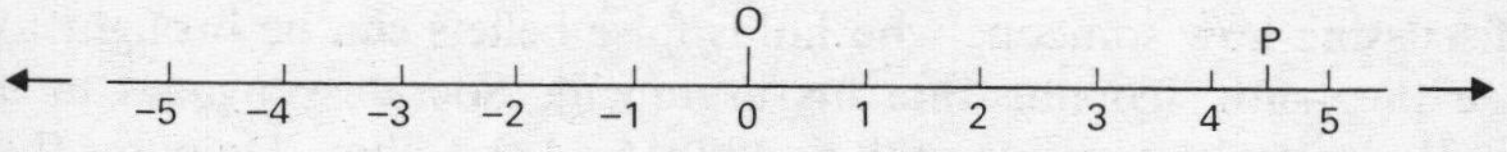

Figure 2.3.

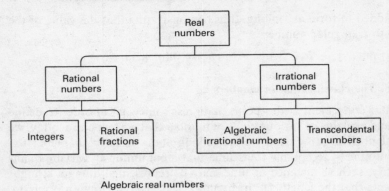

Figure 2.4.

irrational numbers just listed, the first two are algebraic, since they are solutions of the equations $x^2 - 2 = 0$ and $8x^6 - 25 = 0$, while the famous number π is transcendental. The full classification is illustrated in Figure 2.4. Throughout this book the word 'number', without any qualifying adjective, will mean 'positive integer'.

6. The discovery of incommensurability

While the Pythagoreans and their contemporaries were quite happy with rational numbers, they were, in the view of most historians, unable to cope with irrational numbers – or, in geometric terms, with the ratios of two *incommensurable magnitudes*. There came a time, however, probably during the fifth century BC, when the Greek world had to come to terms with the stubborn fact that integers and their ratios were unable to account for some of the fundamental features of simple geometrical figures. The most obvious example is the ratio of the diagonal of a square to its side. In one of Plato's dialogues, the *Meno*, Socrates and Meno (a young aristocrat) are discussing how someone who holds false beliefs can be brought to see the truth. To illustrate his argument, Socrates engages in a mathematical discussion with an untutored slave boy. He poses the problem of finding a square whose area is twice that of a given square. Socrates draws some figures in the sand (probably something like Figure 2.5) and leads the boy, step by step, in the correct geometrical approach.

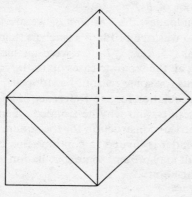

Figure 2.5.

The classic proof of the irrationality of $\sqrt{2}$ is worth looking at in detail. Let us suppose, to begin with, that the theorem is *false*, i.e. that $\sqrt{2}$ can be expressed as a rational fraction, p/q. This is equivalent to saying that the equation $p^2 = 2q^2$ can be satisfied by integral values of p and q, which have no common factors. It follows that p^2, and therefore p, is even, and so $p = 2r$ for some integral value of r. This yields $4r^2 = 2q^2$ or $q^2 = 2r^2$; so q^2, and hence q, is also even. This means that both p and q have a common factor of 2, which contradicts our initial assumption. So the assumption must be false, and the statement that $\sqrt{2}$ is irrational must be true.

The technique of indirect proof – or proof by contradiction (*reductio ad absurdum*) – was much favoured and skilfully exploited by the Greeks; we shall be meeting further examples later. There is an alternative proof of the $\sqrt{2}$ theorem which can be traced back to the Classical period. If p and q are decomposed into their prime factors, then each factor will appear twice in the factorizations of p^2 and q^2. This means that the left side of the equation $p^2 = 2q^2$ will consist of an even number of factors, and the right side of an odd number of factors. Once again, we have established a contradiction. Both these proofs can easily be generalized to establish the irrationality of $\sqrt{P}$, where P is any prime number, and even of higher roots of P (e.g. $\sqrt[3]{P}$, the cube root of P).

In another of Plato's dialogues, the *Theaetetus*, we are told that Theodorus of Cyrene, who was one of Plato's teachers, had proved in about 390 BC that $\sqrt{3}$, $\sqrt{5}$, $\sqrt{6}$, $\sqrt{7}$, ... were all incommensurable with unity, 'taking all the separate cases up to the root of 17 square feet, at which point, for some reason, he stopped.'

There has been much speculation (e.g. in Reference 35) as to how Theodorus obtained his results, and why he stopped when he did. Figure 2.6 shows a geometrical construction for the square roots of successive integers. The reader is invited to continue this construction as far as $\sqrt{17}$; the result may prompt some speculation as to why Theodorus stopped at that point.

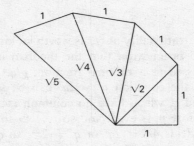

Figure 2.6.

Theaetetus was a prominent figure in Plato's Academy. He died in 369 BC from a combination of battle wounds and dysentery. The dialogue which bears his name was Plato's tribute to his dead friend. In it Theaetetus discusses the nature of incommensurable magnitudes with Socrates and Theodorus, and draws a distinction between what we would now call *quadratic surds* (of the form $\sqrt{p}$) and more complicated irrational numbers. The subject is treated in great detail in Book X of Euclid's *Elements*, which is usually considered to be based on the work of Theaetetus.

The discovery of incommensurability could, indeed, have come about in one of several ways; no evidence of the date and circumstances of the discovery survives. The diagonal-to-side ratio of a square is one obvious possibility; the diagonal-to-side ratio of a regular pentagon, a figure to which the Greeks were much attached, is another. The star pentagon, or pentagram, was in fact the badge of the Pythagorean brotherhood, and also the Greek emblem of health. Figure 2.7 shows a construction which yields an infinite sequence of nested pentagons, a result to which we shall return in the next chapter.

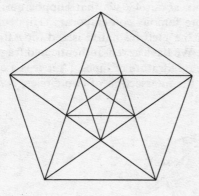

Figure 2.7.

Most, but not all, historians have taken the view that the discovery of incommensurability precipitated a grave crisis in Greek mathematics. Thus the French scholar Paul Tannery, writing 100 years ago, speaks of a 'veritable scandal'. There is the well-known story that the Pythagoreans drowned the colleague who disclosed the awful secret to the outside world. Folklore apart, the consequences were certainly serious. For one thing, the whole theory of the similarity of equiangular polygons was in jeopardy. To take a simple example from the *Meno*, since $\sqrt{2}$ is incommensurable we cannot have a rational side for a square which has twice the area of a square with unit sides. One effect of the 'crisis' was to switch the emphasis of Greek mathematics from the Pythagorean study of number to geometry. Indeed, geometry provided the basis of most of the best mathematics for the next 2000 years. We still speak of x^2 and x^3 as 'x squared' and 'x cubed'.

7. Hippocrates of Chios

The figures of Thales and Pythagoras are shrouded in the mists of legend. We must wait until the second half of the fifth century BC for firm evidence of individual contributions. Those years saw an outburst of creative activity fully worthy of the Golden Age of Periclean Athens. It was in about 430 BC that Hippocrates (not to be confused with his more famous contemporary, Hippocrates of Cos, 'the Father of Medicine') left his native island for Athens and began to study geometry. We have a well-authenticated fragment of his work dealing with the quadrature of lunes. (A *lune* is a crescent-shaped figure formed by the intersection of two circles; *quadrature* means 'to find the area'.)

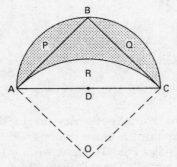

Figure 2.8.

Figure 2.8 shows one of Hippocrates' constructions. A semicircle is circumscribed about a right-angled isosceles triangle ABC. The segment marked R is then constructed as shown so as to be *similar* to the two equal segments marked P and Q (the three segments are similar because each subtends a right angle at its centre). Now, by using the theorem (also attributed to Hippocrates) that the ratio of the areas of two similar segments is equal to the ratio of the squares of their bases, we find that area of segment P + area of segment Q = area of segment R, since

$$AB^2 + BC^2 = AC^2$$

Removing the segments P and Q from the semicircle leaves the triangle ABC; removing the segment R leaves the lune (shown shaded). The area of the lune is therefore equal to the area of the triangle ABC which is itself equal to the square on the radius AD. So far as we know, this is the first quadrature of a curvilinear figure to be rigorously proved. Hippocrates also gave several further examples of lunes whose areas can be evaluated exactly.

8. 'Ruler and compasses' constructions

Greek geometry of the early Classical period was based on – indeed, was circumscribed by – the use of two tools: a straight-edge (or unmarked ruler) for drawing or extending straight lines, and compasses for drawing circles. Such constructions enabled the four basic arithmetic operations and the extraction of square roots to be represented geometrically. The constructions for the addition and subtraction of two magnitudes (represented by lengths along a

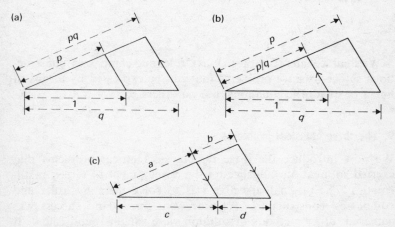

Figure 2.9.

straight line) are trivial. The constructions for multiplication and division, shown in Figures 2.9(a) and (b), are self-explanatory. The same figure, interpreted as in Figure 2.9(c), yields the relation $a : b = c : d$ and provides the construction for 'transferring' ratios from one line to another.

The construction for the extraction of a square root is shown in Figure 2.10. Since the angle ADC in a semicircle is a right angle, we have three similar triangles with the marked angles equal. Hence AB/BD = DB/BC, and so BD = $\sqrt{p}$. The construction also provides a way to obtain a mean proportional or geometric mean, BD, between two lengths AB and BC.

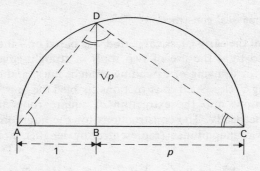

Figure 2.10.

We shall see later (Chapter 7) that 'ruler and compasses' constructions suffice to solve problems that can be expressed by means of linear or quadratic equations, but no others.

9. The three classical problems

We now introduce the three famous geometrical problems that exerted so great an influence on the development of Greek mathematics. The original objective was to solve them by 'ruler and compasses' constructions. As time went on and no success was achieved, other methods of solution were sought – and found. It was indeed the gradual realization (and eventual proof) of the impossibility of 'ruler and compasses' solutions that kept these problems at the centre of the mathematical stage for many centuries.

The first problem, known as the 'duplication of the cube', was to construct a cube whose volume is twice that of a given cube. It is known as the Delian problem, because it was thought to have been posed by the Oracle at Delphi. Its arithmetical formulation is, of course, to find the cube root of 2. Although Hippocrates did not solve the problem, he is credited with seeing that it is equivalent to the problem of inserting two *mean proportionals*, x and y, between two given magnitudes a and b, which we may think of as line segments. Thus if $a:x = x:y = y:b$, we get $x^3 = a^2b$. Taking $a = 1$, $b = 2$ gives the desired result. (We shall use the colon to denote ratios when we wish to emphasize the Greek concept of ratio as a relation between two magnitudes, in contrast to the modern concept of a numerical fraction.)

The Delian problem was brilliantly solved by Archytas in about 400 BC. He did it by finding the point of intersection of three surfaces in three-dimensional space: a cone, a cylinder, and a torus generated by rotating a circle about one of its tangents. It was indeed a most impressive achievement; we must recognize how little was then known about solid geometry and that Archytas had to work out his solution by 'pure' geometry, without the aid of coordinates or equations.

Archytas was born in about 428 BC. Although he lived most of his life in Tarentum in southern Italy, he was in close touch with the philosophers of Plato's Academy. Like so many of his peers, Archytas was a man of parts. As an educationist in the Pythagorean tradition, he is credited with designing the four branches of the *quadrivium* – arithmetic (numbers at rest), geometry (magnitudes at rest), music (numbers in motion) and astronomy (magnitudes in motion) – and so securing the long, continued prominence of mathematics in education. He also wrote on mechanics and the mathematical theory of music. Archytas was elected governor of his city no fewer than seven times, and was admired as an enlightened, just and compassionate ruler, with a special concern for the welfare of slaves and children. His skill in making toys and models elicited the comment from Aristotle that one of Archytas' creations – in fact a rattle – 'was useful to give to children to occupy them from breaking things about the house, for the young are incapable of keeping still.' As commander of the city's defences he was never defeated. Perhaps his most valuable service was that on one occasion he used his influence in high places to save the life of his friend Plato. He was drowned in a shipwreck in about 347 BC.

The second 'classical' problem was 'to trisect an angle': to construct an angle one-third the size of a given angle. The third problem was 'to square the circle'. This problem can take two forms: either to construct a square whose area is equal to that of a given circle, or to find the length of the circumference of a circle of given radius. It was shown eventually that the first two problems are algebraic; in fact each requires the solution of a cubic equation. The third problem is quite different in that it involves the transcendental number π.

10. Hippias of Elis

One of the flourishing schools of fifth-century Athens was the Sophist School. The Sophists supported themselves financially by teaching, but were not popular with the philosophical establishment, who portrayed them as shallow, grasping and venal. One of the most successful of the Sophists was Hippias of Elis; we have accounts of him from both Plato and Xenophon. He is best known for his discovery of a famous curve which is generated in the following manner. In Figure 2.11 the side AB of the square ABCD moves parallel to itself at a constant speed until it reaches the position DC. The side AD rotates uniformly about D until it also coincides with DC, the two motions beginning and ending together. Let XY and DZ be the positions of the two lines at some instant during the motion, P being their point of intersection. The path APQ traced out by the point P, known as the *trisectrix of Hippias*, provides an easy solution to the second 'classical' problem. To trisect the angle PDQ all we have to do is to draw lines X'Y' and X"Y" parallel to AB such that XX' = X'X" = X"D. If these lines cut the trisectrix at P' and P", the lines P'D and P"D will divide the angle PDQ into three equal parts.

This curve can also be used to solve the third problem, that of squaring the circle. It can be proved that the ratio DQ : DC is equal to $2/\pi$, so solving the problem. For this reason the trisectrix is also called the *quadratrix*. The proof is not difficult if modern calculus is used, but the Greeks had to rely on elementary geometry. It is unlikely that Hippias knew of this result; the proof, by contradiction, is usually attributed to Dinostratus, a fourth-century mathematician of the Platonic school. The Greek theorem actually states that DC (in Figure 2.11) is the mean proportional between DQ and the circular arc AC.

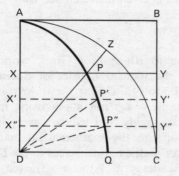

Figure 2.11.

The acute minds of the Greeks realized that there were two logical weaknesses in the arguments presented here, both deriving from the fact that the quadratrix is defined by two simultaneous motions. In Figure 2.11 the horizontal line moves downwards from AB to DC in exactly the same time as the line DZ pivots about D from DA to DC. The two lines start moving together and reach their final position, DC, together. To achieve this we need to know the ratio of the two speeds, which is possible only if we already know the ratio of the side AB to the circular arc AC. But this is what the construction is designed to find out: we seem to be arguing in a circle! The second objection concerns the point Q. The quadratrix is defined by the intersection of two moving lines. When both lines reach the position DC together, they do not intersect at a unique point: one line lies on top of the other. While it is intuitively apparent that the quadratrix is heading for the point Q as shown, does such an argument meet the exacting Greek standards of proof?

In this chapter we have mentioned just one approach to the solution of each of the three 'classical' problems. In fact a variety of other constructions were put forward during Classical and early modern times; these used special curves, mechanical devices, or a marked ruler – none could be constructed using a straight-edge and compasses alone. We shall be meeting some of them in Chapter 4. A good survey of the subject is given in Reference 16.

11. Eudoxus of Cnidus

The intellectual life of the fourth-century Greek world, which opened so tragically with the judicial murder of Socrates, was dominated by two giants: Plato (427–347 BC) and Aristotle (385–322 BC). Although Plato himself was not a practising mathematician, he valued the subject highly and exerted great influence, by inspiration and direction, on the mathematical activities of his time. He frequently raises mathematical issues in his Dialogues and other writings. His Academy at Athens attracted most of the leading thinkers, and soon became the mathematical centre of the Greek world. The most distinguished product of the Academy was Eudoxus (c. 408–355 BC), the greatest mathematician of the Greek Classical period and second only to Archimedes in the whole of Antiquity. The reason he is not better known is that all his works are lost; we have to rely on references by other writers. His work covered a wide field and we shall look briefly at three of his major achievements.

Eudoxus was born at Cnidus, an island off the coast of Asia Minor; he suffered extreme poverty in his youth. After some years of study and travel, he moved to Athens to join Plato and the two men became close friends. We are told that Eudoxus could not afford to live near the Academy, and had to walk in each day from Piraeus, where fish, olive oil and lodging were very cheap. After the execution of Socrates life became difficult in Athens, and Plato and Eudoxus did some travelling together, visiting Egypt among other places. Eudoxus eventually settled down at Cyzicus in Asia Minor, where he founded a distinguished mathematical school. He was also a practising physician and an active legislator.

12. The Eudoxan theory of proportion

We have seen how disturbing was the discovery of the existence of incommensurable magnitudes in geometry. As time went on, more and more were discovered and the problem of their nature became acute. The difficulty was brilliantly resolved by Eudoxus with his theory of proportion. This theory is expounded in Book V of Euclid's *Elements*, and has been the subject of intensive study and discussion for more than 2000 years; indeed, it was not fully understood until the end of the nineteenth century, as we shall see in Chapter 16.

Eudoxus made a sharp distinction between *magnitudes* (of, for example, line segments, areas, volumes, angles or time), which vary continuously, and *numbers*, which jump from one discrete value to another. He was careful not to assign numerical values either to magnitudes or to their ratios, for his objective was to be able to *compare* two ratios – to determine whether they are equal, and if they are not, which is the larger – in a way that is equally valid for ratios of commensurable and incommensurable magnitudes. His theory supplied the Greeks with a logical foundation for dealing with incommensurables and cleared the way for the geometrical advances of the Golden Age and beyond. To the Greeks, the concept of *ratio* was essentially undefined. Eudoxus' point of departure was to assert that two magnitudes of the same kind *have* a ratio if a multiple of either can be found which exceeds the other (i.e. the ratio $a:b$ exists when whole numbers m and n can be found such that $ma > b$ and $nb > a$). Euclid includes this statement among his 'definitions' in Book V. Archimedes, who used it extensively, regarded it as an axiom; he attributed it to Eudoxus, and it is now known as the *Eudoxus–Archimedes axiom*. Its virtue is that it excludes not only zero but also any idea of infinity – either the infinitely large or the infinitely small.

We are now in a position to state the Eudoxan criterion for two ratios to be equal or, as the Greeks would say, for the four magnitudes to be proportional. This is the famous 'Definition 5' of Book V of the *Elements*. Nowadays, like Archimedes, we would regard it as an axiom. Here is the statement expressed in modern terminology for clarity:

If a, b, c and d are four given magnitudes, then the ratio $a:b$ is equal to the ratio $c:d$ if, and only if, given any two whole numbers, m and n, then:

(1) $ma > nb$ implies that $mc > nd$
(2) $ma = nb$ implies that $mc = nd$
(3) $ma < nb$ implies that $mc < nd$.

If the magnitudes are commensurable, (2) is sufficient, both ratios being equal to the rational number n/m. The subtlety of the theory lies in (1) and (3), because (2) *never* holds for incommensurable magnitudes. Out of two inequalities the condition for equality somehow emerges. The four magnitudes need not all be of the same kind: for example, a and b might be the volumes of two spheres, and c and d the cubes of their radii.

Eudoxus goes on to deal with unequal ratios by asserting that if two numbers m and n exist such that $ma > nb$ and $mc < nd$, then the ratio $a:b$ is greater than the ratio $c:d$. We shall see in the next chapter how these three Eudoxan assertions (or axioms) provided a firm foundation from which to extend the Pythagorean treatment of the ratios of whole numbers to deal with incommensurable magnitudes, while avoiding the notion of irrational numbers.

13. The method of exhaustion

It is generally accepted that Eudoxus was the first to formulate an integration procedure, now known as the *method of exhaustion*, which provided a logical justification for all the limiting processes used in Greek mathematics. Archimedes exploited the method to great effect to prove many theorems about the areas and volumes of curvilinear figures, as we shall see in Chapter 4. The basis of the method is contained in the following proposition (Prop. X, 1 of the *Elements*):

If from any magnitude there be subtracted a part, not less than its half, and if from the remainder one again subtracts not less than its half, and if the process of subtraction is continued, ultimately there will remain a magnitude less than any preassigned magnitude of the same kind.

This result, which we may call the 'exhaustion property', is easily proved by *reductio ad absurdum* from the Eudoxus–Archimedes axiom. In modern terminology, it states that, given two magnitudes a and b, with $a > b > 0$, then if $0 \leqslant r_i < \frac{1}{2}$ for all $i = 1, 2, 3, \ldots$, there is a number n such that $ar_1 r_2 \ldots r_n < b$, however small b may be. (In fact this restriction on the r's is unduly severe, for the result remains valid if $0 < r_i \leqslant k < 1$, where k is any fixed magnitude within the prescribed limits.)

To see how the method works, we shall use it to prove the theorem (Prop. XII, 2 of the *Elements*) that 'the areas of circles are to one another as the squares on their diameters'. The theorem itself is attributed to Hippocrates, but we know nothing of his method of proof. An outline of the Eudoxus–Euclid proof is as follows. Let two circles, c and C, have diameters d and D, and areas a and A. We wish to prove that $a : A = d^2 : D^2$. The proof is by contradiction. Let us assume first that $a : A > d^2 : D^2$. Then there is a magnitude $a' < a$ such that $a' : A = d^2 : D^2$. Put $a - a' = \varepsilon > 0$. We must first prove that a circle can be 'exhausted' by a sequence of inscribed polygons. If we inscribe a square in a circle, as in Figure 2.12, the area between the circle and the square (we shall call it the 'excess area') is equal to four times the segment PQR. If we now construct a regular octagon as shown, it is easy to prove that the new excess area (eight times the shaded area) is less than half the previous one. We continue in this way, inscribing polygons of $16, 32, 64, \ldots$ sides; at each step the excess area is reduced by more than half. We must, therefore, eventually reach a stage when the excess area is less than ε. If p_n is the area of the polygon inscribed in circle c at this stage, then $a - p_n < \varepsilon$ and so $p_n > a'$. Let P_n be the area of the corresponding polygon inscribed in the circle C. In the previous proposition (XII, 1) Euclid has proved that $p_n : P_n = d^2 : D^2$. Now, by our initial assumption, this yields $p_n : P_n = a' : A$. We have shown that $p_n > a'$, so it follows that $P_n > A$. Since the polygon P_n is inscribed in the circle C, we have arrived at a contradiction, so our initial assumption that $a : A > d^2 : D^2$ must be false. The alternative assumption, that $a : A < d^2 : D^2$,

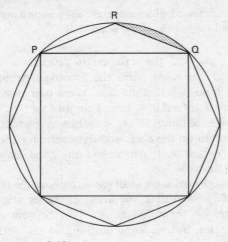

Figure 2.12.

can be disposed of by a similar argument. We conclude, therefore, that $a:A = d^2:D^2$, which was to be proved. Nowadays, of course, we would write $a/d^2 = A/D^2 = \pi/4$. The symbol π was first used in this sense in 1706 by William Jones, but its precise status as a real number was not finally established until 1882, when Ferdinand Lindemann proved π to be transcendental.

Most of the theorems in Book XII of the *Elements*, as well as those in Book V, are thought to be due to Eudoxus. Does this make him the father, or perhaps the grandfather, of the integral calculus? Archimedes in particular pays frequent tribute to him and tells us, for example, that he was the first to give a rigorous proof that the volume of a cone is one-third that of the cylinder which has the same base and the same height (this is Prop. XII, 10 of the *Elements*).

14. Eudoxan astronomy

We must mention one further achievement of this remarkable man, for which he is acclaimed as the 'Father of Mathematical Astronomy'. The object of Greek astronomy was to construct a geometrical model of the Solar System that would, as Plato put it, 'save the

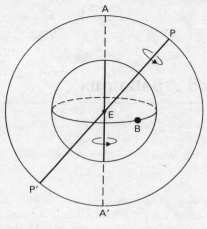

Figure 2.13.

phenomena', i.e. would represent the complex motions of the Sun, Moon and planets as seen from the Earth. No physical interpretation was sought. Eudoxus wrote four astronomical works (all lost), but the essence of his scheme can be reconstructed from later references. With each heavenly body Eudoxus associated a set of rotating spheres, all centred on the Earth, E. The innermost sphere, which carries the body B on its equator, rotates about the axis AEA′, as indicated in Figure 2.13, but the sphere itself is carried along by the rotation of a second sphere about the axis PEP′. We can think of the axis AA′ as an imaginary rod which passes through the poles of the freely rotating first sphere and which is attached to the second sphere at A and A′. The axis of the second sphere is carried around in turn by the rotation of a third sphere, and so on. To reproduce the observed motions adequately. Eudoxus needed three spheres each for the Sun and the Moon, and four for each planet. Each combination was linked to an outermost sphere which rotated once every 24 hours about an axis through the celestial poles. He thus used 27 spheres altogether. We can only marvel at the mathematical skill which enabled him to choose the axial directions, speeds of rotation and radii of so many spheres so as to produce an interlocking dynamic model whose behaviour was in good accord with the observations.

3 Euclid and Apollonius

'Geometry has two great treasures: one is the theorem of Pythagoras; the other, the division of a line in extreme and mean ratio. The first we may name as a measure of gold, the second we may name as a precious jewel.' Johann Kepler

1. Introduction

Most of our knowledge of the Greek mathematics of the Classical age comes from the writings of two men, Euclid and Apollonius, who lived in the third century BC, at the beginning of the Hellenistic period. In 332 BC Alexander the Great founded the city of Alexandria on the Egyptian coast near where Europe, Asia and Africa come together. When the young conqueror died from malaria at Babylon in 323 BC, his empire rapidly fell apart. Three of his generals divided the inheritance between them and one of them, Ptolemy, became ruler of the rich province of Egypt. The Ptolemaic dynasty built up their new capital city until it became – and was to remain for several centuries – the cultural centre of the Greek-speaking world. Two of the Ptolemaic foundations are of special importance. The first is the Museum, a true university, at once a centre of advanced teaching and a research institute which attracted many of the leading natural scientists, astronomers and mathematicians from all over the Greek world. The second is the Library, the largest in Antiquity; at its peak it is said to have contained some 400 000 rolls of papyrus.

2. Euclid: his life and works

Of Euclid's life we know almost nothing, except that he was one of the first teachers at the Museum, as well as one of the greatest. He may well have acquired his early education at Plato's Academy. He was the author of about a dozen treatises, most of which have been lost. These include the *Conics*, soon to be superseded by Apollonius' more comprehensive treatise; the *Porisms*, which may have foreshadowed the analytical geometry of the seventeenth century; and works on *Surface Loci* and *Pseudaria* (*Fallacies*). Fortunately, however, five of his works have survived. They are the *Optics*, an early treatise on perspective; the *Phaenomena*, a work on spherical geometry for the use of astronomers; *On Divisions of Figures* (surviving in Arabic only); and the *Data*. These last two contain a variety of geometrical constructions and pedagogical comments; they were presumably written for the use of students at the Museum. Finally, there is his *magnum opus*, the *Elements*.

3. The *Elements* of Euclid

It is important to realize that the *Elements*, by far the most influential mathematical work ever written (there have been more than 1000 editions since the invention of printing), was compiled primarily as a textbook for students. It is organized in 13 Books and contains no fewer than 467 Propositions. In it is to be found virtually all of the more elementary mathematical knowledge of the Greek Classical period, although some of the later Books are far from elementary. More specialized or advanced topics (e.g. conics or spherical geometry) are excluded. The commentator Proclus, writing in the fifth century AD, remarked that 'the *Elements* bears the same relation to the rest of mathematics as do the letters of the alphabet to the language.' In his elegance, clarity and (for the most part) logical rigour, Euclid was a true Greek of the Golden Age. He presented his results as a sequence of propositions – either *theorems* to be proved, or *problems* to be constructed using straight-edge and compasses only. His mode of exposition has remained a model until recent times. Newton, writing almost 2000 years later, organized his *Principia* on strict Euclidean lines (see Chapter 9).

The first four books of the *Elements* treat the basic properties of rectilinear figures and circles. *Book I* begins with 23 'definitions' (of point, straight line, angle, parallel lines, circle, etc.) followed by 5 'postulates' and 5 'common notions'. Whereas the former are specific to geometry, the latter have a wider relevance – indeed, most people would regard them as self-evident. Most are rules of equality, such as 'If equals be added to equals, the wholes are equal.' Nowadays both postulates and common notions are usually grouped together as *axioms* – that which is to be assumed. The 5th (or parallel) postulate stands out from the others. It states that:

If a straight line falling on two straight lines makes the interior angles on the same side less than two right angles, the two straight lines, if produced indefinitely [this is covered by postulate 2], meet on that side on which the angles are less than two right angles.

After more than 2000 years of intensive effort it was realized, and proved, that this postulate defines a particular kind of geometry that we now call 'Euclidean'. Equally valid geometries can be constructed when the postulate is denied, as we shall see in Chapter 14.

Book I, with 48 propositions, deals with some elementary constructions, congruences of triangles, parallel lines, and the areas of triangles and parallelograms. Many of Euclid's theorems seem to us to be either obvious or trivial: that is because he allows no short cuts in his logical argument. As an example, let us look at Proposition I, 5 (i.e. the fifth proposition of the first Book) which states that the base angles of an isosceles triangle (i.e. a triangle with two equal sides) are equal. Nowadays, we might say that the result follows at once 'from symmetry'. Let us see how Euclid proved it.

At this early stage he has proved only one major theorem (Prop. I, 4), namely the *congruence* (i.e. equality in all respects) of two triangles in which two sides and the included angle are equal, as illustrated in Figure 3.1. Euclid's proof of this theorem is by *super-position*: one triangle is placed on the other so as to achieve a perfect fit. He proves Prop. I, 5 by means of two applications of Prop. I, 4.

He requires some additional construction, as shown in Figure 3.2. The equal sides of the isosceles triangle ABC are extended to D and E. A point F is taken at random on BD, and a point G marked on CE so that AF = AG. (This construction is the subject of Prop. I, 3.) Now join BG and FC. By Prop. I, 4 it follows that the triangles AFC and AGB are congruent, hence FC = BG, ∠ AFC = ∠ AGB and, by subtraction, BF = CG. Thus the triangles BFC and CGB are congruent, so ∠ CBG = ∠ BCF. Since ∠ ABG = ∠ ACG by the first congruence, we have, by subtraction, ∠ ABC = ∠ ACB, which was to be proved.

No wonder that this theorem used to be known as the *pons asinorum*, the bridge that the scholastic asses were unable to cross. A rather simpler demonstration is attributed to the late Greek geometer Pappus, whom we shall be meeting in the next chapter. Let us, he says, consider the triangle ABC in Figure 3.2 as two triangles, ABC and ACB. These are, by Prop. I, 4, congruent, so ∠ ABC = ∠ ACB. Some 16 centuries later an electronic computer, fed with

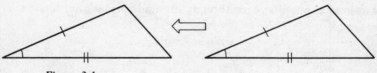

Figure 3.1.

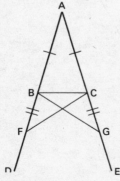

Figure 3.2.

some theorem-solving procedures, 'proved' the result in exactly the same way.

Book I of the *Elements* concludes with Pythagoras' theorem (I, 47) and its converse (I, 48). Euclid proves the former theorem by first establishing that the two shaded triangles in Figure 3.3 are congruent. Since the area of one of them is half that of the square marked L and the area of the other is half that of the rectangle marked R, it follows that these two areas are equal. So, by a similar argument, are the areas marked M and S. The result then follows at once, since $BC^2 = R + S = L + M = AB^2 + AC^2$, which was to be proved.

Euclid proves his theorems and problems (i.e. constructions) by means of an ordered sequence of deductions from the axioms and from propositions already established. In his proof of Pythagoras' theorem, for example, he invokes no less than 24 of the previous 46 propositions. There are many different proofs of this famous theorem. Some, such as the one illustrated in Figure 3.4, provide a direct visual demonstration. It is worth mentioning that the 'parallel postulate' is not invoked until Prop. I, 28. Euclid delayed using it for as long as possible, even though this made his task harder.

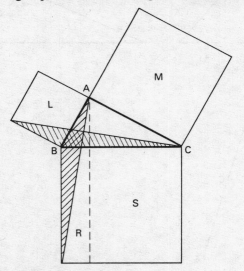

Figure 3.3.

The use of 'logical back referencing' is a characteristic feature of Greek mathematics. Its impact on Thomas Hobbes (1588–1679), the distinguished political philosopher, is worth noting. His friend John Aubrey, whom we shall be meeting again on p. 128, tells us that:

He [Hobbes] was 40 yeares old before he looked on Geometry; which happened accidently. Being in a Gentleman's Library, *Euclid's Elements* lay open, and 'twas the 47*El. libri I.* He read the Proposition. By G—, sayd he, (he would now and then sweare an empaticall Oath by way of emphasis) *this is impossible!* So he reads the demonstration of it, which referred him back to such a Proposition; which proposition he read. That referred him back to another, which he also read. *Et sic deinceps* [and so on] that at last he was demonstratively convinced of that trueth. This made him in love with Geometry.

Aubrey was, of course, able to assume that any educated person would know that the 47th Proposition of the first book of Euclid was the famous theorem of Pythagoras. What a contrast to the situation today!

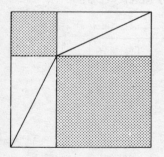 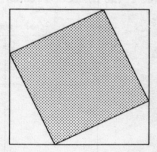

Figure 3.4.

Book II, with 14 propositions, is concerned, on the face of it, with geometrical algebra – with the 'application of areas' to prove algebraic identities such as

$$a^2 = (a - b)(a + b) + b^2$$

We must remember that the Greeks had no symbolic algebra and did not recognize irrational numbers: everything had to be proved geometrically. A number was thought of as a 'line segment', the product of two numbers as a rectangular area, and so forth. Euclid's approach may be illustrated by one example, Prop. II, 6, which is typical of the first 11 propositions of this short book. The algebraic identity to be established can be written as

$$(2a + b)b + a^2 = (a + b)^2$$

Figure 3.5 shows Euclid's construction. The two shaded rectangles have equal areas, from which the result follows at once. Propositions II, 12 and II, 13 generalize Pythagoras' theorem for obtuse- and acute-angled triangles. This generalization is effectively a 'line segment' version of the modern 'cosine formula' in the trigonometry of the triangle. The final proposition shows how to construct a square equal in area to a given rectangle (i.e. to find $c = \sqrt{ab}$), a construction we have already discussed on p. 24 in the context of square root extraction. Some historians interpret the Book II propositions in the context of a pre-Eudoxan treatment of incommensurable magnitudes. (We shall not pursue this contentious topic here; the interested reader is referred to Reference 31.)

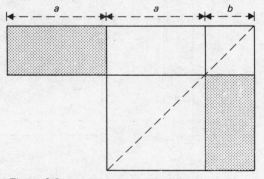

Figure 3.5.

Book III (37 propositions) treats the elementary properties of the circle and its associated chords, segments, angles, tangents and so on.

Book IV (16 propositions) is devoted to the construction of regular polygons, from the triangle to the hexagon, either inscribed in or circumscribed about a circle. We shall return to the construction of the regular pentagon in the next section. Most of the propositions of Books I and II, although probably not their proofs, are thought to be due to the Pythagoreans. The material of Books III and IV, on the other hand, is believed to be largely derived from Hippocrates.

Book V moves to an entirely new level of mathematical sophistication: it expounds the Eudoxan theory of proportion. The treatment is based on the three Eudoxan 'axioms' discussed in the last chapter; they appear here among the definitions. From them, 25 propositions about magnitudes and their ratios are rigorously proved. All the proofs are verbal and are, to our minds, extremely 'wordy'. Here are three examples of Book V propositions in modern algebraic language:

(1) Prop. 4. If $a:b = c:d$, then $ma:nb = mc:nd$.
(2) Prop. 11. If $a:b = c:d$ and $c:d = e:f$, then $a:b = e:f$.
(3) Prop. 17. If $a:b = c:d$, then $(a - b):b = (c - d):d$.

Book VI (33 propositions) treats similar figures, using the results obtained in Book V. The crucial theorem is the first, which states that the areas of triangles (or parallelograms) which have the same altitudes are in the same proportion to one another as their bases. Euclid's proof is a direct application of the Eudoxan criterion embodied in Definition V, 5 (p. 30). From this result, most of the similarity theorems can be proved fairly easily. (An alternative Eudoxan approach to similarity is presented in Appendix 1.) Comparison of the areas of similar polygons is treated in Props. 18–20, while many of the later propositions give geometrical methods for solving quadratic equations.

Books VII, VIII and IX are arithmetical in content, dealing with the properties of the natural numbers and their ratios. The subject is treated geometrically, as is the case throughout the *Elements*. A number is regarded as a line segment and the arguments are presented verbally, with no symbolic aids.

Book VII starts with 19 theorems on ratio and proportion which duplicate the results already proved in Book V, but the treatment in Book VII is restricted to numbers. To cite one example among many, Prop. V, 16 states that 'If four magnitudes be proportional, they will also be proportional alternately', i.e. if $a:b = c:d$, then $a:c = b:d$;

Prop. VII, 13 states that 'If four numbers are proportional, they will also be proportional alternately.' There has been much speculation as to the reasons for such duplication. Perhaps Euclid wished to emphasize the Eudoxan distinction between numbers and magnitudes; perhaps he was influenced by Aristotle, who stressed the dichotomy between the discrete and the continuous; perhaps he did not live long enough to revise his great treatise; perhaps more than one editorial hand was at work. We just do not know. The later propositions of this Book do, however, break new ground. They deal mainly with the primality properties of numbers – either absolute primes or relatively prime pairs (i.e. pairs with no common factors).

Book VIII (27 propositions) is essentially concerned with geometrical progressions, with sequences of numbers in what Euclid calls 'continued proportion', i.e. $a:b = c:d = e:f$. The final theorem states that 'Similar solid numbers have to one another the ratio which a cube number has to a cube number.' What this means is that $(ma.mb.mc):(na.nb.nc) = m^3:n^3$.

Book IX (27 propositions) is a curious mixture. It includes a very elementary and archaic treatment of the properties of odd and even numbers, which almost certainly goes back to the Pythagoreans. By contrast, some of the later propositions are of great importance, and most elegantly proved. Thus, for example, Prop. IX, 20 gives the well-known proof by contradiction that the number of primes is infinite, i.e. that there is no largest prime. We assume, first, that the theorem is false – that there *are* a finite number of primes, $2, 3, 5, 7, \ldots, P$, where P is the largest. Now consider the number $Q = (2 \times 3 \times 5 \times 7 \times \ldots \times P) + 1$. Clearly Q is not divisible by any of the primes up to P, so either Q is a new prime greater than P, or Q is composite and so must be divisible by some prime greater than P. In either case we have established a contradiction, so the initial assumption is false and there is no largest prime number. To convey the flavour of the Greek geometrical approach to a purely arithmetical theorem, we reproduce a translation of Euclid's proof of this proposition in Appendix 2.

Book X (115 propositions) is both the longest and the hardest to understand. It is concerned with the systematic classification and detailed study of certain types of incommensurable magnitudes involving square but not higher roots. Among the forms considered are $a \pm \sqrt{b}$, $\sqrt{a \pm \sqrt{b}}$, $\sqrt{a} \pm \sqrt{b}$ and $\sqrt{\sqrt{a} \pm \sqrt{b}}$, where a and b are commensurable magnitudes. As we have said, Book X is generally held to be based on the work of Theaetetus; its significance remains a contentious issue.

Book XI (39 propositions) introduces the subject of solid geometry. It begins with 28 definitions, of solid angle, the angle between two planes and so on, and such solid figures as the pyramid, prism, sphere, cone and the five regular polyhedra (or Platonic solids).

Book XII (18 propositions) treats the 'measurement' of such figures as circles, cylinders, pyramids, spheres and cones. Their surface areas and volumes are rigorously evaluated by combining the method of exhaustion with a double *reductio ad absurdum* argument. The typical Prop. XII, 2 has been discussed in the last chapter, and indeed all the theorems in this book are usually attributed to Eudoxus.

Book XIII (18 propositions) treats the properties of the five Platonic solids, the main properties of which are summarized in Table 3.1. In the final proposition of the *Elements*, Euclid proves that there are only five Platonic solids, the proof depending on an earlier theorem (Prop. XI, 21) that the sum of the 'face angles' at the vertex of a solid angle must add up to less than four right angles. In view of the mystical as well as the mathematical importance that the Greeks attached to these polyhedra, Book XIII may be seen as a fitting culmination of the whole work. In Plato's dialogue the *Timaeus*, we are told how the Pythagoreans associated the five solids with the basic elements of the Universe – earth, air, fire and water. A special role was allocated to the dodecahedron: 'God used it for the whole; for arranging the constellations on the whole heavens.'

4. The regular pentagon and the golden section

We have already remarked that a study of the properties of the regular pentagon could have led to the discovery of incommensurability, the ratio in this case being that of the diagonal of the pentagon to its side. In Figure 3.6 the two isosceles triangles ABP and DAC are similar, and so AP : AB = DC : DA. Now AP = EP, AB = PB, DC = PB and DA = EB, so we get EP : PB = PB : EB.

Table 3.1. The five regular polyhedra, or Platonic solids

	Tetrahedron	Cube	Octahedron	Dodecahedron	Icosahedron
Number of faces	4	6	8	12	20
Number of vertices	4	8	6	20	12
Number of edges	6	12	12	30	30
Number of edges surrounding each face	3	4	3	5	3
Number of edges meeting at each vertex	3	3	4	3	5
Value of R/r	3	$\sqrt{3} = 1.73\ldots$	$\sqrt{3} = 1.73\ldots$	$\sqrt{3(5-2\sqrt{5})}$ $= 1.26\ldots$	$\sqrt{3(5-2\sqrt{5})}$ $= 1.26\ldots$

R and r are the radii of the circumscribing and inscribed spheres.

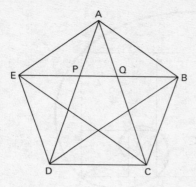

Figure 3.6.

This means that the diagonal AD divides the diagonal EB into two unequal segments (EP and PB) such that the ratio of the lesser segment to the greater is equal to the ratio of the greater segment to the whole diagonal. In fact, any diagonal of the pentagon divides any other intersecting diagonal in this way. This division was known to the Greeks as 'division of a line in mean and extreme ratio'. The modern term, *golden section*, came into use about the time that Kepler (1571–1630) penned the sentence quoted at the head of this chapter. The algebraic formulation of the section is $(a - x)/x = x/a$. This yields the quadratic equation $x^2 + ax = a^2$, whose positive solution is $x/a = \frac{1}{2}(\sqrt{5} - 1) = 0\cdot618\ldots$. The inverse ratio, a/x, is equal to $\frac{1}{2}(\sqrt{5} + 1) = 1\cdot618\ldots$ and may be denoted by g, the *golden number* or *golden ratio*. (Note that $1/g = g - 1$.)

Kepler was not alone in being fascinated by the many striking properties of the golden section; it turns up in the most diverse contexts. If, for example, we have a 'golden rectangle' of sides x and gx as in Figure 3.7, and we remove a square of side x as shown, the remaining rectangle has the same shape as the original one. The process can be continued indefinitely, giving an unending sequence of smaller and smaller 'golden rectangles'. Successive rectangles are obtained by a rotation through a right angle combined with a size reduction of $1:g$. In Figure 3.7 the points A, E and C lie on a straight line, as do the points B, F and D. The two lines intersect at right

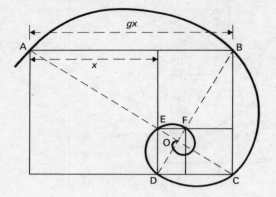

Figure 3.7.

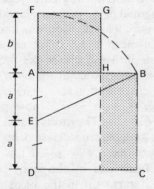

Figure 3.8.

angles at O. In fact, the points A, B, C, D, E and F lie on a curve known as a *logarithmic spiral*, with its 'pole' (or centre) at O.

Euclid shows how to divide a line in golden section in Prop. II, 11 and again in Prop. VI, 20 – another example of duplication. Figure 3.8 shows his construction. ABCD is a square of side $2a$, E is the mid-point of DA, which is extended to F so that $EB = EF = a + b$, and $AH = AF = b$. Euclid proves that $HB.AB = AH^2$, or $HB:AH = AH:AB$, so the point H divides AB in golden section. The first step in his proof is to invoke Prop. II, 6 (discussed in Section 3) to yield

$$FD.FG + AE^2 = EF^2 \quad \text{or} \quad (2a + b)b + a^2 = (a + b)^2$$

Now, since $EF = a + b = EB$ and $EB^2 = AE^2 + AB^2$, we have

$$(a + b)^2 = (2a + b)b + a^2 = a^2 + (2a)^2$$

so $(2a + b)b = 4a^2$, or $FD.FG = AB^2$. Subtracting the rectangle AD.AH from each side leaves

$$AH^2 = HB.BC = HB.AB$$

which was to be proved.

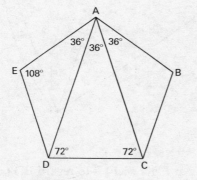

Figure 3.9.

We can now move to Euclid's construction of the regular pentagon (Prop. IV, 11). Figure 3.9 shows some of the angles associated with this figure. If we can construct the triangle ADC it is a simple matter to construct the complete pentagon. The construction of such a triangle (with angles 36°, 72°, 72°) is accomplished in Prop. IV, 10. The golden section proposition (II, 11) is used to construct a point C on a line AB such that $AB.CB = AC^2$, as shown in Figure 3.10. We then draw a circle with centre A and radius AB, mark off the chord BD such that $BD = AC$, join CD and AD, and construct the

circle (shown dashed) which passes through A, C and D. Now, since
$CB.AB = AC^2 = BD^2$, it follows (from a well-known property of
the circle, proved in Prop. III, 37) that BD is a tangent to the dashed
circle at D. Another circle theorem (Prop. III, 32) tells us that the
angles x and z are equal, so $v = x + u = z + u = y$ since the
triangle ABD is isosceles. Since $v = y$, the triangle BCD is isosceles.
This means that $CD = BD = AC$, so the triangle ACD is also
isosceles. Hence $x = u = z$, so each of the base angles of the triangle
ABD is twice the angle at A, and our problem is solved.

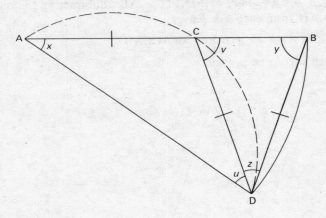

Figure 3.10.

Another way of looking at Figure 3.10 is to regard BD ($= s$, say)
as the side of a regular decagon (10-gon) inscribed in a circle of
radius AB ($= r$, say) and centre A. Then, since the triangle ABD is
similar to the triangle DCB, and $AC = CD = BD = s$, we have
$r/s = s/(r - s)$ – the golden section again – yielding $r/s = g = \frac{1}{2}(\sqrt{5} + 1)$. So we can construct a regular pentagon by first constructing a regular decagon. The last proposition in Book IV shows
how to inscribe a regular 15-gon in a circle.

How far back does knowledge of the golden section extend? There
is an obscure passage in Herodotus which can be interpreted as
saying that the Great Pyramids of Egypt were so constructed that the
area of one of the inclined faces is equal to the square of the height.

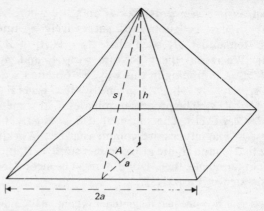

Figure 3.11.

Figure 3.11 represents a pyramid of side $2a$, height h and slant height s. The area of a face is as, so if $as = h^2$ it follows, since $h^2 = s^2 - a^2$, that $s(s - a) = a^2$ or $s/a = g$, the golden ratio. Now, s/a is the secant (p. 8) of the slant angle marked A. If we set $\sec A = g$, we find that $A = 51° 50'$. How does this deduction compare with the facts? The great Pyramid of Cheops at Giza was built around 4750 BC to a height of over 145 metres; it has a slant angle of 51° 52'. The first true pyramid, built slightly earlier at Medumi, has exactly the same slant angle. The other two large pyramids at Giza, built around 4600 BC, have slant angles of 53° 10' and 51° 10'. This close agreement may be no more than a coincidence, or it may be highly significant. The experts differ.

5. The Euclidean algorithm

Some historians argue that the early fourth-century Greeks (i.e. before Eudoxus) were in possession of a technique for handling incommensurable magnitudes, using the so-called *Euclidean algorithm*. Here again, we have a case where Euclid treats the subject twice over: in Props. VIII, 1 and 2 for numbers and in Props. X, 2, 3 and 4 for magnitudes. The procedure, when applied to magnitudes of the same kind, is as follows. Let the two magnitudes (we may think of them as line segments) be A_0 and A_1, where $A_1 < A_0$. Suppose that A_1 can be subtracted successively n_0 times from A_0 to give a remainder A_2 such that $A_0 = n_0 A_1 + A_2$, where $0 \leqslant A_2 < A_1$. We repeat the procedure with A_1 and A_2 to give $A_1 = n_1 A_2 + A_3$. Continuing in this way, we obtain a sequence of remainders $A_1, A_2, \ldots, A_i, \ldots$ given by the general formula $A_{i-2} = n_{i-2} A_{i-1} + A_i$. The process terminates if, at some stage, we get a remainder A_i of zero. This means that A_{i-1} exactly 'measures' A_{i-2}, i.e. the two magnitudes are commensurable. Working backwards, we see that A_0 and A_1 are also commensurable. If, though, the process does not terminate, then A_0 and A_1 must be incommensurable. This result is stated in Prop. X, 2:

If, when the lesser of two unequal magnitudes is continually subtracted in turn from the greater, that which is left over never measures the one before it, the magnitudes will be incommensurable.

When the algorithm is applied to two numbers, the process will always terminate: we eventually reach a remainder, A_i say, of zero. The previous remainder, A_{i-1}, is the largest common factor of A_0 and A_1. If $A_{i-1} = 1$, then A_0 and A_1 are relatively prime.

To see how the algorithm works, let us apply it to the numbers 59 and 16. We have

$$59 = \underline{3} \times 16 + 11$$
$$16 = \underline{1} \times 11 + 5$$
$$11 = \underline{2} \times 5 + 1$$
$$5 = \underline{5} \times 1 + 0$$

The ratio of 59 to 16 is *uniquely* defined by the four underlined numbers, which we may write, for brevity, as [3; 1, 2, 5]. The operation of the Euclidean algorithm can be encapsulated in a single expression; we can write

$$59/16 = 3 + \cfrac{1}{1 + \cfrac{1}{2 + \cfrac{1}{5}}}$$

Nowadays an expression of this form is called a simple *continued fraction*. The general form is

$$x = n_0 + \cfrac{1}{n_1 + \cfrac{1}{n_2 + \cfrac{1}{n_3 + \cfrac{1}{n_4 + \dots}}}}$$

where x is a positive real number and $n_0, n_1, n_2, \dots$ are positive integers. We may write this more succinctly as $[n_0; n_1, n_2, \dots]$. If x is irrational, the sequence of n's will not terminate. If, however, the irrational number is the solution of a quadratic equation with rational coefficients, the terms of the n-sequence will continually recur in the same order, in which case the continued fraction is said to be *periodic*. The simplest periodic continued fraction is given by the golden number, namely $g = \frac{1}{2}(\sqrt{5} + 1) = [1; 1, 1, 1, \dots]$ (This result follows at once from the fact that $g - 1 = 1/g$.) Here are three further examples:

$$\sqrt{2} = [1; 2, 2, 2, 2, \dots]$$
$$\sqrt{3}/\sqrt{2} = [1; 4, 2, 4, 2, \dots]$$
$$\sqrt{7} = [2; 1, 1, 1, 4, 1, 1, 1, 4, 1, \dots]$$

Truncating a continued fraction at successive terms of the n-sequence and working out the value of the resulting expressions gives a sequence of increasingly accurate rational approximations (alternately above and below) to the value of the real number x. These approximations are called *convergents*. Applying the process to $\sqrt{2}$, for example, yields the sequence of convergents 1, 3/2, 7/5, 17/12.

6. Rational approximations

The scarcity of primary source material has engendered a diversity of interpretations of early Greek mathematics and continues to provoke much erudite disputation. Indeed, many scholars do not accept the line of argument presented in the last section, with its emphasis on the Euclidean algorithm as a working tool. There is agreement on one point, however – that the Greeks were certainly able to compute rational approximations. The question at issue is *how* they did it. To focus our discussion of this complex matter, we shall concentrate on two well-known examples of such approximations from the third century BC. The first is due to the Alexandrian astronomer Aristarchus (*c*. 310–230 BC), remembered as one of the first to propose the heliocentric hypothesis – that the Earth and the other planets revolve about a fixed Sun. In one of his books, *On the Sizes and Distances of the Sun and Moon*, he states that: '7921 has to 4050 a ratio greater than that which 88 has to 45'. Now, the continued fraction for the former ratio is [1; 1, 21, 1, 1, 1, 2, 22]. Truncating at the fifth term yields the approximation [1; 1, 21, 1, 1] = 88/45, which we know to be an underestimate. In fact, however, there is a simpler explanation of this result. We have only to observe that

$$\frac{7921}{4050} > \frac{7920}{4050} = \frac{90 \times 88}{90 \times 45} = \frac{88}{45}$$

Were the Greeks able to cancel common factors in this way?

Our second example is Archimedes' statement, in his treatise *The Measurement of the Circle*, that (see p. 81) $265/153 < \sqrt{3} < 1351/780$. The continued fraction for $\sqrt{3}$ is [1; 1, 2, 1, 2, . . .]. The ninth convergent, an underestimate, is 265/153; the twelfth, an overestimate, is 1351/780. An alternative suggestion is that Archimedes used the fact that $\sqrt{27} = 3\sqrt{3} = [5; 5, 10, 5, 10, . . .]$, whose third and fourth convergents are 265/51 and 1351/260. To many, neither of these interpretations is very convincing, and there have been numerous conjectures as to how Archimedes obtained his result. It seems reasonable to look first at the Babylonian method of approximating to $\sqrt{2}$ described in Chapter 1. The method may be generalized as follows. If a is a first approximation to the square root of a given number A, we can write $A = a^2 + b$, where b is a remainder. Then a better approximation to $\sqrt{A}$ is given by $a \pm (b/2a)$. This formula is usually associated with the name of Heron (or Hero) of Alexandria, who lived sometime between 100 BC and AD 100, but it was undoubtedly known much earlier. The result can be extended to give upper and lower bounds thus:

$$a \pm b/2a > \sqrt{a^2 \pm b} > a \pm b/(2a \pm 1)$$

There is yet another approach that is worth looking at. Let us investigate the possible integral solutions (x, y) of the equation

$$x^2 - Ay^2 = B \qquad (3.1)$$

where B is a small integer, either positive or negative, and A is a positive integer (but not a perfect square). The procedure can readily be extended to deal with the case when A is any non-square, positive rational number. If x and y are large compared with B, it is apparent that x/y will yield a good approximation to $\sqrt{A}$. An important special case of equation (3.1) is when $B = 1$. The equation

$$x^2 - Ay^2 = 1 \qquad (3.2)$$

is known – incorrectly – as the 'Pell equation'. (John Pell was a minor seventeenth-century English mathematician who was mistakenly credited by Euler with having investigated this equation.)

We have the evidence of a cuneiform tablet that the algebraic identity

$$(ax + by)^2 - (bx + ay)^2 = (a^2 - b^2)(x^2 - y^2)$$

was known to the Babylonians. A simple extension yields the identity

$$(xx' + Ayy')^2 - A(xy' + x'y)^2 = (x^2 - Ay^2)(x'^2 - Ay'^2) \qquad (3.3)$$

If (x, y) is a solution of equation (3.1) and (x', y') is a solution of equation (3.2), then equation (3.3) tells us that

$$X = xx' + Ayy'; \qquad Y = xy' + x'y \qquad (3.4)$$

is another solution of equation (3.1). We can write equations (3.4) for brevity as

$$(X, Y) = (x, y).(x', y') \qquad (3.5)$$

It follows that if equations (3.1) and (3.2) both have integral solutions, then equation (3.1) has infinitely many. So also does equation (3.2), which is a particular case of equation (3.1). It turns out, however, that for some values of B equation (3.1) has no integral solutions. For example, if $A = 3$ there are no integral solutions of equation (3.1) when $B = 2$ or -1, but infinitely many when $B = 1$ or -2. These solutions provide sequences of rational upper and lower approximations to $\sqrt{3}$.

Solutions of equations (3.1) and (3.2) for small values of A, B and (x, y) can usually be obtained by inspection. Thus for $A = 3$ we have

$$2^2 - 3 \times 1^2 = 1 \quad \text{and} \quad 1^2 - 3 \times 1^2 = -2$$

giving 2/1 (or 2, 1) as an upper bound and 1/1 (or 1, 1) as a lower bound for $\sqrt{3}$. Successive applications of equations (3.4) or equation (3.5) yield increasingly closer approximations: for example,

$$(2, 1).(1, 1) = (5, 3) \qquad \text{(underestimate)}$$
$$(2, 1).(2, 1) = (7, 4) \qquad \text{(overestimate)}$$
$$(2, 1).(7, 4) = (26, 15) \qquad \text{(overestimate)} \qquad (3.6)$$
$$(5, 3).(26, 15) = (265, 153) \qquad \text{(underestimate)}$$
$$(26, 15).(26, 15) = (1351, 780) \qquad \text{(overestimate)}$$

We have obtained not only Archimedes' result, but perhaps also some insight as to why he used such large numbers.

Another feature of interest is that both (x, y) and (x', y') in equations (3.4) may be solutions of equation (3.2), in which case so also is (X, Y). The particular case when $(x, y) = (x', y')$, as in the second and fifth of equations (3.6), yields, from equations (3.4) or equation (3.5),

$$(X, Y) = (x^2 + Ay^2, 2xy) \tag{3.7}$$

This is an overestimate; a corresponding underestimate is

$$(AY, X) = (2Axy, x^2 + Ay^2) \tag{3.8}$$

Applying equation (3.7) to the approximation (2, 1) for $\sqrt{3}$ yields successive overestimates of (7, 4) and (97, 56) with corresponding underestimates of (12, 7) and (168, 97). The numbers soon become excessively large.

Examples of rational approximation are to be found in Greek writings from the fourth century BC onwards. It seems likely that a number of techniques were available at various times, but there is no agreement among scholars as to exactly what they were. The modern concepts of continued fractions and the Pell equation were almost certainly unknown in early Classical times; they have been introduced here for their mathematical interest and to clarify the exposition. We still have a great deal to learn – and there is much that we shall never be able to learn – about the mathematical achievements of this supremely talented people.

7. Apollonius: his life and works

Apollonius, known as 'the Great Geometer', was born in Perga in southern Italy in about 262 BC. He was probably educated at the Museum at Alexandria; he certainly taught there for a time until he moved to the important scientific centre of Pergamum. We know very little of his life, and most of his writings are lost, but we can get a good idea of the contents of some of them from later references.

In his *Plane Loci* Apollonius discussed, among much else, the *locus* of (i.e. the path traced out by) a point P, subject to certain constraints. If, for example, A and B are fixed points and the ratio of PA to PB is held constant, the locus of P turns out to be a circle, now known as the 'circle of Apollonius' (Figure 3.12). His treatise *On Contacts* deals with the famous 'Problem of Apollonius':

Given any three points, lines or circles, or any combination of three of them, construct a circle which passes through the given points and is tangential to the given lines and circles.

There are ten different cases to be considered. Euclid dealt with two of them in the *Elements*; Apollonius probably treated the other eight. Some of his solutions have been lost, but can be reconstructed from information given by the later commentators, notably Pappus of Alexandria (*c.* AD 284–345) – see Chapter 4. The most difficult case, to draw a circle touching three other circles, posed an irresistible challenge to the mathematicians of the sixteenth and seventeenth centuries.

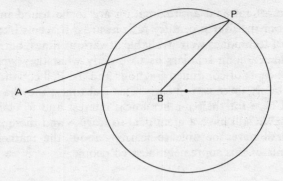

Figure 3.12.

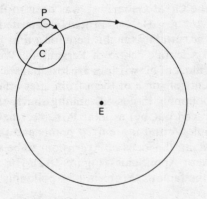

Figure 3.13.

Apollonius was equally distinguished as a geometer and as an astronomer. We have seen that Eudoxus used concentric spheres for his model of the Solar System; Apollonius used circles. In his model the planet P in Figure 3.13 is assumed to move uniformly in a small circle (the 'epicycle') with centre C, which in turn moves uniformly in a larger circle (the 'deferent') centred on the Earth, E. In the course of time the Apollonian circles superseded the Eudoxan spheres, and his epicyclic model was refined by the two greatest astronomers of the later Hellenistic age, Hipparchus in the first century BC and

Ptolemy in the second century AD. None of the astronomers of Antiquity attempted to model the changing distances of the planets, as revealed by changes in apparent brightness; they were concerned only with the angular positions of the heavenly bodies.

8. The *Conics* of Apollonius

Some of the lesser works of Apollonius have survived only in Arabic translations, but it is for his masterpiece, the *Conics*, that he is mainly remembered. This massive treatise is organized in eight books and contains no fewer than 487 propositions, all proved by the rigorous deductive methods characteristic of the Greek masters. The first four books have come down to us from Greek manuscripts of the twelfth and thirteenth centuries; the next three in Arabic translations only. The eighth book is lost, but was partially restored by Edmond Halley in the seventeenth century from indications by Pappus (see Chapter 10).

While the curves known as *conic sections* were known to the fourth-century mathematicians, Apollonius was the first to realize that the three types – *parabola*, *ellipse* and *hyperbola* – could be generated by taking sections of a single circular cone by varying the inclination of the cutting plane. By introducing the 'double napped' cone, he was able to generate the two branches of the hyperbola. Although Apollonius defined his curves as plane sections of a cone, he dispensed with the third dimension as soon as possible. This he did by proving a 'plane' property that is shared by all the conic sections, and on which he could build his vast edifice. We shall present this initial proposition for the ellipse; the other cases are similar. In Figure 3.14 ABC is a triangular section through the axis of a circular cone (which may be oblique), P is any point on an inclined section QPR of the cone, and QR is in the plane ABC, as shown. The lines QR and BC are extended to meet at T. A plane through P parallel to the circular base BC will cut the cone in a circle PDE and the plane QPR in the line PHP'. The line DHE is the diameter of the circle PDE that is perpendicular to PHP'. Then, by similar triangles, we have that $DH/QH = BT/QT$, and also $HE/HR = CT/RT$. Apollonius then uses a property of the circle, illustrated in

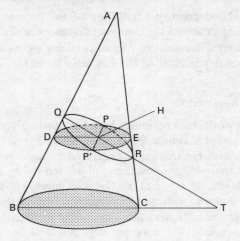

Figure 3.14.

Figure 2.10 on p. 24, namely that $PH^2 = DH.HE$. This yields

$$PH^2 = (QH.BT/QT).(HR.CT/RT) \qquad (3.9)$$

To see what this rather formidable expression means, let us introduce some modern notation (from coordinate geometry – see Chapter 7). Figure 3.15 shows the elliptic section QPRP′ of Figure 3.14. If we put $PH = y$, $QH = x$ and $QR = 2a$, equation (3.9) yields

$$y^2 = kx(2a - x) \qquad (3.10)$$

where the constant k incorporates all the line segments in equation (3.9) whose lengths are independent of the position of P on the ellipse. Now, the well-known equation of an ellipse with axes $2a$ and $2b$, the origin at Q, and the x-axis along QR is

$$\frac{(x - a)^2}{a^2} + \frac{y^2}{b^2} = 1$$

which we can write as

$$y^2 = \frac{2b^2 x}{a} - \frac{b^2 x^2}{a^2} = Lx - \frac{Lx^2}{2a} \qquad (3.11)$$

which clearly has the same form as equation (3.10). *L* is a parameter
of the ellipse known as the *latus rectum*; it is the line segment LSL′
in Figure 3.15, where S is a focus (see p. 63). Apollonius derives
similar results for the hyperbola, namely $y^2 = Lx + (Lx^2/2a)$, and
for the parabola, namely $y^2 = Lx$. For the parabola we may think
of *a* as becoming infinitely large.

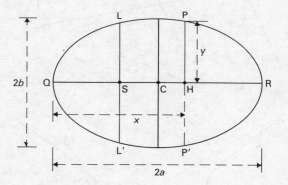

Figure 3.15.

These results provided Apollonius with the starting point for a
comprehensive study of the conic sections, regarded as plane curves.
The earlier books of the *Conics* contain much that was known to
earlier writers (e.g. to Euclid in his lost work on the subject), but new
results soon appear in abundance. In *Book I*, among much else,
Apollonius proves that for the central conics, the ellipse and the
hyperbola, the mid-points of all chords parallel to a diameter, LCM
in Figure 3.16, lie on another diameter, HCK. Since the roles of
the two diameters can be interchanged, they are called *conjugate
diameters*. Apollonius often uses a pair of such diameters as oblique
axes of coordinates. He also proved that a line, such as THT′ in
Figure 3.16, drawn through one end H of a diameter in a direction
parallel to its conjugate diameter is tangential to the conic at H. The
corresponding result for the parabola, which has only one axis of
symmetry, is that the locus of the mid-points of parallel chords is a
straight line parallel to the axis of the parabola.

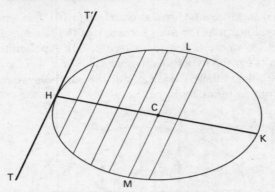

Figure 3.16.

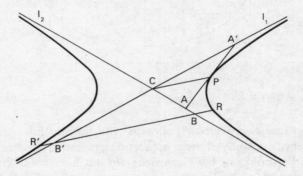

Figure 3.17.

In *Book II* Apollonius explores the distinctive properties of the hyperbola, with its two branches and its pair of *asymptotes*, l_1 and l_2, as illustrated in Figure 3.17. Here are two examples of his many results:

(1)   If the tangent at P cuts the asymptotes in A and A′, then AP = PA′.

(2)   If a chord parallel to CP cuts the asymptotes at B and B′ and the hyperbola at R and R′, then RB = R′B′ and RB.RB′ = CP².

In the latter part of Book II Apollonius investigates the 'tangent properties' of a conic and shows how to draw a tangent to a conic from an external point. Thus, in Figure 3.18, if P is a point on a

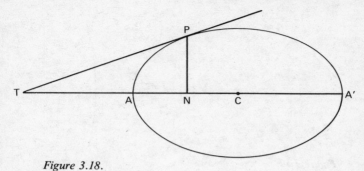

Figure 3.18.

central conic, PN is perpendicular to the axis AA', and the point T is chosen on the extended axis so that T divides AA' externally in the same ratio as N divides it internally, then TP is the tangent to the conic at P.

In *Book III* Apollonius embarks on a detailed investigation (more than 50 propositions) of a famous problem which has engaged the attention of many distinguished mathematicians from Apollonius to Newton. It is now usually known as the 'problem of Pappus', and it will be convenient to defer discussion of it until the next chapter.

Book IV is described by Apollonius as showing 'in how many ways the sections of cones meet one another'. He was fascinated by the double-branched nature of the hyperbola and by the variety of ways in which two hyperbolas can intersect. When taxed about the practical value of such results, he proudly replied:

They are worthy of acceptance for the sake of the demonstrations themselves, in the same way as we accept many other things in mathematics for this and for no other reason.

Book V takes us into new territory: it deals with the maximum or minimum lengths that can be drawn to a conic from various points. The Greeks did not have a very satisfactory definition of the tangent to a curve; presumably that is why Apollonius did not want to define the normal to a curve at a point P as the line through P perpendicular to the tangent at P. Instead he based his approach on the fact that the normal from a point R to a curve is the line through R which

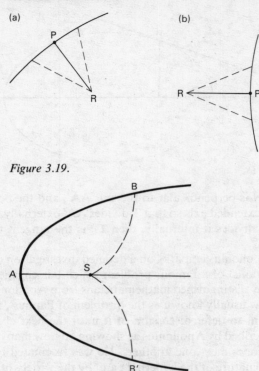

Figure 3.19.

Figure 3.20.

makes the distance from R to the curve (at P, say) a relative maximum or minimum in the sense indicated in Figures 3.19(a) and (b). The tangent at P can then be defined as the line through P perpendicular to PR. He investigates what happens when the point P is in various positions – on the major axis of the conic, for example – and then establishes criteria for deciding how many normals can be drawn from any given point to a conic. Thus in Figure 3.20 the dashed curve (called a *semi-cubical parabola*) separates the region to the right, from any point within which three normals can be drawn to the parabola, from the region to the left, from any point within which only one normal can be drawn. From any point on the dashed

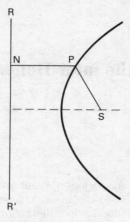

Figure 3.21.

curve itself, two normals can be drawn. There are corresponding results for the central conics.

Book VI treats congruent and similar conics and their segments. *Book VII* returns to the subject of conjugate diameters. *Book VIII*, as we have said, is lost.

It is perhaps surprising, in so comprehensive a treatise, to find no mention of some of the concepts that we regard as fundamental. Nowadays a conic is usually defined (see Figure 3.21) as the locus of a point P which moves so that the ratio of its distances from a fixed point S (the *focus*) and from a fixed line RR′ (the *directrix*) remains constant. We can write PS = e.PN, where e is called the *eccentricity*. If $e < 1$, the conic is an ellipse (a circle if $e = 0$); if $e = 1$, a parabola; if $e > 1$, a hyperbola. There is no mention of the focus/directrix property in the *Conics*; the two foci of a central conic are referred to only indirectly, the two directrices not at all. Once again, we must look to Pappus for the first mention of these terms. Of course, this in no way detracts from the magnitude of Apollonius' achievement. While his methods are entirely synthetic in the Classical Greek tradition, his approach to the subject is strikingly modern. The greatest geometer of Antiquity only just failed to create analytical geometry, for which the world had to wait until modern times.

4   Archimedes and the later Hellenistic period

'It is not possible to find in all geometry more difficult and intricate questions, or more simple and lucid explanations. Some ascribe this to his genius; while others think that incredible effort and toil produced these, to all appearances, easy and unlaboured results.'
Plutarch on Archimedes

1. Archimedes of Syracuse

The position of Archimedes as the most creative and original mathematician of Antiquity has never been in question – indeed, he is usually ranked with Newton and Gauss as one of the supreme mathematical geniuses of all time. He was born in about 287 BC at Syracuse in Sicily, the son of Phideas, the astronomer. He studied for some years at Alexandria, and kept in touch with the Museum when he returned to Syracuse to spend the rest of his days in his native city. We know little of his life, but much – perhaps too much – of the manner of his death in 212 BC. The Greek city of Syracuse had the misfortune to choose the wrong side in the Second Punic War; it was besieged by a Roman army under Marcellus and eventually captured. Archimedes was killed. The historical biographer Plutarch (first century AD), in the chapter 'The life of Marcellus' in his *Lives of the Noble Grecians and Romans*, gives no fewer than three versions of the tragic story. This is what he says:

Nothing afflicted Marcellus so much as the death of Archimedes, who was then, as fate would have it, intent on working out some problem by a diagram, and having fixed his mind alike and his eyes upon the subject of his speculation, he never noticed the incursion of the Romans, nor that the city was taken. In this transport of study and contemplation, a soldier, unexpectedly coming up to him, commanded him to follow to Marcellus; which he declined to do before he had worked out his problem to a demonstration. The soldier, enraged, drew his sword and ran him through. Others write that a Roman soldier, running upon him with a drawn sword, offered to kill him; and that Archimedes, looking back, earnestly besought him to hold his hand a little while that he might not leave what he was then at work upon inconclusive and imperfect: but the soldier, nothing moved by his entreaty, instantly killed him. Others again relate that, as Archimedes was carrying to Marcellus mathematical instruments, dials, spheres, and angles, by which the magnitude of the Sun might be measured to the sight, some soldiers seeing him, and thinking that he carried gold in a vessel, slew him. Certain it is that his death was very afflicting to Marcellus; and that Marcellus ever after regarded him that killed him as a murderer; and that he sought for his kindred and honoured them with signal favours.

Plutarch gives several examples of Archimedes' absent-mindedness (always a popular characteristic of a philosopher). It is, however, to the Roman architect Vitruvius (first century BC) that we owe the story – so attractive to schoolchildren – that when Archimedes discovered the laws of buoyancy in a flash of intuition while sitting in his bath, he ran naked through the streets of Syracuse shouting 'Eureka, eureka' ('I have found it'). This insight enabled him to confirm King Hiero's suspicion that one of his goldsmiths had defrauded him by introducing some silver into a gold crown that he had been commanded to make.

Archimedes' range of interests embraced astronomy, hydraulics, mechanics and general engineering. Indeed, it was for his practical achievements that he was best known to the Syracusian public. He invented the compound pulley and the Archimedean screw – still in use – for raising water; he understood the power of leverage ('Give me a place to stand, and I shall move the Earth'); and he devised a variety of fearsome engines of war for the defence of his beleaguered city. Plutarch gives a vivid account of the state of affairs during the siege:

For the truth was that all the rest of the Syracusians merely provided the manpower to operate Archimedes' inventions, and it was his mind which directed and controlled every manoeuvre. All other weapons were discarded, and it was upon him alone that the city relied for both attack and defence. At last the Romans were reduced to such a state of alarm that if they saw so much as a length of rope or a piece of timber appear over the top of the wall, it was enough to make them cry out, 'Look, Archimedes is aiming one of his machines at us!' and they would turn their backs and run. When Marcellus saw this, he abandoned all attempts to capture the city by assault, and settled down to reduce it by blockade.

In spite of his public services, Archimedes remained firmly within the Greek philosophical tradition in upholding the primacy of abstract thought. It was on his mathematical discoveries and writings that he lavished his affection and devotion. His treatises are unsurpassed in their elegance, lucidity and depth of mathematical content. In one of his prefaces he refers to his habit of sending some of his results to his friends in Alexandria, but without proofs so that they might have the pleasure of discovering them for themselves. He had, however, been annoyed to find that some had adopted his theorems as their own, without bothering to prove them. So, in his last communication, he tells us how he had included two theorems which were false so that 'those who claim to discover everything, but produce no proofs of the same, may be confuted as having actually pretended to discover the impossible.'

2. Archimedes' works

While much of Euclid's *Elements* (and, to a lesser extent, Apollonius' *Conics*) are compilations or extensions of earlier results, every one of Archimedes' works is an original contribution to mathematical knowledge. His writings that are preserved in Greek texts are, in approximate chronological order: *On the Equilibrium of Plane Figures* (2 books), *Quadrature of the Parabola*, *On the Sphere and Cylinder* (2 books), *On Spirals*, *On Conoids and Spheroids*, *On Floating Bodies* (2 books), *The Measurement of the Circle*, *The Sandreckoner* and *The Method*. There are also two treatises and some fragments that have been preserved only in Arabic translations. We know that several of his works are lost. These include two more books on mechanics, *On Levers* and *On Centres of Gravity*; a treatise on optics; and a work entitled *On Sphere-making*, which is believed to describe one of his inventions for simulating the motions of the

Sun, Moon and planets round the Earth. We shall be looking at some of his writings in more detail in later sections.

Archimedes' treatises on mechanics and hydrostatics undoubtedly establish him as the 'Father of Mathematical Physics'. Much of his treatment is severely mathematical; for example, *On Floating Bodies* (*Book II*) contains an exhaustive discussion of the stability and positions of equilibrium of segments of paraboloids of revolution when immersed or floating in fluids in various positions. Several of his books deal with *mensuration*, i.e. with the evaluation of the areas, volumes, centres of gravity, etc., of plane and solid figures. His proofs are indirect, by the method of contradiction; they use an 'exhaustion' procedure followed by a double *reductio ad absurdum* argument. The approach is the same as that taken by Eudoxus and Euclid, as presented in Chapter 2, but Archimedes seldom fails to inject some elegant niceties of his own. All we can do here is to introduce a few topics from his wide-ranging writings.

3. *Quadrature of the Parabola*

One of the main theorems of this early work, which contains 24 propositions, is that the area S of the parabolic segment bounded by the chord AC in Figure 4.1 is equal to four-thirds of the area of the inscribed triangle ABC of greatest area. This is the triangle with base AC, whose vertex B is the point on the segment at the greatest distance from AC. Although the conic sections had been known for at least 100 years, Archimedes was the first to evaluate a curvilinear area associated with a conic. He accomplished a first proof of the theorem in Prop. 16, but was not satisfied with it and developed a second proof in Props. 18–24. He first shows (see Figure 4.1) that the area T of the triangle ABC is equal to four times the sum of the areas of the corresponding 'greatest area' triangles on the bases AB and BC. These triangles, ADB and BEC, are shown shaded in the figure. In the course of his argument he proves that the tangent at B is parallel to AC; that BV (where AV = VC) is a diameter bisecting all chords parallel to AC; and that DV' and EV'' play the same role with respect to the triangles ADB and BEC as BV does for the triangle ABC. (The *Conics* of Apollonius had not yet appeared.) Continuing

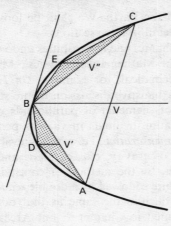

Figure 4.1.

in this way, Archimedes is able to 'exhaust' the parabolic area S by a sequence of triangles, and so prove that, as we would now put it, the area of the segment is given by the sum of the infinite series

$$T + \frac{1}{4}T + \frac{1}{4^2}T + \ldots + \frac{1}{4^n}T + \ldots$$

We know that this sum is equal to $\frac{4}{3}T$, but Archimedes, as a good Greek, could make no use of infinite series. Instead he first proves the exact result that

$$T + \frac{1}{4}T + \frac{1}{4^2}T + \ldots + \frac{1}{4^n}T + \frac{1}{3}\left(\frac{1}{4^n}T\right) = \frac{4}{3}T$$

and then uses a double *reductio ad absurdum* argument to prove that S can be neither greater nor less than $\frac{4}{3}T$.

Archimedes was not able to find the area of a segment of an ellipse or a hyperbola, although he did find the area of a complete ellipse. The reason he failed is that, while the quadrature of a parabola is (in modern terms) an algebraic problem involving polynomials, quadrature of the central conics involves transcendental functions. (Such functions, known as 'elliptic integrals', were first studied in the eighteenth century.)

4. *On the Sphere and Cylinder*

One of Archimedes' favourite works seems to have been the first of his two books dealing with the mensuration of the sphere; at any rate, he asked for a representation of one of his best-known theorems to be carved on his tombstone. His wish was granted. Many years later – in about 75 BC – Cicero found the tomb in a neglected state and restored it. Here is the theorem. If we rotate the plane figure shown in Figure 4.2 (where AD = BD = DC) about AD, we generate a cone inscribed in a hemisphere, which is itself inscribed in a cylinder. The volumes of these three solids are in the ratio 1 : 2 : 3. Now, Eudoxus had already proved, as we noted in Chapter 2, that the volume of the cone is one-third that of the cylinder. Archimedes had to deal with the sphere. In fact he proved two things: first, that the volume of the sphere is two-thirds that of the cylinder; and secondly, that the surface area of the sphere is also two-thirds that of the entire cylinder. Thus in Prop. 33 he proves that 'the surface of any sphere is four times the area of the greatest circle on it' (i.e. $S = 4\pi r^2$). In the next proposition he completes the proof of the double result (the 'tombstone theorem') connecting the sphere and its circumscribing cylinder. Later in the work, in Props. 42 and 43, Archimedes proves that the surface area of a 'spherical cap', namely PABCD in Figure 4.3, is equal to the area of a circle of radius AP, where P is the 'pole' of the cap.

The second book deals with the properties of spherical segments, i.e. the solids enclosed between two parallel planes cutting a sphere. Once again, the theorems on volumes or areas are rigorously proved

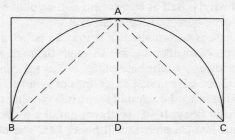

Figure 4.2.

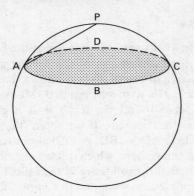

Figure 4.3.

by the indirect Eudoxan method. We shall return to the 'tombstone theorem' in Section 7.

5. *On Conoids and Spheroids*

This work, containing 32 propositions, treats the properties of solids of revolution generated by conics. Archimedes uses the term *conoid* to mean a paraboloid or hyperboloid of revolution; a *spheroid* is generated when an ellipse rotates about either its major or its minor axis. To illustrate the elegance of the Archimedean approach, we shall look at one theorem in some detail. Proposition 21 states that: 'The volume of any segment of a paraboloid of revolution [we will call it a conoid for brevity] is half as large again as the volume of the cone which has the same base and the same axis.'

In Figure 4.4 the base is shown shaded; it may be either a circle or an ellipse. Let us now consider a plane through the axis, AS, of the conoid and perpendicular to the base plane. It cuts the conoid in a parabola BPAC and the base in one of its lines of symmetry, BC. The line EF is the tangent to the parabola which is parallel to BC, P being the point of tangency. If PD is drawn parallel to the axis, then, as we saw in the last chapter, D will bisect BC. The oblique cone referred to in the theorem has vertex P and axis PD. Now, we

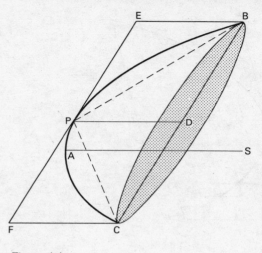

Figure 4.4.

know that the volume of a cone is one-third that of a cylinder on the same base and of the same height. It will be sufficient, therefore, to prove that the volume, $2X$ say, of such a cylinder (let us call it the 'enveloping cylinder') is twice that of the conoid; i.e. that the volume, C, of the conoid is equal to X.

Let us now suppose that the conoid and its enveloping cylinder are sliced into a large number, n, of thin sections by equally spaced cutting planes, all parallel to the base. About each section we inscribe and circumscribe thin cylindrical discs, as shown in Figure 4.5. We now have four solids: the conoid of volume C; the inscribed 'stepped' solid of volume I_n, say; the circumscribing 'stepped' solid of volume C_n; and the enveloping cylinder of volume $2X$. Then, with an obvious notation for the slices, we have

$$I_n = i_1 + i_2 + i_3 + \ldots + i_n$$
$$C_n = c_1 + c_2 + c_3 + \ldots + c_n$$

where $i_1 = 0, i_2 = c_1, i_3 = c_2, \ldots, i_n = c_{n-1}$. This gives

$$C_n - I_n = c_n$$

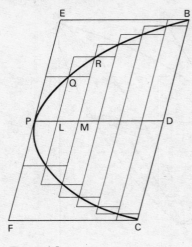

Figure 4.5.

By increasing the number of slices, c_n, and hence $C_n - I_n$, can be made as small as desired. We wish to prove, by contradiction, that $C = X$. There are three possibilities. Let us assume first that $C > X$. Then, by a previous theorem (Prop. 19), we can find a value of n such that

$$C_n - I_n < C - X \quad \text{or} \quad C_n - C < I_n - X$$

Clearly, $C_n > C$, so $I_n > X$.

To Archimedes, a parabola is a curve in which the squares of the ordinates $LQ, MR, \ldots, DB$ in Figure 4.5 – in this case, the radii of the discs, $r_1, r_2, \ldots, r_n$, say – are proportional to the abscissae $PL, PM, \ldots, PD$ – or $h, 2h, \ldots, nh$. Thus we have

$$\frac{r_1^2}{r_n^2} = \frac{h}{nh}, \quad \frac{r_2^2}{r_n^2} = \frac{2h}{nh}, \quad \ldots, \quad \frac{r_{n-1}^2}{r_n^2} = \frac{(n-1)h}{nh}$$

Now

$$2X = n\pi r_n^2 h$$

and

$$I_n = \pi r_1^2 h + \pi r_2^2 h + \ldots + \pi r_{n-1}^2 h$$

Hence

$$\frac{2X}{I_n} = \frac{nr_n^2}{r_1^2 + r_2^2 + \ldots + r_{n-1}^2}$$

since the radius of the base disc of the inscribed 'stepped' solid is r_{n-1}. This yields

$$\frac{2X}{I_n} = \frac{n(nh)}{h + 2h + \ldots + (n-1)h} = \frac{n^2}{\frac{1}{2}n(n-1)} = \frac{2n}{(n-1)} > 2$$

and so $I_n < X$. But we have already established that $I_n > X$, so there is a contradiction. Our assumption that $C > X$ must, therefore, be false. The contrary assumption that $C < X$ also leads, by a similar argument, to a contradiction. We must therefore conclude that $C = X$, which completes the proof.

Once again, we have introduced algebraic symbolism to make the argument easier to follow. Although Archimedes' demonstrations are models of elegance and precision, many of them are tiresomely lengthy and complicated. The inevitable reaction against such classical perfectionism set in at the beginning of the modern era, as we shall see in Chapter 7.

6. The Archimedean spiral

In his treatise *On Spirals* (with 28 propositions), Archimedes investigates the properties of the curve which bears his name. It is defined as the locus of a point P (Figure 4.6) which, starting from an origin O, moves uniformly along the length of a 'half-line' OP, which itself rotates uniformly about O from an initial line OA. Setting OP = r and the angle POA = θ, we have $r = a\theta$ (the *polar equation* of the spiral), since both r and θ are proportional to the time of rotation. With the aid of this curve it is a simple matter to trisect the angle θ. We trisect OP at Q and R, and draw circles of radii OQ and OR, as in Figure 4.6, which intersect the spiral at Q′ and R′. Then the angles R′OA and Q′OA are $\frac{2}{3}\theta$ and $\frac{1}{3}\theta$, respectively.

Archimedes also proved a number of theorems about the area swept out by the spiral. For example, Prop. 24 states that: 'The area bounded by the first turn of the spiral and the initial line [shown

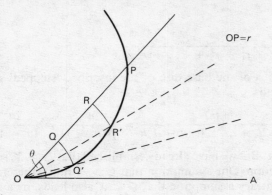

Figure 4.6.

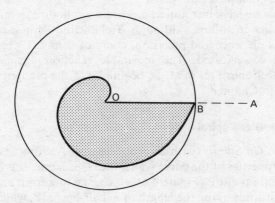

Figure 4.7.

shaded in Figure 4.7] is equal to one-third of the area of the first circle.' The 'first circle' is the circle with radius OB. The shaded area in Figure 4.7 is therefore $\frac{1}{3}\pi.\text{OB}^2 = \frac{4}{3}\pi^3 a^2$, since $\text{OB} = 2\pi a$.

Another of his results is illustrated in Figure 4.8. PQ is the tangent to the spiral at P; Q is the point of intersection of PQ and a line through O perpendicular to OP. Archimedes proved that the length OQ is equal to the circular arc PS, with centre O, which is equivalent to saying that the areas of the triangle OPQ and the sector OPS are

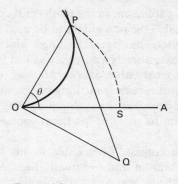

Figure 4.8.

equal. This result has three features of interest. First, it seems likely that Archimedes arrived at it by considering the motion of P as the resultant of a linear and a circular motion. Secondly, the introduction of the tangent foreshadows the 'differential calculus', whereas most of his work deals with problems that are now solved using the techniques of the 'integral calculus'. (These terms will be explained in Chapter 7.) Thirdly, the theorem provides a method of squaring the circle. Thus, if we take θ to be a right angle, we get a triangle and a quarter-circle of equal area. The Archimedean spiral, like the quadratrix, can be used to solve two of the three 'classical' problems.

7. *The Method*

One of the problems posed by the use of 'exhaustion' proofs is that one needs to have a good idea of the answer before the proof can be constructed. The polished, rigorously argued treatises of Archimedes give no hint of how his discoveries were actually made. The matter remained a mystery until the beginning of this century. The works of Archimedes, like those of Euclid, were definitively edited and published by J. L. Heiberg (p. 15). In 1906 his lifetime of devoted scholarship and unremitting labour was fittingly rewarded by the dramatic discovery, in the library of a monastery in Constantinople, of a lost work by Archimedes, now known as *The Method*. The

Greek text had been written on parchment in the tenth century, and had then been washed off to enable the parchment to be used again for devotional writings. Fortunately the original writing – which also contained some Archimedean texts already known – was just visible to Heiberg's expert eyes. The great value of *The Method*, which contains 14 propositions, is that it reveals something of the 'discovery methods' of Archimedes: it describes the preliminary 'mechanical' investigations which led him to some of his finest geometrical results.

The preface to *The Method* consists of a letter to his friend Eratosthenes of Cyrene, the librarian and eventually head of the Museum at Alexandria. Eratosthenes was a remarkable all-rounder in the best Greek tradition. He was renowned for the breadth of his knowledge, and for his excellence as a poet, literary critic, geographer and mathematician. He is probably best known today for his computation of the circumference of the Earth by measuring shadow lengths at two places on the same line of longitude. Here are two extracts from the letter:

I thought fit to write out for you and explain in detail in the same book the peculiarity of a certain method, by which it will be possible for you to get a start to enable you to investigate some of the problems in mathematics by means of mechanics. This procedure is, I am persuaded, no less useful even for the proof of the theorems themselves; for certain things first became clear to me by a mechanical method, although they had to be demonstrated by geometry afterwards because their investigation by the said method did not furnish an actual demonstration. But it is of course easier, when we have previously acquired, by the method, some knowledge of the questions, to supply the proof than it is to find it without any previous knowledge.

I deem it necessary to expound the method partly because I have already spoken of it and do not want to be thought to have uttered vain words, but equally because I am persuaded that it will be of no little service to mathematics; for I apprehend that some, either of my contemporaries or of my successors, will, by means of the method when once established, be able to discover other theorems in addition, which have not yet occurred to me.

Archimedes first applies his method to the quadrature of the parabola; indeed he said as much in his preface to that early work. We have mentioned in Section 3 that he gave a second proof of the theorem because the first, which used a 'mechanical' argument on the lines expounded in *The Method*, did not satisfy him. He then applies this approach to the 'tombstone theorem' discussed in Section 4. He formulates the double result thus:

Any sphere is (in respect of solid content) four times the cone with base equal to a great circle of the sphere and height equal to its radius; and the cylinder with base equal to a great circle of the sphere and height equal to the diameter is one-and-a-half times the sphere.

Let us see how Archimedes tackles this problem in Prop. 2 of *The Method*. He uses a physical argument based on the principle of the lever, which he had established in his treatise *On the Equilibrium of Plane Figures*. In Figure 4.9, ABCD is a circle with centre O and radius r, and X is an arbitrary point on the diameter AC. Other straight lines are drawn as shown, so that EC = CF = AC = $2r$. PQ is a line perpendicular to AC through the arbitrary point X on AC. Let us set AX = TX = x and YX = y. Then

$$AY^2 = x^2 + y^2 = 2rx \tag{4.1}$$

since AY and YC are perpendicular, and we have three similar triangles. CA is then extended to W as shown, so that WA = AC = $2r$. The entire figure is now rotated about the line WAC, so generating a sphere, a cone and a cylinder. The arbitrary line PQ sweeps out a plane which intersects the cylinder in a circle, S say, of radius $2r$; the cone in a circle C_1 of radius x; and the sphere in a circle C_2 of radius y. Then equation (4.1) yields

$$\frac{\pi x^2 + \pi y^2}{\pi(2r)^2} = \frac{x}{2r} = \frac{\text{sum of areas of circles } C_1 \text{ and } C_2}{\text{area of circle S}} \tag{4.2}$$

Archimedes now treats WAC as a lever pivoted at A, and interprets equation (4.2) as an equilibrium condition. If we imagine circular discs whose weights are proportional to their areas, then the circles C_1 and C_2 suspended from W would balance the circle S suspended from X on a lever with fulcrum at A. Furthermore, this condition is satisfied for *any* 'interior' position of the line PQ. Archimedes envisaged the cylinder, sphere and cone as being made up of the circles cut in these solids by every possible plane swept out by PQ.

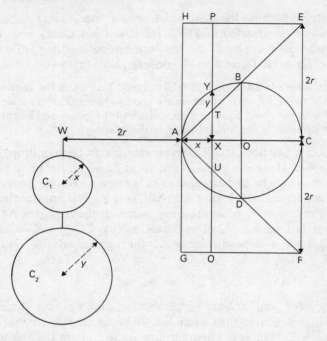

Figure 4.9.

If we now reassemble the circles C_1 and C_2 to form the cone and the sphere and suspend them from W, as indicated in Figure 4.9, they will balance the cylinder in its original position about A as a fulcrum. Since the centre of gravity of the cylinder is at O, we have

$$\frac{\text{volume of sphere} + \text{volume of cone}}{\text{volume of cylinder}} = \frac{\text{AO}}{\text{WA}} = \frac{1}{2}$$

Using the Eudoxan result that the volume of the cylinder HEFG is three times the volume of the cone EAF, we get

$$2 \times \text{volume of sphere} = \text{volume of cone EAF}$$
$$= 8 \times \text{volume of cone BAD}$$

so the volume of the sphere is four times the volume of the cone BAD. This is the first of Archimedes' two results. The second, that the volume of the sphere is two-thirds that of the circumscribing cylinder, follows at once.

What Archimedes is saying here is that the assumption that areas (or volumes) can be envisaged as the aggregate of a large number of line segments (or areas), while useful for discovering theorems, is not acceptable in a rigorous proof. The question was raised again in the seventeenth century when an integration procedure known as the 'method of indivisibles' was put forward – and widely accepted – as a means of short-circuiting the lengthy classical demonstrations. The existence of *The Method* was, of course, unknown at that time.

8. Angle trisection by *neusis*

In his *Book of Lemmas* Archimedes gives a second method of trisecting an angle. Figure 4.10 shows his construction, where AOB is the angle to be trisected. The curve ABC is a semicircle with centre O and radius r. A line PQB is then drawn through B so that P is on the extension of AC, Q lies on the semicircle as shown and the length PQ is equal to r. Then the angle BPA is one-third of the angle BOA. The crucial operation is that of placing a line (of unlimited extent) so that it passes through a fixed point (B in this case), and the ends (P and Q) of a segment of the line of predetermined length (r in this case) lie on two given curves (a straight line and a circle in this case). To do this we have to use a *marked* straight-edge with two marks a prescribed distance apart – not an acceptable Euclidean procedure. This type of construction, known to the Greeks as *neusis*, was used by several mathematicians of the Hellenistic period

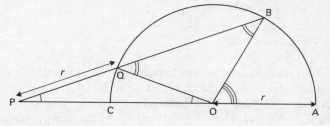

Figure 4.10.

and, indeed, probably earlier. The Greek word itself means 'verging': the line always 'verges towards' (i.e. points to) a given point.

9. Construction of the regular heptagon

About 70 years ago a manuscript was discovered which turned out to be an Arabic translation by Tabit ibn Qorrah of Baghdad (834–901) of a hitherto unknown work by Archimedes. It contains 16 propositions, culminating in the construction of a regular heptagon (7 sides). In Figure 4.11, ABCD is a square with diagonal AC. Here again, we have a neusis; in this case the construction of a line through D that meets AB extended to X so that the areas of the two shaded triangles are equal. Let AC and DX intersect at K, and the line YKL be drawn as shown. The next step is to construct the point E ('above' the square) so that EY = AY and EB = BX. If we now construct the circle through A, E and X, then AE is a side of a regular heptagon inscribed in this circle.

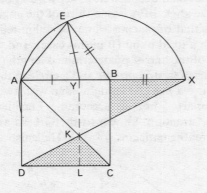

Figure 4.11.

This remarkable result appears as if plucked from the air. Archimedes gives no hint as to how he hit on the construction; we have no '*Method*' information in this case. Regrettably, Archimedes' elegant proof is too long to be presented here (but see Reference 11).

10. Numerical studies

The Measurement of the Circle contains only 3 propositions; it may be incomplete. The first states that the area A of a circle of radius r is equal to that of a triangle of base c, the circumference of the circle, and height r (i.e. $A = \frac{1}{2}rc$). The proof is, once again, by contradiction. It follows that $A : r^2 = c : 2r$; we now call this common ratio π. In Prop. 2 Archimedes shows that if we assume the familiar value of $3\frac{1}{7}$ for π, the area of a circle is to the square on its diameter as 11 is to 14. However, Prop. 3 is the most interesting. It states that the ratio of the circumference of any circle to its diameter is less than $3\frac{1}{7}$ but greater then $3\frac{10}{71}$. Archimedes established these limits by computing the lengths of a sequence of inscribed and circumscribed regular polygons of up to 96 (6×2^4) sides. The calculations are complex, and no working is shown. It is here that Archimedes makes use of the two approximations to $\sqrt{3}$ that we discussed in the last chapter, but he provides no clue as to how he obtained them.

In his late work, *The Sand-reckoner*, Archimedes devised a scheme for writing very large numbers which foreshadows our index notation. He estimated that about 10^{63} grains of sand would be needed to fill the whole Universe. This number he describes as ten million units of the eighth order of numbers. Each order contains 100 million (10^8) numbers or, in his terminology, a *myriad myriad*. (A Greek *myriad* was 10 000, or 10^4.) Numbers of the first order go from 1 to $10^8 - 1$; numbers of the second order from 10^8 to $(10^8)^2 - 1$; and so on. Numbers of the eighth order start at $(10^8)^7$ and Archimedes thought of 10^{63} as $10^7 \times (10^8)^7$. He even extended his system to deal with still larger numbers. All numbers of order up to 10^8 are said to be of the first period. Numbers of the second period would therefore begin with the number $(10^8)^{10^8}$. There is no need to go on!

We may mention, finally, that Archimedes seems to have understood the principle that led eventually (in the seventeenth century) to the invention of *logarithms*: namely, that to find the product of two numbers we add their exponents – or, as he would say, their 'orders'.

11. The later Hellenistic age

Mathematics occupied a central place in Hellenistic culture, which was – in contrast to that of the Classical age – predominantly practical and pragmatic. Over the centuries this led to a marked change in the direction in which mathematics developed. More attention was paid to applications – not only to astronomy, but also to optics, mechanics, hydraulics and geography. Numerical calculations were no longer considered to be unworthy of the attention of an academic mathematician. Geometry became increasingly concerned with problems of mensuration: with obtaining results useful in evaluating lengths, areas and volumes. We have seen how here, as in much else, Archimedes pointed the way forward. Thus, for example, while Euclid was content to prove that (in our notation) the area of a circle could be expressed as kr^2, Archimedes obtained a close approximation to the numerical value of k, so that circular areas could be accurately computed.

This prompts the question of how the Greeks wrote their numbers during the Hellenistic period. The Greeks of the Classical age used the so-called Attic system of numeration, which resembled the primitive Egyptian number-language described in Chapter 1. It was eventually superseded by a new system known as the Ionic or Alexandrian. This system was still non-positional, but used 27 different numerals – the 24 letters of the Greek alphabet, supplemented by 3 obsolete letters – to represent the numbers $1, 2, 3, \ldots, 9$; $10, 20, 30, \ldots, 90$; and $100, 200, 300, \ldots, 900$. Intermediate numbers were written by combining 2 or 3 numerals: thus 149 was written as $\rho\mu\theta$ (the numerals for 100, 40 and 9); 109 as $\rho\theta$. Extra markings were used to denote numbers of 1000 or more. The scheme was reasonably serviceable for integers, but quite inadequate for fractions; various notations were used and there was a good deal of confusion. That is why the Babylonian place-value system was retained by most Hellenistic astronomers, as we shall see in the next section.

The Golden Age of Greek mathematics came to an end with the death of Archimedes in 212 BC, followed by that of Apollonius a few years later. The next four centuries saw substantial advances in astronomy and mechanics, but little of major importance in 'classical' mathematics. The main achievement during this time was the creation of trigonometry by Hipparchus, Menelaus and Ptolemy. From about 200 BC to AD 400 the dominant power in the Western world was Rome – first as an Italian republic and then as a far-flung empire. During all this time the Roman contribution to mathematics was virtually nil. In the words of Alfred North Whitehead, when commenting on the death of Archimedes:

The Romans were a great race, but they were cursed by the sterility which waits upon practicality. They were not dreamers enough to arrive at new points of view, which could give more fundamental control over the forces of nature. No Roman lost his life because he was absorbed in the contemplation of a mathematical diagram.

12. Ptolemy of Alexandria

Claudius Ptolemy, who lived in the second century AD, is the author of the famous treatise best known to us by its Arabic title of the *Almagest* ('the best'). This work played much the same role for mathematical astronomy as Euclid's *Elements* did for elementary geometry. One of Ptolemy's many achievements was to construct a *table of chords* belonging to different angles at $\frac{1}{2}°$ intervals. This is equivalent to a modern sine table. The relation between Ptolemy's *chord* and our *sine* (introduced by the Hindu astronomers) is, as shown in Figure 4.12, that the chord AB of angle θ is given by $2R \sin \frac{1}{2}\theta$. Ptolemy took $R = 60$; like most Hellenistic astronomers, he used the Babylonian number system described in Chapter 1. In computing his table, he made much use of a geometrical result still known as 'Ptolemy's theorem'. It states that, if ABCD is a quadrilateral inscribed in a circle (see Figure 4.13), then AB.CD + BC.DA = AC.BD. (The easiest way to prove this result is to draw AP so that the marked angles are equal, and to exploit the properties of two pairs of similar triangles.) By examining particular cases of the theorem (e.g. where DC is a diameter), Ptolemy was able to derive most of the trigonometrical formulae he needed to build up his table from suitable starting values. (He used formulae such as, in

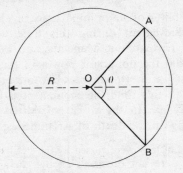

Figure 4.12.

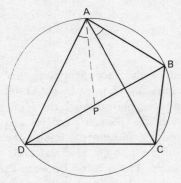

Figure 4.13.

modern notation, $\sin(A \pm B) = \sin A.\cos B \pm \cos A.\sin B$.) The first two entries were chord $\frac{1}{2}° = 0; 31, 25$ and chord $1° = 1; 2, 50$: the former is equivalent to $\sin 15' = 0.004\,361$.

The decline of Greek mathematics during the early Christian era was arrested during the century from 250 to 350 AD (sometimes known as the Greek Silver Age). It was dominated by two very dissimilar figures: Diophantus, known as the 'Father of Higher Arithmetic' (or, as we would now say, of the theory of numbers); and Pappus, the last significant geometer in the classical tradition.

13. Diophantus of Alexandria

As with many others, of the life of Diophantus we know almost nothing, although there is some evidence that he lived to the age of 84. Several of his works are completely lost, but, of a probable thirteen Books of his great treatise, the *Arithmetica*, we now have six in Greek and four more, recently discovered, in Arabic translation. This brilliant and original work stands very much on its own, outside the Greek mainstream. It appeared on the scene too late to revive the declining fortunes of Greek mathematics; its merits were not appreciated until the seventeenth century, as we shall see in Chapter 7.

One of Diophantus' major innovations was to introduce some measure of algebraic symbolism into mathematics. His writings occupy an intermediate position between the old *rhetorical* style, in which everything is written out in words, and the fully *symbolic* style of modern times. In Diophantus we find the mixture of special symbols, verbal abbreviations and ordinary prose that is known as the *syncopated* style. Thus he would write the expression $x^3 + 13x^2 - 5x + 2$ as

$$\mathrm{K}^{\Upsilon}\alpha\Delta^{\Upsilon}\iota\gamma \wedge \zeta\varepsilon\mathrm{M}^0\beta$$

where

K^{Υ} means 'the unknown cubed'*

α is 'one' (the coefficient of x^3) in the Ionic Greek number system (p. 82)

Δ^{Υ} means 'the unknown squared'

$\iota\gamma$ is 'thirteen'

$\wedge$ means 'minus' (addition is indicated, or implied, by putting terms alongside each other)

ζ means 'the unknown' (our x)

ε is 'five'

M^0 is the symbol for unity and indicates that a pure number not involving the unknown follows

β is 'two'.

* Υ is the Greek capital letter upsilon. Δ^{Υ} uses the first two capital letters of *dunamis* ($\Delta\Upsilon\mathrm{NAMI\Sigma}$), and K^{Υ} the first two of *kubos* ($\mathrm{K\Upsilon BO\Sigma}$), meaning 'power' and 'cube', respectively.

The *Arithmetica* exhibits no deductive structure and little logical progression of ideas. It is a fascinating collection of some 150 separate problems, all worked out numerically. No attempt is made to classify them by type; general methods are not presented explicitly but are often apparent from the choice of particular examples. While the algebraic approach suggests Babylonian influence, no direct connection has been established and there are important differences. For one thing, Diophantus is exclusively concerned with *exact* rather than approximate solutions of equations. Such solutions must be either integers or rational fractions. Negative and irrational numbers are not recognized; they are simply ignored. One of the distinctive features of the *Arithmetica* is that it treats both *determinate equations* (e.g. $9x^2 = 4$, with solution $\frac{2}{3}$) and *indeterminate equations* (e.g. $xy = 24$, with an unlimited number of solutions: $(\frac{1}{2}, 48)$, $(1,24)$, $(2, 12)$, etc.). Indeed, the study of the rational solutions of indeterminate equations is now known as 'Diophantine analysis', and the equations themselves are called 'Diophantine equations'. *Book I* consists mainly of problems that lead to determinate equations in two or more unknowns. Where two conditions are to be satisfied by two numbers, Diophantus often chooses the form of the two numbers so that one condition is automatically satisfied, so he then has a single unknown to satisfy the second condition. An example will make the matter clear. The problem is to find two numbers whose sum is 10 and the sum of whose cubes is 370. Diophantus designates the two numbers (in modern notation), not as x and y, but as $5 + x$ and $5 - x$. The second condition yields $(5 + x)^3 + (5 - x)^3 = 370$, which reduces to $x^2 = 4$. Diophantus gives the single answer $x = 2$, so the required numbers are 7 and 3.

Most of the extant books of the *Arithmetica* are concerned with the solution of indeterminate equations in two or three unknowns, of the form

$$y^2 = Ax^2 + Bx + C \quad \text{or} \quad y^2 = Ax^3 + Bx^2 + Cx + D$$

or with sets of such equations. The Diophantine method is to consider only such special values of the coefficients as enable the governing equation, after a suitable substitution, to be reduced to a linear form. Thus he solves the quadratic equation $y^2 = Ax^2 + Bx + C$

when either A or C is the square of a rational number. If $A = a^2$, say, he assumes that $y = ax + m$; if $C = c^2$, he assumes that $y = mx + c$. We can illustrate the method by considering the 'Pell equation', $y^2 = 1 + px^2$ (p. 53). Diophantus assumes that $y = mx + 1$, and obtains $(mx + 1)^2 = 1 + px^2$, yielding $x = 2m/(p - m^2)$. Several particular cases of the Pell equation are treated in the *Arithmetica*, for example $y^2 = 1 + 30x^2$. This equation, like all such equations, has an infinite number of rational solutions, the first three of which are $m = 1, x = 2/29, y = 31/29$; $m = 2, x = 2/13, y = 17/13$; and $m = 3, x = 2/7, y = 13/7$. For Diophantus, however, a single solution is sufficient. The first integral solution ($x = 2, y = 11$) is reached when $m = 5$. The cubic equation $y^2 = Ax^3 + Bx^2 + Cx + D$ is solved when D is the square of a rational number, d^2 say. In this case he assumes that $y = mx + d$, and takes $m = C/2d$ to remove the term in x. Since the d^2 terms also cancel out, we are left with an equation of the form $Ax^3 - Qx^2 = 0$, yielding a rational solution $x = Q/A$.

Another type of indeterminate problem is that of finding two rational numbers a and b such that both $a^2 + b$ and $a + b^2$ are perfect squares. Diophantus uses the same technique as in our first example. He calls the two numbers x and $2x + 1$, thereby satisfying one condition for any value of x. The other condition requires $(2x + 1)^2 + x$ also to be a perfect square. If we choose this to be of the form $(2x - n)^2$, we get a linear equation in x, yielding $x = (n^2 - 1)/(4n - 5)$. Diophantus chooses, as usual, the simplest case of $n = 2$, to give $x = 3/13$ and $2x + 1 = 19/13$. There are, of course, an infinite number of rational solutions when $n = 2, 3, 4, \ldots$

Some *Arithmetica* problems foreshadow modern work in the theory of numbers, and indeed provided the inspiration for Diophantus' most famous pupil, Pierre de Fermat, whom we shall meet in Chapter 7. Thus Problem 8 of Book II is 'to divide a given square number into two squares'. The general solution, in modern notation, is $\{a(m^2 + 1)\}^2 = (2am)^2 + \{a(m^2 - 1)\}^2$. For his example, Diophantus takes $a = 16/5$, $m = \frac{1}{2}$ to obtain $4^2 = (16/5)^2 + (12/5)^2$. The next problem is 'to divide a number which is the sum of two squares into two other squares'. Here he takes $13 = 2^2 + 3^2$ as the given number, and obtains $(18/5)^2 + (1/5)^2$.

We conclude this brief Diophantine survey with a geometrical problem (Problem 18 of Book IV); there are not many such in the *Arithmetica*. We are required to find a rational right-angled triangle such that the line bisecting one of the acute angles is also rational. In Figure 4.14, Diophantus lets $AP = 5x$ and $PC = 3x$, so $AC = 4x$. He then assumes that BC is a multiple of 3, say $3p$, giving $BP = 3(p - x)$. Now, since AP bisects the angle at A, we have (by a well-known result – Prop. VI, 3 of Euclid's *Elements*) $AB/AC = BP/BC$, giving $AB = 4(p - x)$. So $\{4(p - x)\}^2 = (3p)^2 + (4x)^2$, yielding $x = 7p/32$. Once again Diophantus is content with the simplest solution, which is obtained by taking $p = 1$. Multiplying by 32 to get integral solutions, we obtain 28, 96 and 100 for the sides of the triangle ABC, and 35 for the bisector AP.

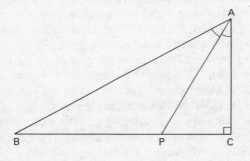

Figure 4.14.

14. Pappus of Alexandria

Virtually all we know of Pappus, the last of the great line of Greek geometers, is that in about AD 320 he composed a large treatise with the title *Synagoge* (Mathematical Collection). The eight of its ten books that have survived are a mine of information. They provide the only record of several important Greek mathematicians whose very existence would otherwise be unknown to us; they contain many alternative proofs and commentaries on the works of the Classical writers; and they include a number of new discoveries and generalizations by Pappus himself. To give a flavour of his geometrical

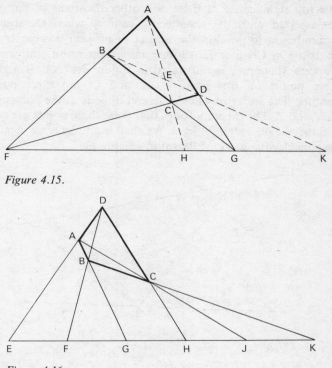

Figure 4.15.

Figure 4.16.

studies we present two theorems on the *complete quadrilateral* (four sides and three diagonals). Figure 4.15 shows such a quadrilateral, ABCD, with its associated lines. Pappus proved that E divides AC internally in the same ratio as H divides it externally. There are similar divisions of the line segments BD and FG. The second theorem (Figure 4.16) concerns the six points in which the six lines comprising a complete quadrilateral ABCD meet an arbitrary straight line EK. The theorem states that if five of these points are fixed, then so is the sixth; and the six points satisfy the condition

$$\frac{EK}{EH} \bigg/ \frac{JK}{JH} = \frac{EK}{EF} \bigg/ \frac{GK}{GF}$$

The full significance of these and other theorems of Pappus was not realized until the seventeenth century when Desargues and others began to develop the subject of projective geometry, to be discussed in Chapter 7. Another important result, still known as 'Pappus' theorem', is illustrated in Figure 4.17. A, B and C are three points on one line; A′, B′ and C′ are three points on another. The six 'cross-joins' intersect in pairs at the points marked X, Y and Z. The theorem states that these three points are collinear (i.e. they lie on a straight line). We shall meet this result again when we consider the work of Pascal in Chapter 7.

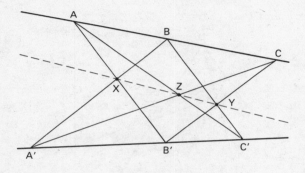

Figure 4.17.

We now, finally, return to the 'problem of Pappus' which we mentioned in the last chapter. Given four lines in a plane (Figure 4.18), the problem is to find the locus of a point P which moves such that the product of the distances from P to two of the lines is proportional to the product of the distances from P to the other two (i.e. $d_1 d_3 = k d_2 d_4$). The distances may be measured obliquely, provided the angles marked a_1, a_2, a_3 and a_4 are kept constant. The locus is in fact a conic section, as can be easily proved using the techniques of modern coordinate geometry. Apollonius demonstrated the result by 'classical' methods, after much labour. An important special case is when two of the lines, say l_1 and l_3, coalesce, so the condition becomes $d_1^2 = k d_2 d_4$. Indeed, the problem was originally called 'the locus of three or four lines'. Pappus' contribution was to

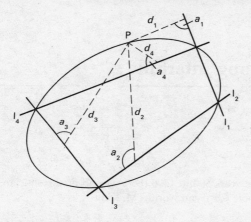

Figure 4.18.

generalize the problem to deal with more than four lines. For the case of six lines (i.e. $d_1 d_3 d_5 = k d_2 d_4 d_6$) he recognized that the locus is a higher-degree curve; it can be expressed in modern terms by a cubic equation in two variables. To Pappus, the locus was defined by the constancy of the ratio between the volumes of two solids. He did not go beyond six lines as 'there is not anything contained by more than three dimensions'. We shall meet this famous problem again in Chapter 7.

5   The long interlude

'From the years 500 to 1400 there was no mathematician of note in the whole Christian world.' M. Kline

1. Introduction

The attempt by Pappus to revive the study of geometry was not successful: the decline continued. In AD 415 Hypatia, the talented daughter of Theon of Alexandria, the Euclidean commentator, was murdered by a Christian mob. This event symbolizes the end, after more than seven centuries, of the dominance of Alexandria as a centre of learning. The Roman Empire in the West finally collapsed in 476; the execution in 524 of Boethius – the influential Roman patrician statesman and author of elementary mathematical text-books – marks the end of classical mathematics in Western Europe. Some scholarly activity lingered on in Athens until 529, when Justinian closed the philosophical schools and the scholars dispersed. Europe had entered its Dark Age. The new climate of thought was well summarized by St Augustine (354–430) when he wrote: 'What-ever knowledge man has acquired outside Holy Writ, if it be harmful, it is there condemned; if it be wholesome, it is there contained.'

Fortunately, however, Europe and the Mediterranean basin were not the whole world. Just as there are often floods somewhere on the planet while other parts are suffering from drought, so did the European darkness coincide with two of the brightest periods (the T'ang and Sung dynasties) in the long history of China. Indeed, during the millennium from 500 to 1500, culture was kept alive by five different civilizations and expressed in five different languages: China (in Mandarin), India (Sanskrit), the Muslim world (Arabic), the Byzantine Empire (Greek) and the successors of the Western Roman Empire (Latin).

2. China

The civilization of China, stretching back to the third millennium BC, remained largely isolated from the West until modern times. That we are now able to appreciate the considerable scientific achievements of this ancient and talented people is due, in large measure, to the scholarly devotion of the Cambridge scientist Joseph Needham. The first volumes of his monumental work *Science and Civilisation in China*, a work still in progress, were published during the 1950s; mathematics is treated in Volume 3 (Reference 39).

Such Chinese mathematical works as have come down to us are in the spirit of the Babylonians rather than the Greeks. They consist of collections of specific problems and present a curious mixture of the primitive and the sophisticated. The Chinese were, however, more advanced than the Babylonians in that they gave general rules, often with formal proofs, as well as solutions of particular examples.

One of the most influential of all Chinese mathematical treatises was the *Chiu-chang suan-shu* (Nine Chapters on the Mathematical Art). It was probably written by Chang Tsang, about 200 BC (early Chinese dates, and indeed many authorships, are uncertain), and was based on surviving fragments of a much earlier work. It contains 246 problems on a variety of topics, from surveying to taxation. Chapter 8 deals with the solution of equations, both determinate and indeterminate; it anticipates by two thousand years the Gaussian method of solving sets of linear equations by the method of successive elimination and back-substitution (see p. 329). The 'Nine Chapters' is also the earliest known mathematical work to discuss the significance of negative numbers. (More information on this and other Chinese mathematical writings can be found in Reference 4.) Although Chang Tsang was most esteemed for his scientific learning, he was also a distinguished soldier and statesman. He began his career as a civil servant and then achieved military distinction in the wars of the first Han Emperor. When peace was restored Chang resumed his official career, becoming Chief Minister in 176 BC, a post he held until his death at a great age in 152 BC.

The Chinese mathematicians were assiduous 'π-chasers'. Liu Hui (third century AD), who reworked and extended the 'Nine Chapters', obtained the value 3·141 59 by calculating the perimeter of a regular polygon of 3072 (6×2^9) sides. A little later Tsu Ch'ing-chih (430–501) spoke of the familiar 22/7 as the 'inexact' value, and 355/113 as the 'accurate' value – it is indeed correct to six decimal places. How did he obtain such a good result? Regrettably his calculations, in which he was helped by his son, are lost. The expansion of π as a continued fraction (p. 51) is [3; 7, 15, 1, 292, . . .]; truncating at successive terms yields 3, 22/7, 333/106 and 355/113. The reason the last approximation is so accurate is that the next term in the expansion (namely 292) is so large. It seems that Tsu carried his calculations even further, as he was able to bracket π between 3·141 592 6 and 3·141 592 7 – an impressive feat of computational stamina.

A galaxy of distinguished Chinese mathematicians flourished during the twelfth, thirteenth and fourteenth centuries (the Sung period). One of the greatest was Chu Shih-chieh, who spent twenty years of his life as a wandering scholar, earning his living by teaching mathematics. He wrote several treatises, the most interesting of which is *Szu-yuen yu-chien* (The Precious Mirror of the Four Elements). It deals mainly with algebra, and includes a numerical method for solving polynomial equations which is equivalent to the nineteenth-century 'Horner's method' of the West. Thus to solve the equation $x^2 + 252x - 5292 = 0$, Chu first establishes that there is a root between 19 and 20. He then uses the transformation $y = x - 19$ to obtain the equation $y^2 + 290y - 143 = 0$, with a root between 0 and 1. His final approximate solution (rather an anticlimax) is $x = 19 + 143/(1 + 290)$. Similarly, to solve the equation $x^3 - 574 = 0$ he sets $y = x - 8$ to give $y^3 + 24y^2 + 192y - 62 = 0$, yielding as an approximate root $x = 8 + 62/(1 + 24 + 192) = 8 + 2/7$. Indeed, 'Horner's method' must have been well known in medieval China; it was used by several Sung mathematicians for the numerical solution of cubic and even quartic equations.

3. India

In the period between the demise of the Greek world and the rise of Islam, it was India that occupied the centre of the mathematical stage. We know that there was some mathematical activity during the first millennium BC, but we have no texts earlier than the fifth century AD. Aryabhata, the author of one of the oldest such texts, was born in 476 AD, the traditional date of the fall of the Roman Empire in the West. Most of the leading Hindu mathematicians of the next 800 years (notably Brahmagupta in the seventh and Bhaskara in the twelfth centuries) had astronomical appointments, and their mathematical writings were largely concerned with the formulae and computational procedures needed in astronomy and mensuration. The Hindu approach, like the Chinese, is more Babylonian than Classical Greek in spirit, with a marked emphasis on equation-solving; we find few proofs, little generalization and an emphasis on specific problems. It must be said, however, that early Hindu work in arithmetic and algebra (they had little taste for geometry) was nourished by the later Alexandrians, notably Diophantus and Ptolemy.

The Indians used negative numbers to represent debts. This led to the acceptance – albeit somewhat hesitantly – of negative coefficients in and negative solutions (or *roots*) of quadratic and higher-degree equations. Irrational roots were also accepted. The logical and philosophical distinctions between rational and irrational numbers that so troubled the Greeks were simply ignored – or bypassed – by the Indian mathematicians, who went ahead and operated on irrational numbers by what were in fact the correct procedures. Such numbers are to be 'reckoned like integers', as Bhaskara put it when he added $\sqrt{3}$ and $\sqrt{12}$ to give $3\sqrt{3}$.

The most significant achievement of Hindu mathematics during the European Dark Ages, and certainly the best known, was the creation of the decimal positional number system that is universally used today. We need not go into the long history of Indian numeration; suffice it to say that their method of counting was purely decimal, and that separate symbols (i.e. numerals) for the numbers from 1 to 9 had been in use since the time of the Buddhist Emperor

Asoka (third century BC). There was also a notation for representing successive powers of ten. Final evolution into a mature place-value system probably took place in the sixth century AD. This last and most difficult step was the promotion to full membership of a tenth numeral: a round symbol for *zero* – or *sunya* (which means 'empty'), as the Hindus called it. Confusion about the status of this mysterious numeral persisted for centuries, and as late as the fifteenth century it was described as 'a symbol that merely causes trouble and lack of clarity'. How, it was asked, could a symbol which means 'nothing', when placed after another numeral, enhance its value tenfold?

4. The Muslim world

The prophet Mohammed made his fateful journey from Mecca to Medina (the Hegira) in 622. Less than a century later the conquests of his militant followers extended from India to Spain and included North Africa, southern Italy and large parts of western Asia. In 755 the Islamic Empire split into a western kingdom with its capital at Cordova, and an eastern kingdom centred on Baghdad. Both kingdoms rapidly developed a rich culture, absorbing intellectual nourishment from the Greek, Jewish, Persian and Indian worlds. The ninth century was the Golden Age of Arab mathematics. Baghdad, with its 'House of Wisdom', fine library and lavishly equipped astronomical observatory, became the new Alexandria.

The most influential mathematician of the Arab Golden Age was Mohammed ibn Musa al-Khwarizmi (*c.* 780–850). He was born in Khwarizm, the modern Kiva, a town situated south of the Aral Sea in Soviet Central Asia. He was summoned to Baghdad and appointed court astronomer. Two of his books, one on arithmetic and one on algebra, were destined to play important roles in the history of mathematics. The Arabic text of his arithmetical treatise is lost, but a twelfth-century Latin translation by John of Seville, with the title *Algoritmi de numero indorum*, is extant. The work itself is largely based on an Arabic translation of Brahmagupta, and gives a full account of the Hindu system of numeration. Such was its influence in Europe that the new system and its use became known as *algorism*, and was even thought to be of Arabic origin. The title of al-Khwarizmi's second and most important book, *Hisab al-jabr w'al-muqabala* (which roughly means 'the science of equations'), has given us the word *algebra*. *Al-jabr* means transposing a quantity from one side of an equation to the other, while *muqabala* signifies the simplification of the resulting expression.

The first six chapters, in which Babylonian and Hindu influences are apparent, deal with the six recognized types of linear and quadratic equations. We list them below in modern notation, with their rhetorical forms as used by Al-Khwarizmi (*a*, *b* and *c* are positive integers). The treatment is by means of numerical examples.

(1) $ax^2 = bx$: squares equal to roots
(2) $ax^2 = b$: squares equal to numbers
(3) $ax = b$: roots equal to numbers
(4) $ax^2 + bx = c$: squares and roots equal to numbers
(5) $ax^2 + b = cx$: squares and numbers equal to roots
(6) $ax^2 = bx + c$: squares equal to roots and numbers

Chapter 7 sees a sudden change:

We have said enough, so far as numbers are concerned, about the six types of equations. Now, however, it is necessary that we should demonstrate geometrically the truth of the same problems which we have explained in numbers.

The demonstrations which follow are reminiscent of those in Book II of Euclid's *Elements*, but there are substantial differences. To illustrate al-Khwarizmi's approach we consider his treatment of the 'type 4' equation, typified by $x^2 + 10x = 39$. Around a central square (Figure 5.1) of side x units he places four equal rectangles of width $2\frac{1}{2}$ units, thereby adding an area of $10x$ units. The total area is now $x^2 + 10x$, which we are told is equal to 39 units. To complete the large square he adds four corner squares each of side $2\frac{1}{2}$, giving a further added area of 25 units. Now the side of the large square is $x + 5$ units, and its area is $39 + 25 = 64$ units, so its side is 8 units. The result, $x = 3$, follows at once, and checks (or perhaps proves) the answer he has already found by algebraic means. The second half of the *Al-jabr* contains a wide variety of illustrative problems, together with some general rules for performing algebraic

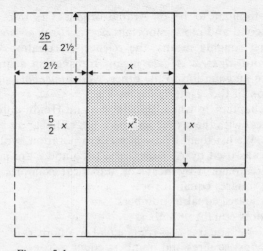

Figure 5.1.

manipulations. Although al-Khwarizmi is often hailed as the 'Father of Algebra', in two respects he looked backwards rather than forwards: he rejected both algebraic symbolism (his treatment is entirely rhetorical) and the Hindu acceptance of negative roots and negative coefficients of equations.

The three centuries of Arab pre-eminence saw many useful – but not outstanding – advances in algebra, number theory, trigonometry, optics and, to a lesser extent, geometry. We end this section with a brief look at one of the last significant mathematicians of the Muslim era, better known in the West for his poetry than for his mathematics: Omar Khayyam (*c*. 1050–1123). Omar held the post of official astronomer at the Caliph's court at Merv. Called upon to advise on the reform of the calendar, he produced, in the words of Edward Gibbon, 'a computation of time which surpasses the Julian, and approaches the accuracy of the Gregorian style.' In his book *Algebra*, he went beyond al-Khwarizmi to investigate cubic and even some quartic equations. He believed – mistakenly, as we shall see in the next chapter – that the general cubic equation could not be solved algebraically, but that geometrical methods, involving the use of conic sections, were necessary.

Let us see how Omar treats the general cubic equation. We shall
follow his method (taken from Reference 2) in some detail, but shall
use some modern symbolism for clarity. Since Omar's treatment is
geometrical we must preserve dimensionality, so let us express the
equation to be solved as

$$x^3 + ax^2 + b^2x = c^3 \tag{5.1}$$

The coefficients a, b and c are assumed to be positive, so ensuring
that there is a single positive root, x. Omar defines his problem
as: 'a solid cube plus squares plus edges equal to a number'.
His construction is shown in Figure 5.2. He begins his exposition
thus:

We draw BH to represent the side of a square equal to the given sum of the
edges, and construct a solid whose base is the square of BH, and which
equals the given number. Let its height BG be perpendicular to BH.

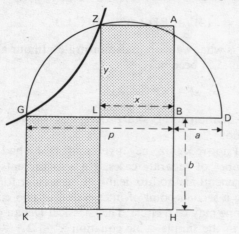

Figure 5.2.

So BH = b and BG = c^3/b^2 = p (say). We then draw BD = a
as shown, construct a semicircle on GD as diameter, and complete
the rectangle BGKH. The next step is to draw a rectangular hyperbola through the point G with the lines HB and HK as asymptotes.

(The hyperbola is called 'rectangular' because the asymptotes are perpendicular to each other.) The hyperbola intersects the circle at G and at another point, Z say. Draw ZLT, ZA and AB as shown, and set LB = x and LZ = y.

A basic property of a rectangular hyperbola is that the product of the distances from any point on the curve to each of the asymptotes remains constant. Two such points are G and Z, so we have GB.GK = ZA.ZT (or $pb = x(y + b)$) – an equality of two areas. Subtracting the area LB.BH from both sides gives ZL.LB = GL.GK = GL.BH (or $xy = (p - x)b$) – the two shaded areas in Figure 5.2. Hence ZL/GL = BH/LB. Now, from the semicircle property that we established on p. 24, we see that GL.LD = ZL^2 (or $(p - x)(x + a) = y^2$), so ZL^2/GL^2 = BH^2/LB^2 = LD/GL, i.e.   BH^2.GL = LB^2.LD = LB^3 + LB^2.BD   (or $b^2(p - x) = x^3 + ax^2$). (This, to Omar, represents the equality in volume of two solids.) We now add BH^2.LB to both sides to give

$$BH^2.GB = LB^3 + BD.LB^2 + BH^2.LB \qquad (5.2)$$

This, says Omar, 'is what we wish to demonstrate'. In our algebraic notation, equation (5.2) becomes

$$b^2p = c^3 = x^3 + ax^2 + b^2x$$

which is indeed equation (5.1), and proves that LB ($= x$) is the desired positive root.

Since Omar did not recognize negative numbers, he had to consider a large number of separate cases. He also presents several variations of his general method to deal with particular forms, but all involve finding a second point of intersection of two conics, of which, as above, one may be a circle. The Classical Delian problem (p. 25) gives rise to the simple cubic equation $x^3 - 2 = 0$, which was solved in the fourth century BC by Menaechmus – one of Alexander the Great's teachers – by obtaining the point of intersection either of two parabolas ($x^2 = y$ and $y^2 = 2x$) or of a parabola and a hyperbola ($x^2 = y$ and $xy = 2$).

The Arabic period saw a marked narrowing of the gap between algebra and geometry. We have seen one example of this trend in the work of al-Khwarizmi. Omar summed up the situation when he wrote:

Whoever thinks algebra is a trick in obtaining unknowns has thought it in vain. No attention should be paid to the fact that algebra and geometry are different in appearance. Algebras are geometric facts which are proved.

The Muslim contribution viewed as a whole exhibits a nice blending of Greek, Babylonian and Hindu influences. The debt of the West is twofold. First, the Islamic scholars collected, preserved and translated the Classical Greek mathematical texts. Secondly, they adopted the fully developed Hindu system of numeration, which was in due time transmitted to the West and eventually to the whole world. Although the new system had become known in the West by the year 1000, it took several centuries to displace the Roman number-language in Western Europe and the Ionic Greek number-language in the Byzantine Empire. We must remember that most arithmetical calculations were performed on an abacus or a counting-frame; only the results needed to be recorded on paper, and for this the Roman or Greek numerals were quite adequate. Indeed, both the medieval abacus and the counting-frame were direct physical analogues of the Hindu decimal system of numeration, with the symbol for zero corresponding to the empty column. The battle between the 'abacists' (the supporters of the old Roman numbers) and the 'algorists' (who advocated the new system of al-Khwarizmi) continued for centuries. As late as 1299 the city of Florence issued an edict prohibiting the commercial use of the Hindu–Arabic numerals; they were thought to be too easy to falsify on accounts.

5. Leonardo of Pisa

One of the first, and certainly the most influential, of the medieval textbooks on the new arithmetic was the *Liber abbaci* (often misspelt as *abaci*), which was completed in 1202 and revised in 1228. Its author, Leonardo of Pisa (*c.* 1175–1250), is better known as Fibonacci ('son of good nature'). His father was a Pisan merchant who also served as a customs officer in North Africa. The young Leonardo travelled widely and learnt the

Arabic methods from a Muslim teacher. The first chapter of his book opens with the sentence:

These are the nine figures of the Indians: 9 8 7 6 5 4 3 2 1. With these nine figures, and with this sign 0 which in Arabic is called *zephirum*, any number can be written, as will below be demonstrated.

He then explains how to use the new numerals to perform the basic arithmetical operations, including the extraction of square roots. Then follow a number of problems, the most famous of which asks:

How many pairs of rabbits will be produced each month, beginning with a single pair, if every month each 'productive' pair bears a new pair which becomes productive from the second month on?

(Rabbits do not produce young during their first month of life.) Let us number the months $1, 2, 3, \ldots$, and let u_n be the number of pairs alive at the beginning of the nth month. Deaths are supposed not to occur. The number of pairs alive at the beginning of month $n + 1$ will then be the sum of the u_n pairs alive at the beginning of month n, and the pairs born during the nth month. These come from pairs over one month old, that is from the u_{n-1} pairs alive at the beginning of month $n - 1$. This yields the recurrence relation $u_{n+1} = u_n + u_{n-1}$, where clearly $u_1 = u_2 = 1$, and thence the *Fibonacci sequence* $1, 1, 2, 3, 5, 8, 13, 21, \ldots$. There are many more realistic examples of biological growth involving this famous sequence: the branching of trees, the reproduction of bees, the patterns of petals on the flowers of dahlias and other plants, and much else.

An explicit formula for the nth term of the sequence was obtained in the nineteenth century. It may be written as

$$u_n = (1/\sqrt{5})[g^n - (1/g)^n]$$

where $g = \frac{1}{2}(1 + \sqrt{5})$ is the golden number once again. Since $1/g$ is less than 1, the second term gets steadily smaller as n increases. This means that u_n tends to $g^n/\sqrt{5}$, and the ratio of successive Fibonacci numbers tends to $g = 1 \cdot 618 \ldots$. The convergence is very rapid: thus if $n = 8$, we have $u_9/u_8 = 1 \cdot 6190 \ldots$.

Another of Leonardo's feats was to compute a remarkably accurate approximation to the positive root of the cubic equation

$$x^3 + 2x^2 + 10x = 20$$

He expressed his answer as a sexagesimal fraction: 1; 22, 7, 42, 33, 4, 40. We do not know how he obtained his result, but it retained its position for some 300 years as the most accurate European approximation to an irrational root of an algebraic equation. It is interesting to find Leonardo using the Babylonian number system for a theoretical calculation while stressing the advantages of the new system for commercial affairs.

Here is another example of Leonardo's very personal mathematical style. He arranged the consecutive *odd* numbers in a triangular pattern, as shown in Table 5.1, and observed that the *n* numbers in the *n*th row always form an arithmetical progression of mean value n^2, and sum $n.n^2 = n^3$. Hence the sum S_n of *all* the numbers in the first *n* consecutive rows is given by $S_n = 1^3 + 2^3 + 3^3 + \ldots + n^3$.

Table 5.1. Arrangement of the odd numbers used by Fibonacci to find an expression for the sum of the first *n* cubes

										Row total
									1	1^3
								3	5	2^3
							7	9	11	3^3
						13	15	17	19	4^3
					21	23	25	27	29	5^3
				31	33	35	37	39	41	6^3
			43	45	47	49	51	53	55	7^3
		57	59	61	63	65	67	69	71	8^3
	73	75	77	79	81	83	85	87	89	9^3
91	93	95	97	99	101	103	105	107	109	10^3
⋮	⋮	⋮	⋮	⋮	⋮	⋮	⋮	⋮	⋮	⋮

We know (p. 16) that the sum of the first p odd numbers is equal to p^2. Now, the total number of entries (all odd numbers) in the first n rows of the table is $1 + 2 + 3 + \ldots + n = \frac{1}{2}n(n + 1) = T_n$, say, so we have $S_n = (T_n)^2$, i.e.

$$1^3 + 2^3 + 3^3 + \ldots + n^3 = (1 + 2 + 3 + \ldots + n)^2$$
$$= [\tfrac{1}{2}n(n + 1)]^2$$

Leonardo wrote several other books on a variety of mathematical topics; many were too advanced to be understood by his contemporaries. He was undoubtedly the most creative mathematician of the medieval Christian world. As a final example of his powers, we may mention a problem that was posed at one of the mathematical contests held at the court of the Emperor Frederick II: 'To find a rational number such that if 5 is either added to or subtracted from the square of the number, the result in each case will also be the square of a rational number.' Leonardo solved the problem; the answer is $3\frac{5}{12}$. A short discussion of the generalized form of this problem is given in Appendix 6.

6. An architectural note

One of the glories of medieval Europe was its architecture. A master mason would need to know a lot about practical geometry, both

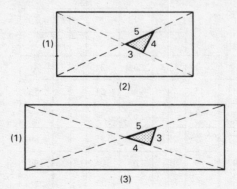

Figure 5.3.

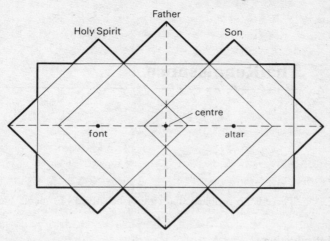

Figure 5.4.

plane and solid. The familiar 45° and 30°/60° set squares would almost certainly be found in his bag of tools. Perhaps he also carried another 'square' in the shape of a Pythagorean triangle with sides in the ratio 5 : 4 : 3. Figure 5.3 illustrates how such a tool could be used to delineate a double and a triple square. Figure 5.4 shows a common medieval church layout based on a double square and three diagonally orientated squares symbolizing the three persons of the Holy Trinity.

6   The Renaissance

'There is no problem that cannot be solved.' François Viète

1. Introduction

The European Renaissance – the rebirth of humanist learning and culture in the West – began in Italy around 1370 and was substantially complete by 1600 (or perhaps 1603 for the English). The revival started rather later in mathematics and science than in literature and the visual arts, in the sixteenth rather than the fifteenth century. While substantial contributions to geometry, trigonometry, mechanics and astronomy were made during this period, the main progress was in algebra. In this chapter we look briefly at the contributions of some of the leading algebraists of the Renaissance.

It was during the twelfth century that Europe began to be receptive to the culture of its Muslim neighbours. Translations were made of the Classical writers, first from Arabic to Latin, and then increasingly from the original Greek into various European languages. Thus, a Latin translation of Euclid's *Elements* from the Arabic was produced by Adelard of Bath in 1142. At this time there were three main cultural bridges from the Islamic to the Christian world: Spain, Sicily, and the Byzantine Empire centred on Constantinople. The first two were the more important, as the third was often blocked by the wars of the Crusades. The revival gathered momentum during the thirteenth century; in 1269 William of Moerbecke brought out a Latin translation of the main works of Archimedes.

(The original Greek manuscript from which he worked was discovered in the Vatican library as recently as 1884.) Western mathematics was beginning to stagger to its feet.

In 1453 Constantinople fell to the Turks; the Byzantine Empire, after an unbroken existence of more than 1000 years, came to an end. In Western Europe, mathematical activity continued to expand, and by the end of the fifteenth century most of the surviving Greek and Arabic mathematical works were available in Latin translations, and increasingly in printed book form. As yet, however, few scholars could both read Greek and appreciate the more advanced mathematical works of Antiquity – those of Archimedes and Apollonius, for example. Nevertheless, the humanist tide was now flowing so strongly that it could not be halted.

The Renaissance mathematicians, with their algebraic emphasis, derived more inspiration from Arabic and medieval mathematics than from the much richer inheritance of Classical Greece. To ensure that their subject was sufficiently healthy to sustain further growth, the Renaissance algebraists needed to do two things: first, to extend the number system, to come to terms with negative and then with complex numbers (p. 116) – irrational numbers had been accepted for centuries; and secondly, to develop an efficient symbolic notation. Most of ancient and Arabic mathematics was rhetorical – everything was written out in words. Diophantus was the great exception: he introduced what is known as syncopated algebra, where equations are written partly in words and partly in symbols (p. 85). However, the process needed to be taken a lot further. Indeed, the full development of algebraic symbolism was to entail much effort by many people over a long period.

2. Nicolas Chuquet

The year 1484 saw the appearance of a manuscript treatise entitled *Triparty en la science des nombres*. Although the work was not printed until the nineteenth century, it had considerable contemporary influence. We know very little about its author, Nicolas Chuquet, except that he was born in Paris, qualified in medicine, practised in Lyons and died about 1500. The *Triparty*, as its name implies, is in three parts. The first explains the Hindu–Arabic number system and how to perform the standard arithmetical operations with such numbers. The status of zero still remained a source of confusion. Chuquet tells us that 'the 10th figure does not have or signify a value, and it is called cipher or nothing or figure of no value.' The second part, which deals with square and higher roots of numbers, uses a good deal of syncopation. Thus, for example, Chuquet writes

$$\sqrt{18 - \sqrt{150}} \quad \text{as} \quad \text{R})^2.18.\bar{m}.\text{R})^2 150$$

The most important part of the *Triparty* is the third, which considers the *Règle des premiers*, the rule of the unknown thing – or, to us, algebra. One of Chuquet's achievements was to devise an exponential notation in which the power of the unknown is indicated by an exponent. Thus he wrote $6x$ and $12x^3$ as $.6.^1$ and $.12.^3$. Zero and negative exponents were included in the scheme, so that $4x^0$ becomes $.4.^0$, and $7x^{-2}$ becomes $.7.^{2.\bar{m}}$ ($\bar{p}$ and $\bar{m}$ denote 'plus' and 'minus'). Here is a quotation which shows how well he understood the laws of indices:

He who wants to multiply $.12.^3$ by $.10.^5$ must first multiply $.12.$ by $.10.$, obtaining $.120.$, then must add the denominations together, which are $.3.$ and $.5.$, giving $.8.$. Hence the multiplication gives $.120.^8$. Similarly, he who would multiply $.8.^3$ by $.7.^{1.\bar{m}}$ will find it convenient first to multiply $.8.$ by $.7.$. He obtains $.56.$, then he must add the denominations, and will take $.3.\bar{p}$ with $.1.\bar{m}$ and obtain $.2.$. Hence the multiplication gives $.56.^2$ and in this way we must understand other problems.

Chuquet also produced what is, in effect, a small table of logarithms to base 2. The second half of the third part of the *Triparty* deals with the solution of equations. Here again, the treatment contains several novel features. Thus he writes $.4.^1$ *egaux a* $\bar{m}.2.^0$ (in our notation, $4x = -2$). This seems to be the first time an isolated negative number appears in an algebraic equation. Chuquet recognized that an equation with positive coefficients and exponents can yield, as we would say, an imaginary solution. He dismissed such cases with the remark: 'Tel nombre est ineperible'.

3. Luca Pacioli and Michael Stifel

The best-known algebraic work of the late fifteenth century was the *Summa de arithmetica, geometrica, proportioni et proportionalita*, which was published in Italy some ten years after the *Triparty*. The author, the friar Luca Pacioli (*c*. 1445–1514), was a man of many accomplishments: among other things, he is generally regarded as the inventor of double-entry book-keeping. Interest in algebra grew steadily during the second half of the fifteenth century; the *Algebra* of al-Khwarizmi had been translated into Italian by 1464 at least; the first printed book on arithmetic was published at Treviso, near Venice, in 1478. Pacioli borrowed freely from other authors and made good use of the available material. The *Summa* was widely read, and completely overshadowed the more original *Triparty*.

The development of symbolic notation continued steadily throughout the sixteenth century. Indeed, as early as 1489 Johann Widman, a lecturer at Leipzig, published a commercial arithmetic in which the familiar + and − signs first appeared in print. In Northern, now largely Protestant, Europe, the growth of trade and commerce led to the publication of many books on algebra and commercial arithmetic. The German *Rechenmeisters* took the lead in the move to replace the old methods of computation, using counters and Roman numeration, by pen-and-paper calculations with the Hindu–Arabic numerals. At the same time, the algebraists continued to improve notation and extend the range of their subject. Thus, to cite one example among many, the Lutheran preacher Michael Stifel (*c*.1487–1567), in his influential *Arithmetica integra* of 1544, was one of the first to allow negative coefficients in equations, although he called them *numeri absurdi*. This enabled him to reduce al-Khwarizmi's five types of quadratic equation to a single form. However, Stifel did not recognize negative numbers as roots of an equation, and he spoke of irrational numbers as 'being hidden under a cloud of infinitude'.

4. Girolamo Cardano

We saw in Chapter 1 that the gentle art of solving equations goes back to at least 2000 B.C. The Babylonians of the 'Old' period were able to solve quadratic equations (with positive coefficients) by 'completing the square'. In the West, matters remained at the quadratic stage until the sixteenth century when a major and quite unexpected breakthrough occurred. This was the publication at Nuremberg in 1545 of a treatise entitled *Artis magnae sive de regulis algebraicis* (The Art of Solving Algebraic Equations), usually called the *Ars magna*, which explained how to solve not only the cubic (in certain cases) but the quartic equation as well. The author, Girolamo Cardano – or, in English form, Jerome Cardan – was a larger-than-life Renaissance character with an immense range of interests, activities and eccentricities. (See Illustrations 1 and 2.) In his old age he published his *De vita propria* (The Book of My Life) in which he revealed the most intimate and outrageous details of his extraordinary life.

Cardano was born a sickly infant in 1501 in Pavia, the illegitimate son of a lawyer. His mother, he tells us, tried and failed to induce an abortion. After a wretched childhood plagued by ill-health and extreme poverty, he studied medicine at Pavia and Padua. After qualifying as a doctor Cardano moved to Milan, where in 1534 he was appointed Professor of Mathematics at the Academia Palatina. There is little doubt that his disabilities and youthful hardships permanently soured his passionate nature. He gambled heavily, cheated at games, and was often malicious and contemptuous of his contemporaries. His main mathematical works – and indeed most of his voluminous writings – were composed in intervals between practising medicine, astrology, debauchery and miscellaneous rogueries. His family life, too, was not uneventful. His elder son was beheaded in 1560 for uxoricide, while Cardano himself, in a fit of rage, cut off the ears of his scoundrelly younger son.

By the mid-century Cardano had become famous throughout Europe as a physician, mathematician, humanist and astrologer. In 1552 he accepted an invitation to travel to Scotland to treat the Archbishop of St Andrews for asthma and dropsy, and was handsomely rewarded in money and gifts by the grateful prelate. Cardano then visited England where he cast the horoscope of the invalid boy-king, Edward VI. He predicted a long, albeit disease-ridden life for the young king, who died a year later. Cardano managed to talk

Illustration 1. Girolamo Cardano, 1501–76 (Mary Evans Picture Library)

himself out of that fiasco, and his fame continued to grow. On his return to Italy, he obtained a Chair at the University of Bologna, but in 1570 was imprisoned on a charge of heresy – his offence was to have cast the horoscope of Jesus Christ. Cardano was in fact released, but the scandal was such that he had to leave Bologna. He went to live in Rome where the Pope granted him a life pension, partly for his medical skills and partly to secure his services as Astrologer to the Papal Court. Finally, we have the story that, having foretold the date of his own demise – 21 September 1576, Cardano felt obliged to starve himself to death to maintain his reputation.

Cardano was indeed a quintessential product of the Renaissance: in his own person he bridged the gap between the medieval and modern worlds. He was a firm believer in the occult and in 'natural magic', and wrote many volumes on palmistry, dreams, ghosts, portents, angels and demons; yet his searching scientific and medical studies are entirely free from mysticism and the supernatural. The volume and range of his writings was immense: although much is probably lost, the extant material has been estimated to fill about 7000 pages.

5. Tartaglia and the strange story of the cubic equation

The *Ars magna*, as we have said, astonished the mathematical world by giving algebraic solutions of both cubic and quartic equations. It is completely in character that Cardano was not the original discoverer in either case, as indeed he admitted in the book. The story is a tortuous one, and we can give only a brief outline here.

It seems that the first to discover how to solve a cubic equation was Scipione del Ferro (1465–1526), a professor of mathematics at Bologna. He did not publish his findings, but in about 1510 he disclosed his method to one of his students, Antonio Maria Fior.

The next character in the story is Niccolò Fontana, always known by his nickname of Tartaglia (the stammerer), who was born in about 1500 in Brescia into a very poor family. When he was 12 years old he received severe head wounds from the sabre of a French soldier during the sack of Brescia. Let Tartaglia tell the story:

In the Cathedral, in front of my mother, I was given five mortal wounds, three on my head (each of them exposing the brain) and two on my face. If my beard did not hide my scars, I should look like a monster. One wound went right across my mouth and my teeth, breaking my jaw and my palate in two. This wound not only stopped me from talking, except with my throat the way magpies do, but I could not even eat . . . Things were so bad that I had to be fed with liquids, and even that was very difficult.

For some time afterwards, he tells us, 'I could hardly talk and always stuttered.' That was how he got his nickname, which he decided to adopt. His father died when Niccolò was a child; at the age of 14 he went to a writing school and 'began to write the alphabet just as far as the letter K.' Tartaglia explains this curious situation thus:

The terms of the contract with that particular teacher were to pay a third in advance, and another third when one knew how to write the alphabet as far as the letter K, and the rest when one knew how to write the whole alphabet. Since at that time I did not have the money to pay him and since I wanted to learn, I tried to get hold of one of his complete alphabets and the model letters he'd written out; after that I didn't go back any more. Since then I've learned on my own. I've had no other tutor, but only the constant company of a daughter of poverty called Industry. I've always worked from the books of men who are dead.

Tartaglia taught himself not only writing but mathematics as well. He taught the subject for a living in a number of Italian cities, often without getting the salary due to him, and also found time to write large treatises on arithmetic, geometry, algebra and the works of Archimedes. By about 1500 Tartaglia had learned – or had discovered – how to solve certain types of cubic equation. How much he knew of del Ferro's work is not clear. When Tartaglia's claim became known a mathematical contest was arranged in 1535 between Tartaglia and Fior in which each contestant was to set 30 questions, to be solved by the other within a prescribed time. In the event, Tartaglia solved all Fior's questions, but Fior could not solve a single one of Tartaglia's. It seems that Fior was able to deal only with equations of the type $x^3 + px = q$, with p and q positive; Tartaglia, on the other hand, had discovered how to solve some other types, including equations of the form $x^3 + rx^2 = q$.

At this point Cardano comes into the story. When he heard of Tartaglia's triumph, he invited him to Milan. At first Tartaglia vehemently refused, but in 1539 he was prevailed upon to make the journey by the prospect of paid employment in the city. Tartaglia eventually consented to reveal to Cardano his method of solution, which he expressed in 25 lines of rather obscure verse. A rough translation of the first nine lines is:

When the cube and the things themselves add up to some discrete number, take two others different from the first, choosing them so that their product is always equal to the cube of one-third of the third thing, and the difference between their cube roots will give you the main thing.

These lines give the rule for solving cubics of the form $x^3 + px = q$ ($p, q > 0$), as we shall see shortly. The next nine lines deal with equations of the form $x^3 = px + q$; the next three invite the reader to generalize the argument. The last four lines date the discovery to 1534, when Tartaglia was in Venice ('the city surrounded by the sea'). He hoped to publish his method, and exacted a solemn oath from Cardano that he would never reveal the secret. When the *Ars magna* was published and Cardano's breach of faith was manifest, Tartaglia was – not surprisingly – very angry, and a malicious and sordid controversy ensued. It is difficult to get at the truth of the matter, but we do know that Tartaglia himself was not above publishing other people's work and passing it off as his own. In the *Ars magna* Cardano does indeed acknowledge his debt to Tartaglia, thus:

Niccolò Tartaglia communicated it [i.e. his method of solution] to us when we asked for it, but suppressed the demonstration. With this aid we sought the demonstration, and found it, though with great difficulty, in the manner which we shall now set out.

6. Cardano's published treatment of the cubic equation

Cardano's discussion of the cubic in the *Ars magna* is mainly rhetorical. In this he followed his mentor, al-Khwarizmi; the book makes tedious reading today. Cardano did not accept the existence of negative numbers in equations and so had to consider no less than 13 different types, from

$$x^3 + px = q \quad \text{to} \quad x^3 + rx^2 + q = px \quad (p, q, r > 0)$$

HIERONYMI CAR

DANI, PRÆSTANTISSIMI MATHE

MATICI, PHILOSOPHI, AC MEDICI,

ARTIS MAGNÆ,

SIVE DE REGVLIS ALGEBRAICIS,

Lib.unus. Qui & totius operis de Arithmetica, quod

OPVS PERFECTVM

inscripsit,est in ordine Decimus.

HAbes in hoc libro,studiose Lector,Regulas Algebraicas (Itali, de la Cossa uocant) nouis adinuentionibus,ac demonstrationibus ab Authore ita locupletatas,ut pro pauculis antea uulgo tritis,iam septuaginta euaserint. Neç solum, ubi unus numerus alteri,aut duo uni,uerum etiam,ubi duo duobus, aut tres uni æquales fuerint,nodum explicant. Hunc aut librum ideo seorsim edere placuit,ut hoc abstrusissimo, & planè inexhausto totius Arithmeticæ thesauro in lucem eruto, & quasi in theatro quodam omnibus ad spectandum exposito, Lectores incitarétur,ut reliquos Operis Perfecti libros, qui per Tomos edentur,tanto auidius amplectantur,ac minore fastidio perdiscant.

Illustration 2. Title page of Cardano's *Artis magnae sive de regulis algebraicis,* usually called the *Ars magna* (1545)

a chapter being devoted to each type. All his equations have numerical coefficients, but it is clear that he thinks of each of them as representing its general type. Thus, in his first example, when he says 'let the cube and six times the side be equal to 20' (i.e. $x^3 + 6x = 20$), he is clearly thinking of the general case of, as he puts it, 'cube and the thing equal to a number'. His discussion of this equation covers several pages of rhetorical prose. He first gives a geometrical demonstration of his method, and then a rule for solving the general case.

His procedure, in modern symbolism, is to set $x = u - v$, where $uv = p/3 = 2$. Substituting in the cubic equation gives $u^3 - v^3 = 20$, and hence $u^6 = 20u^3 + 8$, yielding $u^3 = \sqrt{108} + 10$. Since $u^3 - v^3 = 20$, we have $v^3 = \sqrt{108} - 10$, which yields

$$x = \sqrt[3]{\sqrt{108} + 10} - \sqrt[3]{\sqrt{108} - 10}$$

Cardano then moves on to his next case, 'cube equal to thing and number', or $x^3 = px + q$ ($p, q > 0$). He proceeds in a similar fashion but sets $x = u + v$ instead of $u - v$. Equations of this type can cause trouble, however. Consider, for example, the equation $x^3 = 15x + 4$, which yields the 'Cardano' solution of

$$x = \sqrt[3]{2 + \sqrt{-121}} + \sqrt[3]{2 - \sqrt{-121}}$$

Such a situation was called a *casus irreducibilis*. Cardano knew that a negative number has no square root, but he also knew that $x = 4$ is a solution of the equation. He was not able to make sense of this paradox; he had, as it were, blundered into the minefield of complex numbers.

A *complex number* has the form $a \pm \sqrt{-b}$, where $b > 0$. (If $a = 0$ such a number is, regrettably, called an imaginary number.) The two complex numbers $a + \sqrt{-b}$ and $a - \sqrt{-b}$ are said to form a *conjugate pair*; they are to be thought of as the two roots of the quadratic equation $(x - a)^2 + b = 0$. In a polynomial equation with real coefficients, *complex roots* must occur in such conjugate pairs – both the sum and the product of two complex conjugates being real numbers. This extension of the concept of number enables us to assert that any polynomial equation of the nth degree (with real coefficients) has exactly n roots, either real or as complex conjugate pairs. Of course, some of the roots may be equal. Thus the roots of the equation

$$x^4 - 6x^3 + 16x^2 - 18x + 7 = 0$$

are $1, 1, 2 + \sqrt{-3}$ and $2 - \sqrt{-3}$.

In a later chapter of *Ars magna*, 'On the rule of postulating a negative', Cardano does address the problem posed by the existence of complex roots of quadratic equations. In his example he wishes to divide the number 10 into two parts such that their product is 40. The solution is, of course, $5 \pm \sqrt{-15}$, which Cardano wrote as '5p: Rm: 15' and '5m: Rm: 15'. He concludes his discussion with the comment that his result in this case is as subtle as it is useless. Reverting to modern notation, the general solution of the cubic equation $x^3 + px + q = 0$ (where now p and q can be either positive or negative) is given by

$$\sqrt[3]{-\frac{q}{2} + \sqrt{\frac{q^2}{4} + \frac{p^3}{27}}} + \sqrt[3]{-\frac{q}{2} - \sqrt{\frac{q^2}{4} + \frac{p^3}{27}}}$$

Thus Cardano's procedure fails when $27q^2 + 4p^3$ is negative. Now this is just the condition for all three roots to be real: the *casus irreducibilis* is the least complex! When this condition is not satisfied, one root is real and the other two form a complex conjugate pair. Thus we have the paradoxical result that the Cardano formula contains complex quantities when all the roots are real, and does not do so when two of the roots are complex. Clearly, the challenge presented by the existence of complex numbers could not be ignored much longer.

7. Ferrari's solution of the quartic equation

The *Ars magna* also presents solutions of several types of quartic equation. The method in this case, as Cardano puts it, is 'due to Luigi Ferrari, who invented it at my request'. Ferrari (1522–60) started at the age of 18 as Cardano's servant, then became his secretary, and eventually Professor of Mathematics at Bologna. Once again, the restriction to positive coefficients means that no less than 20 different cases must be considered in turn. One example must suffice to illustrate the method. We will consider, he says, the case of 'square-square and square and number equal to side'. Cardano's chosen example is $x^4 + 6x^2 + 36 = 60x$ (he knew how to remove the x^3 term). We shall modernize Cardano's argument for clarity. Six steps are needed, as follows:

(1)   Add terms in x^2 and x^0 as necessary to make the left side of the equation (as written) a perfect square. This gives

$$x^4 + 12x^2 + 36 = (x^2 + 6)^2 = 6x^2 + 60x$$

(2)   Introduce a new unknown, y, so that the left side remains a perfect square, thus:

$$(x^2 + 6 + y)^2 = (2y + 6)x^2 + 60x + (y^2 + 12y)$$

(3)   Choose y so that the right side is also a perfect square. This is the crucial step. The condition for this (well known before Cardano's time) is that

$$60^2 = 4(2y + 6)(y^2 + 12y)$$

(4)   This yields a cubic in y which can be solved by the procedure already discussed. (Note that the equation contains a y^2 term, which must be removed.)

(5)   Substitute the value of y obtained in step (4) into the equation in step (2), and take the (positive) square root of both sides.

(6)   This leads to a quadratic equation for x which is easily solved. We now know, of course, that step (5) leads to two quadratic equations, and so to four roots, but Ferrari and Cardano obtained only a single real root.

8. François Viète

The breakthrough of being able to solve equations of the third and fourth degrees gave a powerful stimulus to the further development of algebra. The quintic equation resisted all attempts to solve it, and it was not until the 1820s that Abel and Galois proved the impossibility of doing so. We must pass over the many sixteenth-century algebraic contributions – by Bombelli, Harriot, Recorde and Stevin

among others – and come to the most significant mathematical figure
of the later Renaissance, François Viète (or Franciscus Vieta, the
Latin form of his name).

Viète was born at Fontenay in 1540, the son of a lawyer. Following his father's profession he studied law at Poitiers, and was
called to the bar in 1560. Four years later he abandoned legal
practice and entered the service of the house of Soubise. The
head of the family, Jean de Parthenay, was a prominent soldier in the
Huguenot cause. One of Viète's duties was to be tutor to Jean's
daughter, Catherine. He wrote several textbooks for her, and in 1591
he dedicated his most important mathematical work to his former
pupil. When Jean died in 1567, Viète stayed with the family and
moved to the Huguenot stronghold of La Rochelle as secretary to
Jean's widow. In 1571 he became an Advocate in the Parlement de
Paris, and rose rapidly in the service of the state. He was made a
Privy Councillor and was entrusted with several special missions by
King Henry III, but in 1584 he fell foul of the Catholic faction at
Court and was relieved of his duties. Much of Viète's best mathematical work was done during the next five years of enforced leisure. In
1589, on the accession of Henry IV of Navarre, he returned to royal
favour and public service.

A few years later, the Netherlands Ambassador remarked to the
King that France had no one capable of solving a problem that
Adriaan Van Roomen had proposed to the mathematicians of the
world. It required the solution of a polynomial equation of the 45th
degree, of the form $f(x) = A$ (more details are given in Appendix 3).
The King summoned Viète and told him of the challenge. Now,
Viète had in fact been working on multiple-angle formulae for the
sine and cosine, and he realized that Van Roomen's problem was, in
essence: 'Given the chord of an arc of a circle, find the chord of the
45th part of that arc.' The story goes that Viète was able to give two
pencilled solutions to the King within a few minutes, and that he
found the full set of 23 positive solutions later. Viète in turn
proposed a problem to Van Roomen – in fact the famous Apollonian
problem of drawing a circle to touch three given circles (p. 55).
Van Roomen solved it by the use of conic sections, whereupon Viète
sent back a solution which used Euclidean methods only. This so

impressed Van Roomen that he travelled to meet Viète, and the two men became firm friends.

At this time France was at war with Spain, and the Spaniards were using a cipher containing more than 500 characters; it was changed periodically and was believed to be unbreakable. The King gave some intercepted dispatches to Viète and asked him to decipher them. Viète succeed in finding the key, which the French used for the next two years to their great advantage. Philip II was so convinced that his cipher could not be broken that he complained to the Pope that the French were using sorcery against him, 'contrary to the practice of the Christian faith'.

9. Viète's new algebra

Viète's mathematical writings cover a wide range, but undoubtedly his most important work is the *In artem analyticam isagoge* (Introduction to the Analytical Art) which was published in Tours in 1591. Viète had a philosophic cast of mind, and was a great admirer of the Greeks (he wrote a famous essay on Apollonius). His aim was to uncover and restore the algebraic (or as he preferred to say, the analytical) relationships that were, he believed, hidden behind the geometrical presentations of the Greek masters. Foreshadowing Leibniz, Viète believed this to be a necessary element in the search for a universal science. Success in this enterprise would make everything possible; the triumphant quotation at the head of this chapter is, in fact, the final sentence of the *Isagoge*. The first chapter of the *Isagoge* takes up the distinction made by Pappus between analysis and synthesis. Roughly speaking, we may say that in analysis we assume the truth of what we wish to prove, and deduce a chain of consequences until we reach something that is accepted as being true. In synthesis, this process is reversed: we start from known truths and make a sequence of deductions until we arrive at the desired result. Most of the Greek mathematics that had survived was synthetic. Where analysis was used (e.g. by Apollonius), it was followed by a synthetic demonstration. (The term 'analysis' seems to have been introduced, very late, by Theon of Alexandria.) Diophantus, in his *Arithmetica*, was once again the great exception to this generalization.

Viète sought to improve on Diophantus by introducing a fully symbolic algebra. This he called *logistica speciosa*, as opposed to *logistica numerosa* – or, as we would say, algebra as opposed to arithmetic.

The notion of using letters systematically, not just to represent unknown quantities but also as general coefficients, is Viète's most important contribution to algebraic theory. Before his time there was no accepted scheme for writing a single equation to represent a whole class (or 'species', as he called it) of equations. Viète used capital vowels $(A, E, I, \ldots)$ for unknowns and consonants $(B, C, D, F, \ldots)$ for known quantities, assumed to be positive. (Our present convention of denoting unknown quantities by letters near the end of the alphabet is due to Descartes.) Viète's algebra had a firm geometrical foundation; he kept his equations homogeneous, the dimensions of the quantities being indicated where necessary. Thus he would write the equation $3BA^2 - DA + A^3 = Z$ as

*B*3 *in A quad* − *D plano in A* + *A cubo aequator Z solido*

The constants B, D and Z have dimensions 1, 2 and 3, respectively; *in* means 'multiplied by'. In fact, Viète's main objective was to use algebra (i.e. the analytic art) to help solve geometric problems, an approach that was soon to be taken further by Fermat and Descartes. Viète realized that algebra would need considerable development if it was to replace the geometrical methods of the Greeks, and he initiated several steps in this direction.

In his use of symbols, Viète was a transitional figure. He used the German + and − signs, and introduced square brackets and braces into his equations, but his equations still contained many words and verbal abbreviations. Indeed, by retaining *A quadrature* and *A cubus* for A^2 and A^3, he failed to reach the position achieved by Chuquet a century earlier.

10. Viète's other mathematical contributions

Up to about 1450, trigonometry was largely spherical; surveying relied on the simple geometrical techniques of the Romans. During the next 100 years plane trigonometry began to develop again from where Ptolemy and the Arabs had left it. The formulae for the solution of plane and spherical triangles were tidied up and a number of good trigonometrical tables were constructed. Viète wrote several trigonometrical works, the first being his *Canon mathematicus* of 1579. Many of our basic trigonometrical results are due to him, including the set of four identities typified by

$$\sin A + \sin B = 2 \sin \tfrac{1}{2}(A + B) \cos \tfrac{1}{2}(A - B)$$

Thus Viète can be said to have foreshadowed Napier's discovery of logarithms, whereby the operations of multiplication and division can be performed by additions and subtractions only.

We have already mentioned Viète's mastery of the multiple-angle sine and cosine formulae. By considering regular polygons of 4, 8, 16, ... sides inscribed in a unit circle, he was able to express π as an infinite product:

$$\frac{2}{\pi} = \sqrt{\tfrac{1}{2}} \times \sqrt{\tfrac{1}{2}(1 + \sqrt{\tfrac{1}{2}})} \times \sqrt{\tfrac{1}{2}(1 + \sqrt{\tfrac{1}{2}(1 + \sqrt{\tfrac{1}{2}})})} \times \dots$$

(more details are given in Appendix 4). This seems to be the first exact formula for this elusive constant; the many infinite-series formulae had to wait until the seventeenth and eighteenth centuries.

Viète's trigonometrical skills enabled him to deal with Cardano's *casus irreducibilis* of the cubic equation with three real roots. He starts from the identity (in modern notation)

$$\cos 3A = 4 \cos^3 A - 3 \cos A$$

$$\text{or} \quad z^3 - \tfrac{3}{4}z - \tfrac{1}{4}\cos 3A = 0, \quad \text{where } z = \cos A$$

Suppose, as in Section 6, that the cubic to be solved is given by $x^3 = px + q$ ($p, q > 0$). If we introduce an arbitrary constant n, setting $x = nz$, then

$$z^3 - p/n^2 . z - q/n^3 = 0$$

We can now match coefficients in the two forms, $p/n^2 = \tfrac{3}{4}$ and $q/n^3 = \tfrac{1}{4}\cos 3A$, so that $n = \sqrt{4p/3}$. With this value of n, we can select a value of A so that

$$\cos 3A = 4q/n^3 = \tfrac{1}{2}q/\sqrt{p^3/27}$$

Since $p > 0$, n is real, and the condition (mentioned earlier) for three real roots ensures that $|\cos 3A| < 1$. The notation $|x|$ denotes x if $x \geqslant 0$ and $-x$ if $x < 0$. So $|x|$ is always positive (or zero); it is called the *modulus* or *absolute value* of x, and may also be written mod x. We can therefore find $3A$, and hence A, from a cosine table. Viète obtained only one root, namely $n \cos A$; the two others are given by $n \cos(A + 120°)$ and $n \cos(A + 240°)$. Applying the method to the example of Section 6, namely $x^3 = 15x + 4$, we find that $n = 2\sqrt{5}$ and $\cos 3A = 2/5\sqrt{5}$. It turns out that $\cos A = 2/\sqrt{5}$, so $n \cos A = 4$, as expected. A little trigonometrical manipulation yields

$$\cos(A + 120°) = -(2 + \sqrt{3})/2\sqrt{5}$$
$$\cos(A + 240°) = -(2 - \sqrt{3})/2\sqrt{5}$$

so the two other real roots are $-2 \pm \sqrt{3}$. Viète's trigonometrical treatment of the cubic equation is contained in a paper 'On the review and correction of equations', written in 1591 but not published until 1615. A complete discussion of the cubic equation had to wait for Euler in 1732.

Viète also discusses cubic and quartic equations in the *Isagoge*. For the cubic, he is able to simplify Cardano's treatment. Starting, once more, with the equation $x^3 + px + q = 0$, he introduces the transformation $x = z - p/3z$ to give

$$z^3 - \frac{p^3}{27z^3} + q = 0$$

This quadratic in z^3 is then solved to give

$$z^3 = -q/2 \pm \sqrt{p^3/27 + q^2/4}$$

Viète uses only the positive value of the square root, as did Cardano, and so obtains the value of the single real root. His treatment of the quartic equation is essentially the same as – but rather more direct than – Ferrari's.

When Viète died in 1603 the transition to modern mathematics was substantially complete. The scene was set for the great triumphs of the seventeenth century: logarithms, and coordinate and projective geometry in the first half; the calculus and universal gravitation in the second.

7 Descartes, Fermat and Pascal

'As long as algebra and geometry proceeded along separate paths, their advance was slow and their applications were limited. But when these sciences joined company, they drew from each other fresh vitality and thenceforward marched on at a rapid pace towards perfection.' Joseph Louis Lagrange

1. Introduction

At the beginning of the seventeenth century we enter what is, mathematically speaking, the modern world. For the first time we feel at home with the approach, the style and the symbolism of most of the leading mathematical writers. Although some of the most distinguished seventeenth-century scientists – Kepler and Newton, no less – still had one foot in the modern and the other in the medieval world, the decisive transition had been achieved by the 1630s. The middle years of the century also saw a great increase in the number of talented mathematicians whose names have survived. We cannot do justice to them all; in this chapter we shall focus attention on three Frenchmen of genius who bestrode the world of mathematics in the immediate pre-Newtonian years.

2. René Descartes

René Descartes (see Illustration 3) – philosopher, soldier, gentleman and mathematician – was born at La Haye in Touraine on 31 March 1596. His father was a successful lawyer, a councillor of the Parlement

Illustration 3. René Descartes, 1596–1650 (Mansell Collection)

of Rennes, Brittany. René was born into an old family whose sons were destined for careers as 'independent gentlemen' in the service of France; he never needed to earn his living. His mother died a few days after he was born, but his devoted father did all he could to make up for the loss. Because of his fragile health René was kept at home until the age of eight, when he was sent to the Jesuit College of La Flèche. He soon attracted the attention of the rector, who allowed his talented pupil to lie in bed in the mornings until he felt able to join his class. The habit remained with him throughout his life. He maintained in later years that the long, quiet mornings of bedtime meditation were the prime source of his best philosophical and mathematical thinking. Socrates, we are told, was able to cerebrate while standing for hours in the snow on military service; Descartes's mind worked properly only when he was warm and horizontal.

After eight years of rigorous Jesuit training in the classics, and in the conduct befitting a gentleman, Descartes left the college and went to Paris. He was already in a highly sceptical frame of mind, questioning much of what he had learned from the good fathers. Having decided that he must see the world and learn something of its ways, he immersed himself in the many pleasures afforded by the great city. However, he soon became bored with his vapid social life, and retired to a quiet lodging for two years of study and reflection. He also found time to take a law degree. In 1617, surprisingly, he enlisted in the Dutch army at Breda. Holland was then at peace and Descartes was able to enjoy two more years of meditation, while at the same time learning the art of soldiering from the brilliant Prince Maurice of Orange. With the outbreak of the Thirty Years' War, Descartes transferred his allegiance to the Bavarian army. One night in November 1619, when his regiment was resting in winter quarters, he had a traumatic dream-experience which convinced him yet again of the futility of his present way of life, and strengthened his resolve to become a philosopher. He tells of how he awoke with the intense conviction that he had been presented with a magic key which would enable him to unlock the treasure-house of nature. Descartes is not very explicit as to the nature of the key, but it is widely believed to have been an infallible 'method' for exploring natural phenomena by controlled experiment and rigorous mathematical reasoning. Some people, on the other hand, took the view that the key was more specific – indeed, nothing less than the application of algebraic techniques to geometry – and proclaim 10 November 1619 as the true birthday of analytical geometry and, by extension, of modern mathematics. Be that as it may, Descartes published nothing on the subject for another eighteen years. Instead, he went on with his soldiering and took part in the battle for Prague in 1620, but in the following year, completely disillusioned with the war and its terrible consequences, he abandoned his military career. After some travelling in Northern Europe and Italy he returned to Paris in 1625. There he spent three comparatively happy and peaceful years, although we are told that his convivial friends of former days would keep calling on him, even in the mornings during his thinking hours in bed. It seems, however, that the military life still retained some attraction, for in 1628 Descartes embarked on his last bout of soldiering. He joined the forces of the Duke of Savoy, who was besieging the Huguenot stronghold of La Rochelle, where Viète had lived some years earlier. Descartes was offered a Lieutenant Generalship, which he declined. Once again he returned to Paris, but soon decided to move to the more relaxed and tolerant atmosphere of the Netherlands. He lived there for some twenty years, moving around a good deal and enjoying that country's relative freedom of expression and publication. Even in Holland, however, he was attacked by Protestant bigots and was saved from persecution on more than one occasion only by the personal intervention of the Prince of Orange.

By 1634 Descartes's long-awaited philosophical treatise, *Le Monde*, was nearly ready for the press. At this point he took fright and withheld publication. He was stunned when he heard how Galileo had been forced, on his knees, to renounce a heresy that Descartes had espoused for years as a matter of course. He knew, as a good Catholic, that the Pope was infallible and, as a good scientist, that Copernicus was right. What had gone wrong? Finally, however, in 1637, he yielded to the entreaties of his friends and agreed to the publication, not of *Le Monde*, but of his greatest work, *Discours de la méthode*, with its three appendices. Almost at once he became famous and much sought after throughout Europe. He might well have accepted a court appointment in England but for the outbreak of the civil war. In 1646, at the age of 50, he was unfortunate enough to attract the interest of Queen Christina of Sweden, already established at the age of 19 as a capable and masterful ruler, with a passionate thirst for knowledge and a remorseless determination to get her own way. She decided that she must have the great man at her court. Descartes held out for three years, but eventually succumbed when Christina sent one of her admirals with a ship to fetch him. His reception in Stockholm was boisterous and extravagant. He lived in the house of the French Ambassador in the full glare of court publicity. Christina demanded daily lessons from her new philosopher, and insisted that five o'clock in the morning was the proper time for a busy monarch to study science and philosophy – in the ice-cold court library. The Swedish winter of 1649–50 was one of the severest on record. Christina seemed to be impervious to cold, but it proved too much for Descartes. He died on 11 February 1650, at the age of 54 – a grim sacrifice to the vanity and insensitivity of a headstrong young woman.

Seventeen years later, Descartes's bones were returned to Paris and placed in what is now the Panthéon. Even then, a proposed funeral oration was hastily forbidden by order of the Crown; a few years before he died the Church had put his works on their Index of prohibited books. It is ironic that the man who 'desired only tranquillity and repose' (as he put it) had to seek them in military camps and a rumbustious court. Descartes did indeed live in troubled times: the old order was rapidly passing away; the new was not yet established.

Descartes never married. John Aubrey (1626–97), the unreliable but entertaining author of *Brief Lives*, put the matter rather more robustly:

He was too wise a man to encumber himself with a Wife; but as he was a man, he had the desires and appetites of a man; he therefore kept a good-conditioned handsome woman that he liked, and by whom he had some Children (I think 2 or 3).

In fact there is firm evidence for only one natural child, a daughter whose death at the age of five was one of the great sorrows of his life.

3. Descartes's philosophical position

Descartes is generally regarded as the first modern philosopher; his outlook was profoundly affected by the new physics and astronomy. He was not prepared to build on the foundations laid by Plato and Aristotle, but sought instead to construct a complete new philosophical structure. His labours were so successful that his ideas – not only in philosophy but also in biology, physics and cosmology – came to dominate the thinking of his Continental successors until well into the eighteenth century.

Descartes's approach was to doubt everything that could possibly be doubted, starting with the evidence of the senses. What then, if anything, remains? Descartes's answer, in his own words, is:

While I wanted to think everything false, it must necessarily be that I who thought was something; and remarking that this truth, *I think, therefore I am*, was so solid and so certain that all the most extravagant suppositions of the sceptics were incapable of upsetting it, I judged that I could receive it without scruple as the first principle of the philosophy that I sought.

(The well-known phrase '*cogito ergo sum*' has become Descartes's philosophical trademark.) It was on this thinnest of foundations that Descartes sought to rebuild the edifice of knowledge. What emerged was a stark, metaphysical structure embodying a full-blooded dualism of mind and matter, of soul and body. For Descartes, the bodies of men and animals are machines. An animal is an automaton, but man is different: he has a soul which can make contact with the 'vital spirits' and so bridge the gap that separates the two modes of existence. God has so arranged matters that events in the two worlds can proceed independently along parallel paths, and yet remain in step. Physical phenomena are to be explained entirely in terms of matter and motion: indeed, the Universe consists solely of matter in ceaseless motion. Descartes also put forward an elaborate and extremely influential cosmological theory of vortices to explain how the Earth and the other planets are kept moving in their paths around the Sun. It is all very odd!

In his famous *Discours de la méthode pour bien conduire sa raison, et chercher la vérité dans les sciences* (Discourse on Method), Descartes set out his programme of philosophic inquiry. As we have seen, he called for a critical and systematic use of what we would now call the 'scientific method'. His emphasis on certainty, clarity and order led him to proclaim:

I was especially delighted with the Mathematics, on account of the certitude and evidence of their reasonings: but I had not as yet a precise knowledge of their true use; and thinking that they but contributed to the advancement of the mechanical arts, I was astonished that foundations, so strong and solid, should have had no loftier superstructure reared on them.

Descartes then proceeded to add to the superstructure some not inconsiderable bricks of his own.

Attached to the *Discours* are three appendices in which Descartes sought to apply his general method to three specific areas of knowledge: optics, meteorology and mathematics. *La Dioptrique* contains the first publication of Snell's law of optical refraction; *Les Météores* includes the first credible explanation of the rainbow; *La Géométrie* embodies his outstanding contributions to mathematics. It is important to remember that Descartes was only incidentally a mathematician; *La Géométrie* was his only mathematical work, apart from a few fragments and some references in his letters. But what a work! It pointed mathematics in a new direction which, broadly speaking, has been maintained ever since. The extent of his influence was undoubtedly much enhanced by his excellent literary style. His books were written in French, not Latin, and were addressed to the general educated public rather than to a small specialist elite or a coterie of devoted followers.

4. *La Géométrie*

Descartes's reputation as a mathematician rests, above all, on his invention of coordinate geometry. (The modern term 'analytic geometry' did not appear until the nineteenth century.) This reputation may be well deserved, but the matter is not as simple as it might first appear. His prime objective was to unify the hitherto largely separate disciplines of algebra and geometry, in particular to enlist the aid of algebra to solve geometrical problems. The scene is set in the opening sentence:

Any problem in geometry can easily be reduced to such terms that a knowledge of the lengths of certain straight lines is sufficient for its construction.

He then points out that such lines need not, in many cases, be drawn on paper: 'It is sufficient to designate each by a single letter.' After discussing a few simple examples, he presents his method in these words:

If, then, we wish to solve any problem, we first suppose the solution already effected, and give names to all the lines that seem needful for its construction – to those that are unknown as well as to those that are known. Then, making no distinction between known and unknown lines, we must unravel the difficulty in any way that shows most naturally the relations between these lines, until we find it possible to express a single quantity in two ways. This will constitute an equation.

He goes on to explain that we must find as many equations as there are unknown 'lines', and then combine them so as to leave us with a single unknown line expressed in terms of known lines by an algebraic equation. For Descartes, the final step is to construct the unknown line geometrically, using the fact that its length satisfies an algebraic equation. (Descartes uses the term 'line' in a broad sense: it could be represented by x, x^2, x^3, etc., in contrast to Viète's usage.)

An example should make the procedure clear. Suppose the geometrical problem requires us to find an unknown length, denoted by x, which, after any necessary algebraic manipulation, is found to satisfy the quadratic equation $x^2 = ax + b^2$, where a and b are known lengths (and so greater than zero). Now, we know that the 'algebraic' solution of this equation is $x = a/2 + \sqrt{a^2/4 + b^2}$. Descartes's construction for x is shown in Figure 7.1. ABC is a right-angled triangle with $AB = b$ and $AC = a/2$. We construct a circle as shown. The extended line BC cuts the circle at P and Q; BP is then the desired value of x. Descartes dismisses the other (negative) root, which is given in magnitude by QB, as 'false'.

It is perhaps worth pointing out that Descartes's remark that 'we first suppose the solution already effected' derives directly from Viète's 'analytic art', and indeed goes back to Pappus.

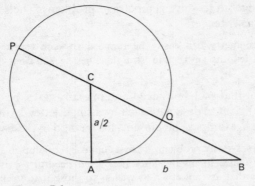

Figure 7.1.

La Géométrie is organized in three Books. In Book I Descartes first discusses *determinate* Euclidean problems, and goes on to consider *indeterminate* problems in which many different lengths can serve as possible answers. The end-points of these lengths, when drawn out using a suitable coordinate scheme, will delineate a curve. 'It is required,' he says, 'to discover and trace the curve containing all such points.' The crucial idea here is that the curve is described by an indeterminate equation (in x and y, say) which expresses the unknown length y in terms of arbitrary lengths denoted by x. However, once a value of x is specified the value of y (or a set of values which are the solutions of an equation in y) is determined. So the curve, which may be expressed in modern notation as $f(x, y) = 0$, can be constructed point by point.

To illustrate the power of his method, Descartes considers the famous 'three or four lines' problem of Pappus (p. 90). In Figure 7.2 the four lines are marked l_1, l_2, l_3 and l_4, as they were in Figure 4.18; the four lines PN_1, PN_2, PN_3 and PN_4 from the variable point P are shown dashed. It is required to find the locus of P satisfying the condition

$$PN_1.PN_3 = k.PN_2.PN_4 \qquad (7.1)$$

Descartes denotes AN_1 by x and PN_1 by y, and then obtains, by simple geometry, the values of PN_2, PN_3 and PN_4 in terms of known

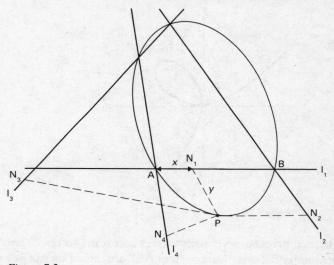

Figure 7.2.

lengths. The condition (7.1) yields a second-degree equation of the form

$$y^2 = ay + bxy + cx + dx^2 \tag{7.2}$$

where a, b, c and d are algebraic expressions involving known quantities only. Now, for any value of x, we have a quadratic equation for y which can be solved algebraically. The length y can therefore be constructed by the Euclidean methods he has already discussed. As x is varied, the point P traces a curve whose equation is given by equation (7.2): it is, of course, a conic. What Descartes has done (although he does not say so explicitly) is to set up an oblique coordinate system with A as origin and AB as the x-axis. The y-axis is a line through A making a fixed angle with AB – in fact, the angle between l_1 and l_2.

Once Descartes has the concept of the 'equation of a curve', he develops it in several directions. He considers, for example, two curves, each represented by an equation, and finds their point (or points) of intersection by solving the two equations simultaneously.

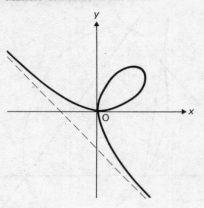

Figure 7.3.

He then proceeds to classify curves according to the degree of their defining equations. 'Geometric curves' are those that can be expressed by an algebraic equation (of finite degree) in x and y. One of his favourite geometric curves is the *folium of Descartes* (Figure 7.3), whose equation is

$$x^3 + y^3 - 3axy = 0$$

Curves, such as the Archimedean spiral or the quadratrix, which are not 'geometric', he calls 'mechanical'. (The modern term, 'transcendental', was introduced by Leibniz some fifty years later.) Descartes then seeks to generalize the problem of Pappus to more than four lines. He knew that for five or six lines the locus is a cubic curve; for seven or eight lines it is a quartic, and so on. One curve that appears several times in *La Géométrie* is a five-line Pappus locus where four of the lines are parallel and equally spaced, as shown in Figure 7.4. The locus is the cubic curve given by

$$(a + x)(a - x)(2a - x) = kxy$$

Newton, in his detailed study of cubic curves, called it the *Cartesian parabola* or *trident*, as we shall see in the next chapter.

In both *La Dioptrique* and Book III of *La Géométrie*, Descartes investigates some of the practical problems that arise in the design

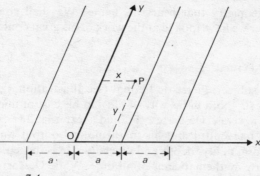

Figure 7.4.

of optical instruments. One example must suffice here – that of finding the surface of separation between two media such that a pencil of light rays starting from a point in the first medium would strike the surface, refract into the second medium, and then converge once again to a point. Descartes discovered that the required surface is that generated by the revolution of a closed curve belonging to a family of curves now known as the *ovals of Descartes*. Such a curve may be defined as the locus of the point P such that $(m.\text{SP} + n.\text{S}'\text{P})$ is constant, where S and S′ are two fixed points, and m and n are real numbers.

In Book III of *La Géométrie* Descartes returns to the topic treated in Book I, namely the geometrical construction of the roots of determinate algebraic equations, but he now considers equations of higher degree than the second. (We have already met such curves in Figures 7.3 and 7.4.) Descartes tells us that: 'We should always choose with care the simplest curve that can be used in the solution of the problem.' To this end, it is necessary to know quite a lot about algebraic equations, in particular, for example, whether or not a given equation is reducible to one of lower degree. Much of Book III is concerned with the theory of algebraic polynomial equations. It treats such questions as how to remove a specified term of an equation, how to increase (or decrease) all the roots by a prescribed amount, and how to determine the number of positive or negative roots.

In this section we have sought to give an impression of Descartes's mathematical achievement. But was he the prime creator of the branch of geometry that bears his name? We shall consider this question after we have looked at the work of his great contemporary.

5. Pierre de Fermat

Our second subject, Pierre de Fermat (see Illustration 4), affords a sharp contrast – both in his way of life and his cast of mind – to his better-known rival, Descartes. Fermat's external life was quiet, orderly and uneventful; but his mathematical output was remarkable for its originality, depth and diversity. He is indeed one of the greatest 'pure' mathematicians of all time. While Descartes focused his mathematical thinking on the single subject of analytical geometry, Fermat made outstanding contributions to no less than four branches of mathematics: analytical geometry, the calculus, the theory of numbers and the mathematical theory of probability. His investigations into the last of these are embodied in an exchange of letters with Pascal, and is discussed at the end of the chapter.

Fermat was born at Beaumont de Lomagne in August 1601. His father was a leather-merchant; his mother came from a family of public service lawyers, a tradition carried on by her son. After the usual local schooling Pierre continued his studies at Toulouse, where he qualified as an advocate. In 1631 he married his mother's cousin (she was to present him with three sons and two daughters) and was installed as a Commissioner of Registry. In 1648 he was promoted to the post of King's Councillor in the Parlement of Toulouse, a position he filled with quiet competence and scrupulous integrity until his death in 1665 in his 65th year.

In addition to his considerable legal and administrative abilities, Fermat was a brilliant classical scholar and philologist with a wide knowledge of the main European languages and literatures. His verse compositions in Latin, French and Spanish were much admired. Above all, however, he devoted most of his leisure time to his favourite recreation, mathematics. He indulged his intellectual passion for the best of reasons – sheer love of the subject. Throughout his life Fermat was a hard-working and efficient public servant; how, then, did he find the time and energy to produce so much first-class mathematics? The answer (apart from his orderly habits

Illustration 4. Pierre de Fermat, 1601–65 (Mary Evans Picture Library)

and equable temperament) must be sought in the fact that parliamentary councillors were expected to hold themselves somewhat aloof from the community and to abstain from many of its social activities. Not being a professional mathematician, Fermat had little incentive to publish, and indeed most of his mathematical writings did not appear in print until after his death. Some, however, were circulated in manuscript within his lifetime, and he also revealed many of his discoveries, usually in abbreviated form, in letters to his wide circle of mathematical correspondents.

6. Fermat and coordinate geometry

Fermat was a cultural conservative – again unlike Descartes – who looked for inspiration to the classical masters, especially Apollonius and Diophantus, and to his immediate mentor, Viète. One of his early projects (1629) was to reconstruct a lost work of Apollonius, *Plane Loci* (p. 55), based, needless to say, on information given by Pappus. This led him to the discovery of the fundamental principle of analytical geometry, which he expressed thus:

Whenever in a final equation two unknown quantities are found, we have a locus, the extremity of one of these describing a line, straight or curved.

This profound statement was written at least a year before the appearance of Descartes's *La Géométrie*. Fermat developed his ideas further in a short treatise entitled *Ad locus planos et solidos isagoge* (Introduction to Loci Consisting of Straight Lines and Curves of the Second Degree). Fermat begins with the simplest case, the linear equation, which he writes as $ax + by = c^2$; he retains Viète's homogeneity. He sets up a pair of rectangular coordinate axes and draws the locus as a straight line in the first quadrant only. (Both Fermat and Descartes admitted only positive values of the coordinates.)

As an example of the power of his analytical approach, we may cite his discovery that:

Given any number of fixed lines in a plane, the locus of a point such that the sum of any multiples of the segments drawn at given angles from the point to the given lines is constant, is a straight line.

Having established that every equation of the first degree represents a straight line, he proceeds to investigate equations of the second degree in x and y. By applying the appropriate translation and/or rotation of the coordinate axes, he is able to reduce the general equation of the second degree to one of the familiar 'standard forms' for the parabola, the ellipse (with the circle as a special case) and the hyperbola, including both branches (with a pair of straight lines as another special case). Fermat concludes his treatise with a discussion of the proposition:

Given any number of fixed lines, the locus of a point such that the sum of the squares of the segments drawn at given angles from the point to the lines is constant, is a solid locus [i.e. a conic section].

The 'problem of Pappus' is disposed of, for good measure, in an appendix.

Ad locus was not published until 1679, some years after Fermat's death: that is why Descartes is widely acclaimed as the sole inventor of coordinate geometry. By the 1630s both men were in possession of the crucial idea of using algebraic equations to represent curves and to investigate their properties. It must be said, however, that Fermat's exposition is undoubtedly the more systematic and direct; it is also much easier to understand. He is primarily concerned with the concept of 'locus' – the path traced out by a point whose coordinates satisfy an indeterminate equation. Descartes, in contrast, places the emphasis on the use of determinate equations to solve geometrical construction problems; he had little interest in plotting curves from their equations and rarely used rectangular coordinate axes. Indeed, a recent historian of mathematics has gone so far as to say that 'there is a little in *La Géométrie* that resembles what is usually thought of today as analytic geometry.' Without going as far as that, there is no doubt that Fermat's approach is much closer to the modern treatment of the subject. However, Fermat retained the cumbersome 'syncopated' style of Viète, whereas Descartes's symbolic conventions are entirely modern, virtually those we use today.

7. Towards the calculus

Nowadays we think of the calculus (more precisely, the *infinitesimal calculus*) as a part of *analysis* – that large area of mathematics which is concerned with infinite processes, and in particular with the limiting values that are approached by varying quantities as certain 'infinitely small' quantities get smaller and smaller. One of the central concepts of analysis is that of a *function*, which expresses a relation between two variable quantities. We say that a variable y is a function of another variable x (known as the independent variable) if, given any value of x within a prescribed range, there is some rule which determines a *unique* value of y. We commonly write the relationship as $y = f(x)$. Simple examples are $y = ax^2 + b$ and $y = \sin x$, where x is any real number; and $y = \sqrt{x}$ when both x and y are positive. The scope of the function concept has been progressively enlarged during the last 400 years. The nineteenth-century analysts were prepared to accept such seemingly bizarre functions as '$y = 1$ when x is a rational number; $y = 0$ when x is irrational'. At this stage of our story, however, we can assume that all functions are 'well behaved' in the sense of being continuous and having, at most, only a few sudden changes of direction.

The calculus was created in the seventeenth century primarily to deal with problems posed by contemporary developments in science, most of which fell into one of four groups:

(1)   to find the instantaneous rate of change of a quantity, e.g. the velocity or acceleration of a projectile whose position varies with time;

(2)   to find the tangent and normal to a curve at any point on it;

(3)   to find the maximum or minimum values of a function;

(4)   to find the length of a specified portion of a curve (known as *rectification*), and also the areas enclosed by curves, the volumes bounded by curved surfaces, and the centres of gravity of such areas and volumes.

The first three of these are closely related and form part of the *differential calculus*; the fourth is the domain of the *integral calculus*. Fermat made major contributions to both.

In 1637, as an extension of his work on loci, Fermat produced a manuscript (again, not published during his lifetime) entitled *Methodus ad disquirendam maximam et minimam* (Method of Finding Maxima and Minima). His point of departure was the observation that the value of a continuous function changes very slowly near a maximum or minimum value of the function. Kepler, in 1615, had the same idea: 'Near a maximum the decrements on both sides are in the beginning only imperceptible.' If the function is plotted in the usual way, the slope of the corresponding curve at either a maximum or minimum point is zero, as illustrated in Figure 7.5.

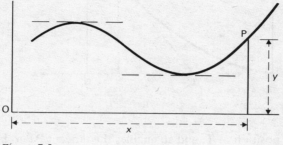

Figure 7.5.

Fermat first applied his method to the problem of dividing a line segment, AB in Figure 7.6, at a point C so that the rectangle AC.CB has maximum area. Setting $AB = a$, $AC = x$ and $AC' = x + e$, where CC' is small, he argues that AC.CB and AC'.C'B can be taken to be nearly equal. This yields

$$x(a - x) \approx (x + e)(a - x - e) \quad \text{or} \quad e(a - 2x) - e^2 \approx 0$$

(The symbol $\approx$ means 'nearly equals'.) At this stage Fermat first divides through by e and then sets $e = 0$, to give $x = a/2$. So to maximize the area AC.CB, AB is bisected at C.

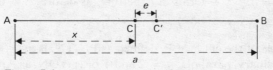

Figure 7.6.

Fermat then treats the related problem of finding the tangent PT to a curve at a point P on the curve. P' is a neighbouring point, and the construction is completed as shown in Figure 7.7. Fermat's objective is to find the length TN (now called the *subtangent*). This

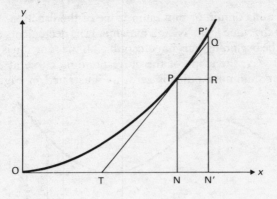

Figure 7.7.

gives the position of T, and so enables the tangent PT to be drawn. His argument goes like this. The triangles PQR and TPN are similar, so TN/PN = PR/QR. Now, since P and P′ are very close together, P′R is nearly equal to QR, so we can write $t/y \approx e/d$, where we have set TN = t, ON = x, PN = y, NN′ = PR = e and P′R = d. Fermat applied his 'tangent' method to many different curves. Here, as an illustration, is his treatment of the 'generalized parabola' whose equation is $y = x^n$, where n is a positive integer. Since P′, with coordinates ($x + e$, $y + d$), is also on the curve, we have

$$y + d = (x + e)^n = x^n + nx^{n-1}e$$

$$+ \text{ terms in } e^2 \text{ and higher powers of } e$$

giving $d/e = nx^{n-1}$ + terms in e. Setting, as before, $e = 0$, we get

$$t = ye/d = x^n/nx^{n-1} = x/n$$

and the *slope* of the curve at P, defined as PN/TN, is $ny/x = nx^{n-1}$. For the 'ordinary' parabola, $n = 2$ and T bisects ON, as Apollonius had proved some centuries earlier. Although Fermat does not attempt a rigorous justification of his method, it is operationally identical with the modern differentiation procedure. This is why Laplace was able to salute Fermat as 'the true inventor of the differential calculus'.

Descartes's approach to the tangent problem is quite different. He follows Apollonius (p. 61) in first constructing the normal to the

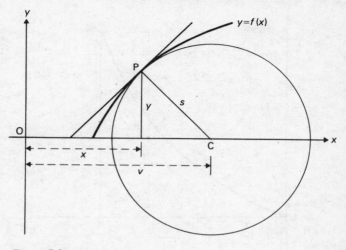

Figure 7.8.

curve at a point P and then defining the tangent as a line through P perpendicular to the normal at P. If we can draw a circle with centre C on the x-axis which touches the curve at P, as shown in Figure 7.8, then, says Descartes, CP will be the normal at P. The touching condition requires that a certain equation (in x) has two equal roots, corresponding to the coalescence of two points of intersection of the curve and the circle. This condition enables the position of the point C, and so the radius r, of the circle to be determined. The method, essentially algebraic, can only be applied in simple cases, but it avoids the use of infinitely small quantities.

To illustrate Fermat's approach to quadrature (or integration), let us see how he found the area under the 'generalized parabola' $y = x^n$. We wish to find the area OPN under the curve, where ON $= a$ and PN $= a^n$, as shown in Figure 7.9. Fermat divides the length ON into a large number of subintervals such that the x-values (or *abscissae*, as they are called) of successive points of division – working backwards from N to O – are in decreasing geometric progression, namely $a, ar, ar^2, \ldots$, where $r < 1$. The corresponding y-values (or *ordinates*) are constructed as shown, and the desired

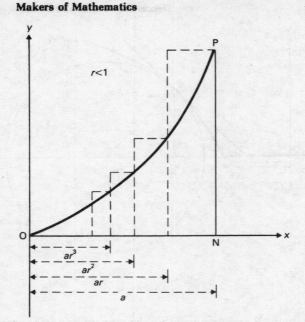

Figure 7.9.

area is taken to be approximately equal to the sum of the rectangular areas so formed. These areas are

$$a^n(a - ar), (ar)^n(ar - ar^2), (ar^2)^n(ar^2 - ar^3), \ldots$$

We now have, in fact, a geometric progression with first term $a^{n+1}(1 - r)$ and common ratio r^{n+1}. The total area of p such rectangles is therefore

$$\frac{a^{n+1}(1 - r)}{1 - r^{n+1}} \left\{ 1 - r^p(n + 1) \right\} =$$

$$\frac{a^{n+1}}{1 + r + r^2 + \ldots + r^n} \left\{ 1 - r^p(n + 1) \right\}$$

As r approaches 1 (from below), the rectangles become narrower and more numerous and their sum approaches the desired area under the curve. As P increases indefinitely the quantity r^p becomes smaller and smaller and may ultimately be neglected. The final step in the limiting process is to set $r = 1$ to give the area OPN as $a^{n+1}/(n + 1)$.

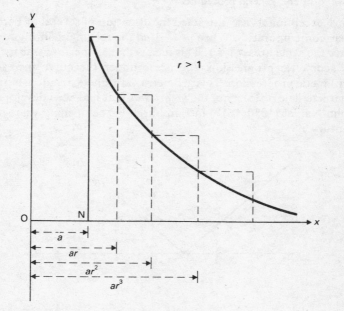

Figure 7.10.

So far we have assumed that n is a positive integer. The results can easily be generalized to cover the case when n is a rational fraction, p/q. When n is a negative integer we have a family of 'generalized hyperbolas' given by $y = x^{-m} = 1/x^m$, with $m > 0$. Fermat's procedure here is to take $r > 1$ and then work forwards from $x = a = $ ON, as shown in Figure 7.10, to obtain the area under the curve 'from a to infinity' as the limit of the sum of a large number of rectangles as r approaches 1 from above. The desired area is $1/(m - 1)a^{m-1}$. Clearly the procedure fails when $m = 1$, i.e. when $xy = 1$ (a rectangular hyperbola with the x- and y-axes as asymptotes). Fermat was, of course, aware of this and states the position thus:

I say that all these infinite hyperbolas except the one of Apollonius, or the first, may be squared by the method of geometric progression according to a uniform and general procedure.

This 'exceptional case' attracted the attention of several of Fermat's near-contemporaries. When $m = 1$ *all* the rectangles in Figure 7.10 have the same area, $r - 1$. This means that as the abscissae increase in geometric progression, the areas under the curve increase in arithmetic progression. This geometric–arithmetic relationship is the characteristic property of the *logarithms* that had been developed by John Napier (1550–1617) in Edinburgh in the opening years of the century.

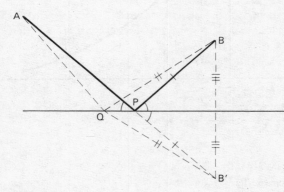

Figure 7.11.

Although primarily a 'pure' mathematician, Fermat was always ready to apply his mind to physical problems. In two letters written in 1657 and 1662 he enunciated his 'principle of least time'. This states that a ray of light travelling from a point A to another point B, and being reflected and refracted in any manner in the course of its journey, will take the quickest path from A to B. Figure 7.11 illustrates a single reflection. Clearly the shortest – and hence the fastest – path from A to B via the reflecting surface (i.e. AP + PB) requires P to be so positioned that the marked angles are equal. Any

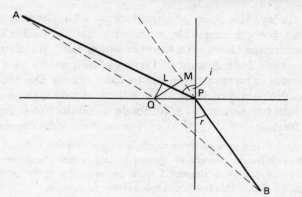

Figure 7.12.

other path (e.g. AQ + QB) is obviously longer. Figure 7.12 shows a single refraction. The ray from A to B is 'bent' when passing from one medium to the other. APB is the actual path; AQB is a slightly displaced path. If we assume that AP − AQ ≈ LP = PQ sin i and QB − PB ≈ MP = PQ sin r (where i and r are the angles of incidence and refraction of the ray APB), then the time difference between rays travelling from A to B along the two paths may be taken as

$$PQ \left(\frac{\sin i}{v_1} - \frac{\sin r}{v_2} \right)$$

where v_1 and v_2 are the velocities of light in the two media. Since, by his principle, the time from A to B is a minimum, Fermat is able to set the time difference equal to zero. The result, known as Snell's law, follows at once. It is usual to write $\mu = v_1/v_2$ and to express Snell's law as sin $i = \mu$ sin r. We shall see in Chapter 15 how Fermat's optical principle was generalized in the nineteenth century to enable the behaviour of a variety of physical systems to be interpreted.

8. The theory of numbers

There can be little doubt that Fermat's favourite mathematical recreation was to explore the properties of the whole (or natural) numbers. He is universally acclaimed as the founder of modern number theory. His point of departure was Diophantus, whose works had become accessible again with Claude Bachet's 1621 publication of the Greek text, together with a Latin translation, of the *Arithmetica*. Indeed, some of Fermat's finest results were noted in the margins of his copy of the Bachet Latin edition. One such note relates to Problem 8 of Book II (p. 87): 'to divide a given square number into two squares', i.e. to obtain a solution, in integers or rational fractions, of $x^2 + y^2 = a^2$. Fermat's marginal comment is:

On the contrary, it is impossible to separate a cube into two cubes, a fourth power into two fourth powers, or, generally, any power above the second into powers of the same degree: I have discovered a truly marvellous demonstration which this margin is too narrow to contain.

Alas, Fermat never found a strip of paper wide enough to contain his proof! What he is asserting, in modern terminology, is that a rational solution of $x^n + y^n = a^n$ is impossible when $n > 2$. This result is known as 'Fermat's last theorem'; no problem in the whole of mathematics has attracted more attention and generated more labour. The result has been proved for a vast number of particular values of n (in fact, for all n less than 4 million), but – until perhaps very recently, as reported in Appendix 8 – a completely general proof still eludes us. Fermat also made a number of conjectures, some of which have turned out to be false. We shall look at one of these in Chapter 12, and see how Euler was able to produce a counterexample. The point to stress, however, is that in no case where Fermat claimed to have a *proof* (rather than a mere conjecture) has he been shown to be wrong.

Another of Fermat's major discoveries, known simply as 'Fermat's theorem', is that if p is a prime number and a and p are relatively prime (i.e. have no common factors), then $a^p - a$ is divisible by p (e.g. if $a = 4$ and $p = 3$, then $a^p - a = 60 = 3 \times 20$). This theorem was to be of fundamental importance in later developments. Here is a simple proof. Consider the expansion of the multinomial expression

$$(x_1 + x_2 + \ldots + x_a)^p$$

where a and p satisfy the conditions of the theorem. If we multiply out, we get a terms of the form $(x_i)^p$, with all the other terms containing the factor p. Setting all the x_i equal to 1, we get, with an obvious notation, $a^p = a + pF(p)$, and the result follows at once.

To indicate the range of Fermat's discoveries, here are four more of his theorems:

(1) No prime number of the form $4n + 3$ is the sum of two squares. Every prime number of the form $4n + 1$ is the sum of two squares, and can be so expressed in only one way.

(2) The 'Pell equation' (pp. 53 and 87), $x^2 - Ay^2 = 1$ (where A is a positive integer which is not a perfect square), has an unlimited number of integral solutions. Fermat also asserted that he could determine the values of A and B for which the equation $x^2 - Ay^2 = B$ is solvable in integers, and could solve the equation for these values.

(3) The solution $y = 3$, $x = 5$ is the *only* rational solution of the Diophantine equation $y^3 = x^2 + 2$.

(4) Fermat generalized the familiar $3:4:5$ result for right-angled triangles with integral sides by asserting that a prime number of the form $4n + 1$ is the hypotenuse of one and only one right-angled triangle with integral sides; that the number $(4n + 1)^2$ is the hypotenuse of two and only two such triangles; and so on. In general, the number $(4n + 1)^s$ is the hypotenuse of s and only s such triangles, for any positive integral value of s. Taking $n = 1$, the hypotenuse 5 gives the triangle $(5, 4, 3)$; 5^2 gives $(25, 20, 15)$ and $(25, 24, 7)$; 5^3 gives $(125, 100, 75)$, $(125, 120, 35)$ and $(125, 117, 44)$. One essentially 'new' triangle emerges at each step in the sequence.

Fermat rarely gave details of his proofs. They had to be found – often only after considerable labour – by his successors, notably Euler and Legendre. Fermat published virtually nothing on number theory during his lifetime, and even today we have to rely for most of our information on his marginal notes and letters. We do know, however, that he made some of his discoveries (e.g. theorem (1) above) by using a technique of his own devising known as the 'method of infinite descent'. It is not at all easy to apply; there are no general rules as to how to proceed. It seems that Fermat first employed the method to prove a number of negative propositions, for example that there are no rational solutions of the equation $a^3 + b^3 = c^3$. Positive theorems gave him more trouble, as he explains in a letter written in 1659:

For a long time I was unable to apply my method to affirmative propositions because the twist and the trick for getting at them is much more troublesome than that which I use for negative propositions.

He mentions his difficulty in proving the second part of the theorem (1) above:

I found myself in a fine torment. But at last a meditation many times repeated gave me the light I lacked. The course of my reasoning in affirmative propositions is such: if an arbitrarily chosen prime of the form $4n + 1$ is not a sum of two squares, I prove that there will be another of the same nature, less than the one chosen, and next a third still less, and so on. Making an infinite descent in this way we finally arrive at the number 5, the least of all numbers of this kind. It follows that 5 is not a sum of two squares. But it is. Therefore we must infer by a *reductio ad absurdum* that all numbers of the form $4n + 1$ are sums of two squares.

Since we have so few of Fermat's proofs, we conclude this section by looking at Euler's 1770 'descent' proof of a negative proposition – the special case of Fermat's last theorem when $n = 4$. In fact Euler proved a stronger theorem, namely that the equation $a^4 + b^4 = c^2$ (not c^4) has no integral solutions, and therefore no rational solutions. The proof is by contradiction, so we start by assuming that there *are* two numbers a and b such that $a^4 + b^4$ is a square number. We may assume that any common factors have been divided out, so that a and b are relatively prime. This means, in particular, that they cannot both be even. If, on the other hand, they are both odd, we can set $a = 2r + 1$ and $b = 2s + 1$. On multiplying out the two binomial expressions we see that $a^4 + b^4$ must be an even number of the form $4A + 2$. Now, an even square number must clearly be divisible by 4, and $4A + 2$ is not. So a and b cannot both be odd: they must, therefore, be of opposite parity. Let us assume that a is odd. We can then write

$$a^2 = p^2 - q^2, \qquad b^2 = 2pq$$

where p and q are relatively prime and of opposite parity. Now, if p is even and q is odd, $p^2 - q^2$ is of the form $4C - 1$, whereas all odd square numbers are of the form $4D + 1$. Thus $p^2 - q^2$ can be a square number only if p is odd. Since $2pq$ is a square number, both p and $2q$ must be squares since they are relatively prime. Further, since $p^2 - q^2$ is a square, it is necessary that

$$p = m^2 + n^2 \quad \text{and} \quad q = 2mn$$

where again m and n are relatively prime integers of opposite parity. But since $2q = 4mn$ is a square, so also is mn. Both m and n must therefore be squares, so we can set $m = x^2$ and $n = y^2$, to give

$$p = x^4 + y^4$$

We have already established that p is a square. Thus it follows that if $a^4 + b^4$ is a square, then so also is $x^4 + y^4$, where x and y are clearly much smaller than a and b. Continuing in this way we obtain a sequence of smaller and smaller numbers, u and v say, such that $u^4 + v^4$ is a square. We must eventually reach the smallest such pair of numbers. But we know that there is no such pair, from which it follows that there is no pair of numbers, however large, having the specified property. So our initial assumption must be false, and the theorem is proved by contradiction. We have got there at last; as we remarked earlier, 'method of descent' proofs do not come easily.

9. Blaise Pascal

The last of our three subjects, Blaise Pascal, comes across to us over the years as a tragic figure. Blessed with a rare talent for mathematics and an excellent literary style, his short life was blighted by chronic illness, both physical and mental. His outstanding intellectual powers were exercised mainly on sterile theological speculations occasioned by the sectarian religious controversies of his day.

Pascal (see Illustration 5) was born in Clermont in the Auvergne on 19 June 1623; his mother died four years later. His father, Étienne, was a lawyer and a keen amateur mathematician. (It is his name, not his son's, that is attached to the kidney-shaped curve known as the *limaçon of Pascal*.) The family was completed by two beautiful and talented sisters, Gilberte and Jacqueline. At the age of 7 Blaise moved with his family to Paris. His precocious talents were already apparent, and Étienne, mindful of his son's poor physique, tried to slow down the pace of his studies. However, Blaise's passion for mathematics overcame the parental anxieties, and his youthful geometrical feats have become legendary. At the age of 14 he was allowed to accompany his father to the weekly meetings of the Paris mathematical circle organized by Father Mersenne. During this time Étienne was having trouble with the authorities, who wanted him to relax his severe standards of financial rectitude. He incurred the displeasure of Cardinal Richelieu, and thought it prudent to go into hiding for a time. The story is that Jacqueline's brilliant acting and dancing so captivated the Cardinal that he restored her father to favour, and even placed him in a comfortable job at Rouen. It was here, at the age of 16, that Blaise produced a short paper in which he enunciated one of the most beautiful theorems in mathematics (to be discussed in the next section).

Two years later the versatile youth invented and built one of the earliest genuine calculating machines. He conceived the design, we are told, to assist his father who, at the time, was engaged in reorganizing the collection of the local taxes. Pascal's machine is essentially an adding device; subtraction is performed by the addition of complements. What makes it a true calculating machine is that the 'carry' in addition, from one place to the next more significant place, is dealt with mechanically – by means of a ratchet that is gradually raised as the number being registered on the 'results wheel' approaches 9. Several of Pascal's machines were made and worked well. They must have included those refinements in design that make all the difference between a paper conception and a working instrument.

By this time, alas, Pascal's health was deteriorating; for the rest of his life he suffered from acute dyspepsia, chronic insomnia and recurring bouts of severe depression. In 1646, at the age of 22, he experienced his first religious conversion. His family had embraced the austere and intolerant creed of Jansenism, and Blaise was convinced that God had specially called him to attack the Jesuits in this life and prepare them for their predestined fate of eternal damnation in the next. His mathematical activities virtually ceased, but in 1648 his early interest in physics was briefly rekindled. He made a number

Illustration 5. Blaise Pascal, 1623–62 (Mansell Collection)

of experiments to find out how atmospheric pressure varied with height and developed the necessary theory to explain his findings.

Blaise and Jacqueline then returned to Paris, to be joined later by their father, now once more in good standing as a state councillor. There the family received a formal visit from Descartes, but the two men were so mutually antipathetic that the visit was not a success. At this point, in 1648, Jacqueline declared her intention to become a nun and to move into the Jansenist convent of Port Royal, near Paris. Étienne was strongly opposed to the project. Blocked by parental opposition, Jacqueline directed her religious fervour towards her unfortunate brother who, in her view, was not so thoroughly converted as he should be. In 1650 Étienne died and Jacqueline was free to enter Port Royal.

In November 1654, Pascal experienced another conversion; this time it was for life. The story is that he was driving a four-in-hand when the two leading horses plunged over the parapet of the bridge at Neuilly. Fortunately the traces broke and Pascal survived unhurt. He took the incident as a divine warning of the danger of plunging over the moral precipice, a fear that had haunted his morbid imagination for years. He was indeed a miserable sinner who had been snatched from the very mouth of hell. Thereafter the fear of the precipice never left him. Supported by his sister, he turned his back on the world and moved into Port Royal himself – to contemplate, as he put it, 'the greatness and misery of man'. It was while at Port Royal that Pascal wrote his *Lettres provinciales* and his *Pensées*: both are now literary classics. The *Pensées* are a collection of notes for a theological treatise that was never completed. The *Lettres* (18 in all) were written in defence of a prominent Jansenist who was facing a charge of heresy laid by the Jesuits.

During this time, the mathematical powers of the neurotic invalid remained as strong as ever and he fell from grace once more, to great effect. He was lying awake one night in 1658, racked by severe toothache. To distract his mind, he began to think hard about the *cycloid*, an elegant curve which attracted considerable attention throughout the seventeenth century. Much to Pascal's surprise, the pain stopped. He took this as a sign from Heaven that he could address his mind to the cycloid without endangering his immortal

soul, and then spent eight concentrated and highly productive days
on what proved to be his last mathematical effort. Later in the year
his illness worsened; four years later, in August 1662, his tortured
existence came to an end in convulsions. He was not yet 40. Pascal
has been called the great 'might have been' of mathematics. Apart
from his literary productions, he will probably be best remembered
for two of his many achievements: the 'mystic hexagram' theorem in
geometry and the creation, with Fermat, of the mathematical theory
of probability.

10. Pascal's theorem and projective geometry

While attending the meetings of the 'Mersenne Academy' (the
forerunner of the Académie Royale des Sciences), the young Pascal
met Desargues, who encouraged him to apply the new methods of
projective geometry to the study of the conic sections. The result was
spectacular. In 1640, at the age of 16, Pascal astonished the world by
publishing a short tract entitled *Essai pour les coniques* in which he
announced his famous theorem of the *hexagrammum mysticum*. It is
now known simply as Pascal's theorem, and states that the three
'meets' (i.e. the points of intersection) of opposite sides of a hexagon inscribed in a conic are collinear, as indicated in Figure 7.13.
(Figure 7.14 shows another configuration.) Pascal's formulation
was to say that, in either Figure 7.13 or Figure 7.14, if AB and

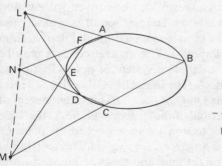

Figure 7.13.

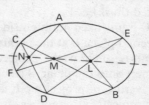

Figure 7.14.

DE intersect at L, and BC and EF intersect at M, then LM, CD and FA form a *pencil* of rays emanating from the point N. The power of the theorem is attested by the large number of corollaries that may be deduced from it. Many of them are obtained by letting two or more of the six vertices coincide. Here, by way of illustration, are two examples.

(1)   Figure 7.15 shows how to construct the tangent to a conic at a prescribed point A. Comparing it with Figure 7.13, we see that the points A and B have been made to coincide. We obtain the points M and N as before. If DE meets MN at L, then LA is the tangent to the conic at A.

(2)   If we let A and B, and also D and E, coincide as shown in Figure 7.16, we see that the tangents at A and D meet (at L) on the line MN, which is a diagonal of the complete quadrilateral ACDF. This yields the general result that if a quadrilateral is inscribed in a conic, the tangents at its vertices meet in pairs on the sides of the 'diagonal triangle', i.e. the triangle formed by the three diagonals of a complete quadrilateral. Thus, in Figure 7.16 MN is one side of the diagonal triangle of the quadrilateral ACDF. In the special case when the conic (in this case a hyperbola) degenerates into a pair of straight lines, Pascal's theorem reduces to the 'theorem of Pappus' discussed in Chapter 4 (p. 90 and Figure 4.17).

Pascal did not give an explicit proof of his theorem. He says only that since it is true of a circle, it must, by *projection and section*, also be true of all conics. The 'circle' result can be proved without too much difficulty by classical methods. The method of projection and section is indeed implied in the term 'conic section' and derives from the Greeks, as we saw in Chapter 3. The general concept may be stated thus. Let $P_1, P_2, P_3, \ldots$ denote a system of points in a plane A; O is any fixed point outside A. The lines $OP_1, OP_2, OP_3, \ldots$ meet a second plane B at the points $Q_1, Q_2, Q_3, \ldots$. This latter system of points is said to be the 'projection' of the system in plane A (the 'base plane') onto the plane B (the 'section') with respect to the point O, the 'vertex' of projection. It is apparent, for example, that a straight line projects into another straight line, and that the point of intersection of two lines projects into the point of intersection of their

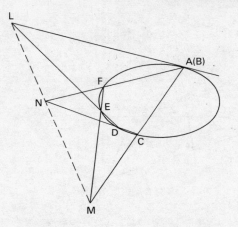

Figure 7.15.

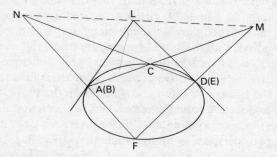

Figure 7.16.

projections. A conic remains a conic under projection; indeed, any
conic can be projected into a circle, and, conversely, the projection
of a circle is a conic. Projective geometry is concerned with those
geometrical properties that are unaltered by the operation of projection and section. The 'metrical' theorems of geometry, those which
involve absolute lengths, areas or angles, congruence, similarity, etc.,
have no place in projective geometry.

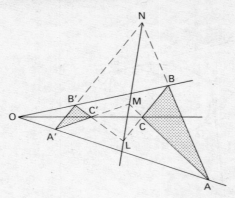

Figure 7.17.

The founder of projective geometry was Girard Desargues (1593–1661), an architect and military engineer from Lyons. He spent some years in Paris where, as we have seen, he met the young Pascal. His major work, published in 1639, virtually created a new branch of mathematics, but it used curious terminology and was difficult to understand. Indeed, with the notable exceptions of the three subjects of this chapter, it was almost entirely ignored by his contemporaries. His basic theorem, still called Desargues's theorem, is illustrated in Figure 7.17. The triangle A′B′C′ is obtained from the triangle ABC by projection and section from the vertex, O. The theorem states that the three pairs of corresponding sides such as AB and A′B′ (suitably extended) of two 'perspective triangles', as they are called, meet in three collinear points (L, M and N in the figure). The converse theorem is also true. Both theorems are valid when the two triangles are in the same or in different planes. In the latter case the proof is very simple since LMN is the line of intersection of the planes in which the two triangles lie. In his essay on conics, Pascal handsomely acknowledged his debt to Desargues when he said of him: 'I should like to say that I owe the little that I have found on this subject to his writings.'

To ensure the full generality of Pascal's theorem, we must allow for the possibility that one or more pairs of opposite sides of an inscribed hexagon may be parallel (as, for example, in a regular hexagon). To deal with this case, Desargues introduced the concept of the *line at infinity*. Any two parallel lines may be regarded as meeting at a point on this fictitious line, the position of the point along the line being determined by the direction of the parallel lines. This yields yet another corollary to Pascal's theorem: if two pairs of opposite sides of a hexagon in a conic are parallel, then so is the third pair. Most modern proofs of Pascal's theorem involve ideas such as the *cross-ratio* of four points, which although familiar to Desargues, and even to Pappus, did not move to the centre of the mathematical stage until the nineteenth century. These later developments are beyond the scope of this book.

11. The cycloid

No transcendental curve was studied more intensively during the seventeenth century than the *cycloid* – the curve traced out by a marked point on the circumference of a circle as it rolls along a fixed straight line, as shown in Figure 7.18. When the circle (of radius a) has rolled through an angle θ, the marked point, initially at A, has moved to P, where the circular arc PN = AN = $a\theta$. The angle θ must be measured in *radians* (or *circular measure*), i.e. as the ratio of the length of a circular arc to its radius. (Thus, for example, a right angle = $\pi/2$ radians.) Taking coordinate axes as shown, the

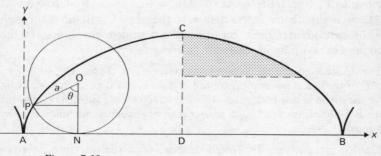

Figure 7.18.

Figure 7.19.

coordinates of the point P are

$$x = a\theta - a \sin \theta, \qquad y = a - a \cos \theta$$

Clearly, AB $= 2\pi a$ and CD $= 2a$. The complete curve consists of an infinite number of repeated arches, a single arch being traced out by one revolution of the rolling circle. Adjacent arches are connected by *cusps*, as at A and B in Figure 7.18.

In 1630 Galileo Galilei (1564–1642) had called attention to the graceful shape of the cycloid and suggested that arches of bridges should be built in this form. (A bridge across the Cam in the grounds of Trinity College, Cambridge, has cycloidal arches.) In his *Discorsi a due nuove scienze* (Dialogues Concerning Two New Sciences) of 1638 – the book that established dynamics as a mathematical science – Galileo presented his readers with an apparent paradox. During one complete revolution of the rolling circle, the point B in Figure 7.19 will move to the position E; the centre will move from A to D; and the point C inside the circle on the radius AB will move to F, where BE $=$ AD $=$ CF $= 2\pi a$ and AB $= a$. Now, if AC $= b$, say, then b is less than a, so the point C will move through a distance greater than $2\pi b$ during a complete revolution of the circle. Galileo puts his question in these terms:

The circumference of the smaller circle never leaves the line CF, so that no part of the line is left untouched, nor is there ever a time when some point on the circle is not in contact with the straight line. How, now, can the smaller circle traverse a length greater than its circumference unless it go by jumps?

Galileo gives a very thorough discussion of the question, starting with an analysis of the motion of a regular hexagon which 'rolls' in a bumpy motion along a straight line. The circle is then presented as the limiting case of a polygon with an infinitely large number of sides. It is, of course, the transition from polygon to circle that contains the essence of the paradox. Galileo sums up in these words:

These difficulties are real; and they are not the only ones. But let us remember that we are dealing with infinities and indivisibles, both of which transcend our finite understanding, the former on account of their magnitude, the latter because of their smallness. In spite of this, men cannot refrain from discussing them, even though it must be done in a roundabout way.

The quadrature of the cycloidal arch was evaluated independently in about 1630 by Gilles de Roberval (1602–73) in Paris and by Evangelista Torricelli (1608–47), a pupil of Galileo, in Florence. Roberval's argument, as presented in his *Traité des indivisibles* of 1634, goes like this. He constructs a second curve, which he calls the 'companion to the cycloid', and which is the locus of the point Q in Figure 7.20, such that $AN = a\theta$ and $QN = a - a\cos\theta$. Now, $PQ = a\sin\theta$ is a chord of the semicircle MPN. Invoking the fashionable 'method of indivisibles' (p. 79), Roberval argues that the area (shown shaded) between the cycloid and its companion can be thought of as being made up of an infinite number of lines, such as PQ (or, if preferred, of very thin rectangles). It follows that the shaded area = sum of the 'lines' PQ = sum of the 'chords' of the semicircle = area of the semicircle = $\frac{1}{2}\pi a^2$. Using a similar argument, but with vertical lines, he asserts that the area AQCD is equal to the area CQAB, since any line such as RS in the first area can be

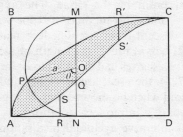

Figure 7.20.

paired with a corresponding line R'S' in the second area, where AR = CR'. Each of these areas is therefore equal to half that of the rectangle ABCD, i.e. to πa^2, the area of the rolling circle. So:

area of a cycloidal arch = 2 (area of semicircle + area of circle)

$$= 3\pi a^2$$

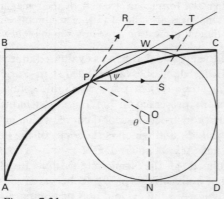

Figure 7.21.

Roberval went on to construct a tangent to the cycloid at any point P on the curve. He thought of the motion of P as the resultant of two motions: one of translation, represented in Figure 7.21 by PS, and one due to the rotation of the generating circle, represented by PR. The two motions are equal in magnitude, so the required tangent PT is in the direction that bisects the angle RPS and passes through W, the uppermost point of the circle. The angle ψ in Figure 7.21 is equal to $90° - \theta/2$. This method of Roberval's is, of course, in the direct line of descent from Archimedes' treatment of his spiral (p. 73). Finally, Roberval found the length of a cycloidal arch (it is $8a$), something that Descartes had declared to be impossible. Christopher Wren (1632–1723), the great architect, had indeed succeeded in rectifying the cycloid when he showed that, in Figure 7.21, the arc PC = 2PW ($= 4a \cos \theta/2$, in modern notation).

During his eight days of intensive cycloidal meditation, Pascal discovered a number of new theorems. Among other things, he calculated the area of the segment of a cycloid shown shaded in Figure 7.18, the 'centroid' of such a segment, and the volumes of the solids of revolution generated when such a segment is rotated about either CD or DB. Then, under a pseudonym, he announced some of his findings as a challenge to other mathematicians and offered two prizes for the best demonstrations. In the event only two replies were received, and neither was considered prizeworthy. Pascal then published his own solutions, together with some further results, in a series of *Lettres de Dettonville*. Another product of Pascal's later years was the *Traité des sinus du quart de cercle*. In this work he came very close to discovering the calculus, as Leibniz handsomely acknowledged when he read the text some years later.

We cannot leave the cycloid without mentioning two remarkable properties that were discovered by Christiaan Huygens (1629–95), the leading mathematical physicist of his time. In 1658 he discovered that the cycloidal arch is a true *tautochrone*: a 'frictionless particle', if allowed to slide from rest down an inverted cycloidal arch, will take the same time to reach the bottom, no matter where it starts from. This is exactly what is needed for a pendulum to keep perfect time, regardless of the size of swing. The problem was: how could the bob of a pendulum be constrained to oscillate in a cycloidal rather than in a circular path? Huygens then made this second discovery. In Figure 7.22, PQ and PR are two equal inverted cycloidal half-arches, with a cusp at P as shown. Huygens was able to prove

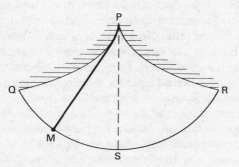

Figure 7.22.

that the bob, at M, of a pendulum PM of length equal to the arc PQ (or PR) will oscillate, if constrained by PQ and PR, in an arc QSR of an equal cycloid. Such a pendulum oscillating between cycloidal jaws will be *isochronous*: the period of oscillation will be independent of the angle of swing. Although the horological advantages turned out to be modest, Huygens' discoveries proved to be of great mathematical importance, but we cannot pursue the subject here.

Our discussion of the cycloid in this section illustrates two characteristic features of the mid-seventeenth-century scene. First we have the ingenuity and variety of the special methods that were devised before the fully developed calculus was available to provide a battery of general operational procedures; and secondly, the number of first-class mathematicians who were investigating the properties and applications of the 'higher curves' – albeit without bothering too much about the strength of the logical foundations of their discoveries – was considerable.

12. The mathematical theory of probability

In 1654 the Chevalier de Méré, an expert gambler, consulted Pascal on some problems connected with games of chance. How, for example, should the stake be divided between two players of equal skill if their game were to be interrupted prematurely – when, for instance, the first player to throw eight 'heads' wins, and the game is stopped when one player has thrown three heads and the other five? Pascal wrote to Fermat on the matter, and the ensuing correspondence (a set of eight letters, the first of which is lost) can be said to mark the birth of the mathematical theory of probability. Initially, the two men used different approaches to obtain the same results. However, as the correspondence developed, attention was increasingly focused on 'combinatorial methods' for investigating the number of different ways in which a specified outcome can occur, on the tacit assumption that all the various possibilities are equally likely. Thus, if we throw two dice, there are 36 equally likely outcomes. One yields two 6's, 10 yield one 6, and 25 no 6's. So the probabilities of these outcomes may be taken to be 1/36, 10/36 and

Table 7.1. Pascal's triangle

Column number	0	1	2	3	4	5	6	7	...
Sum of column entries	2^0	2^1	2^2	2^3	2^4	2^5	2^6	2^7	...
	1	1	1	1	1	1	1	1	...
		1	2	3	4	5	6	7	...
			1	3	6	10	15	21	...
				1	4	10	20	35	...
					1	5	15	35	...
						1	6	21	...
							1	7	...
								1	...
									...

25/36: the three terms in the expansion of the binomial expression $(1/6 + 5/6)^2$.

Some years earlier Pascal had written a treatise, not published until after his death, entitled *Traité du triangle arithmétique*. Table 7.1 shows his triangle, which although known in China many centuries earlier, and to Cardano and others in the sixteenth century, is now always known as Pascal's triangle. The numbers in the nth column (counting from 0) are the coefficients of successive terms in the expansion of $(a + b)^n$; they sum to 2^n. The number in the rth position in the nth column (again counting from 0) is the number of ways in which r objects can be selected from a collection of n objects. It is called the number of *combinations* of n things taken r at a time, is written as

$$^nC_r \quad \text{or} \quad \binom{n}{r}$$

and is equal to

$$\frac{n(n - 1)(n - 2)\ldots(n - r + 1)}{1 \times 2 \times 3 \times \ldots \times r}$$

The denominator of this fraction is usually written as $r!$ and called 'r factorial'. Since ^nC_r must be a positive whole number, it follows that the product of *any r* consecutive positive integers is divisible by $r!$ – an important result in the theory of numbers.

Pascal made a thorough study of the properties of his triangle, in the course of which he gave the first clear exposition of the principle of *mathematical induction*. This principle, which asserts the validity of reasoning by recurrence, is now regarded as a fundamental axiom of modern mathematics. It may be stated thus: 'If a property is true of the number 1, and is also true of the number $n + 1$ provided it is true of n, then it will be true of all whole numbers.' Here is a simple example of a proof by induction. If we assume (p. 16) that

$$S(n) = 1 + 3 + 5 + \ldots + (2n - 1) = n^2$$

then

$$S(n + 1) = n^2 + (2n + 1) = (n + 1)^2$$

Since $S(1) = 1^2$, the general result is established. Many properties of the numbers in Pascal's triangle can be conveniently proved by induction. Such proofs rely heavily on a theorem, sometimes known as Pascal's recurrence law, which asserts that $^nC_r + {}^nC_{r-1} = {}^{n+1}C_r$. This law (and other results in what is now called 'combinatorial mathematics') was skilfully exploited by the two correspondents. The build-up of their reasoning – to answer objections and to deal with ever more difficult problems, as the exposition passes from one man to the other and back again – makes compelling reading, but we must refer the interested reader to the letters themselves. (There are translations in Reference 8.)

8   Newton

'. . . the statue stood
Of Newton with his prism and silent face,
The marble index of a mind forever
Voyaging through strange seas of Thought, alone.'
William Wordsworth, *The Prelude*

1. Introduction

With Isaac Newton (see Illustration 6) we reach the second member
of the triumvirate of mathematical 'supremos'. To some, indeed, he
is revered as the noblest intellect the human race has produced, 'he
who in genius surpassed the human kind'. But how can we compare
a Newton with a Shakespeare, a Mozart with a Rembrandt? Until
recently Newton's fame has rested mainly on his contributions to
natural philosophy as set out in his two great books: the *Principia*,
published in 1687, and the *Opticks*, published in 1704. A few of his
mathematical works found a publisher during the eighteenth century, but the great bulk of his writings remained unpublished – and
so almost unknown – until recent times. At long last, almost two
and a half centuries after his death, the Newtonian archives have
been opened, sifted and published. Now, for the first time, we are
able to appreciate the variety and depth of his mathematical achievements. For this happy state of affairs we are indebted to Professor
D. T. Whiteside of Cambridge. He edited the scholarly, definitive
The Mathematical Papers of Isaac Newton (Reference 10), published
in eight large volumes between 1967 and 1981, which brings together
in a fully annotated and critical edition – in the original Latin

and in English translation – all Newton's significant mathematical manuscripts, whether previously published or not.

There has, indeed, been a most welcome flow of scholarly Newtonian publications during the past thirty years. They include the seven volumes of the Royal Society's edition of Newton's correspondence (1959–81); a new two-volume Latin edition of the *Principia*, with variant readings (1972); a definitive biography of Newton by Professor R. S. Westfall (1980, Reference 42); and a detailed study of the famous dispute between Newton and Leibniz (1980, Reference 34). Newton's masterpiece is, of course, the *Principia*. It is not an easy book to read – indeed, it is not much read – and it should certainly not be disposed of in a few paragraphs. It has therefore been accorded a chapter of its own (Chapter 9).

2. His life

Isaac Newton was born in 1642 in the family manor house at Woolsthorpe in Lincolnshire. Galileo had died a few months earlier, and the English Civil War was in its opening stages. Newton's forebears were prosperous Lincolnshire yeomen. His grandfather, Robert, purchased the manor of Woolsthorpe in 1623 and so became a man of some local importance. Robert Newton settled the estate on his eldest son Isaac in 1641, who was thus enabled to marry his betrothed, Hannah Ayscough, in April 1642. It is surprising that none of Newton's paternal ancestors was able to sign his own name: their wills bear only their marks. It was, in fact, through the Ayscough marriage that the Newtons made their first contact with formal learning. Hannah's brother, William, an MA of Cambridge, became rector of a neighbouring parish in 1642. It is indeed fortunate that the young Isaac was brought up by the Ayscoughs rather than by the Newtons, who might well have left him as illiterate as they were themselves.

Isaac Newton senior – 'a wild, extravagant and weak man' – died in October 1642; his famous son was born posthumously on Christmas Day of that year. He was a premature baby, 'so tiny that no one expected him to survive', and his life hung in the balance for a week. (He was to survive to the age of 84.) The security of Newton's childhood was further disrupted at the age of 3 when his mother

Illustration 6. Isaac Newton, 1642–1727 (National Portrait Gallery)

married an elderly clergyman, Barnabas Smith, the rector of the next parish, and bore him three children. Isaac was left at Woolsthorpe with his grandmother Ayscough. He never recorded any affectionate recollections of either his grandmother or his stepfather. In a psycho-analytical study of Newton by Professor F. Manuel, published in

1968, this maternal remarriage is presented as the crucial event of Newton's life.

In 1653 Barnabas Smith died and his widow returned to Woolsthorpe with her three young children. Two years later Isaac was sent off to the grammar school at Grantham, where he learnt Latin thoroughly, and a smattering of Greek, but virtually no mathematics. He lodged with the town apothecary and developed a talent for making mechanical devices and toys. When he was 17 years of age his mother called him back to Woolsthorpe to learn to manage the estate. He hated it. Fortunately there were two men of influence who had taken the boy's measure – the Reverend William Ayscough and the schoolmaster, Mr Stokes. They prevailed upon Mrs Smith to send Isaac back to school to prepare for university. So Newton returned to Grantham and was boarded by Mr Stokes in his own house. (The Woolsthorpe servants, we are told, rejoiced at parting with him, declaring he was fit for nothing but the 'versity'.) In June 1662 Newton set out for Cambridge and his brave new world. 'He purchased a lock for his desk, a quart bottle and ink to fill it, a pound of candles and a chamber pot, and was ready for whatever Cambridge might offer.' (In later, more affluent days he acquired several silver specimens of the last of these items.)

The teaching given to undergraduates at Cambridge in the 1660s had remained virtually unchanged since the Middle Ages; there was certainly nothing to stimulate a budding mathematician. During his first term Newton picked up a book on astrology, but was unable to understand some of the mathematics it contained. So he bought a copy of Euclid's *Elements*, but was disappointed to find that many of the propositions were 'obvious'. He went on to teach himself mathematics by reading the works of Kepler, Viète and Wallis, among others, but above all a Latin translation of *La Géométrie* of Descartes, which he found heavy going. He also attended the lectures of Isaac Barrow, the first Lucasian Professor and the only mathematician of distinction in the University. (We shall meet Barrow again in Chapter 10.)

Soon after Newton had taken his BA degree in May 1665, the Cambridge colleges were shut down because of an outbreak of plague. They remained closed for most of the next two years.

Newton returned to Woolsthorpe, to enjoy the priceless gifts of leisure and quiet. The years 1665 and 1666 are generally regarded as his *anni mirabiles*: during some 20 months of intense creativity, Newton laid the foundations of all his major discoveries – on infinite series, the calculus, optics and gravitation. As Whiteside puts it, 'Never did one man build up so great a store of mathematical expertise, much of it his own discovery, in so short a time.' Newton himself, writing some fifty years later, concludes a chronological account of his early achievements thus:

All this was in the two plague years of 1665–1666. For in those days I was in the prime of my age for invention and minded Mathematics and Philosophy more than at any time since.

During the next ten years he extended his researches, publishing nothing but communicating some of his results privately. He became a respected member of the small scientific community of Restoration England. This was also the time of his rapid academic advancement. He was elected to a Minor, and then to a Major Fellowship at Trinity College. In October 1669, on the retirement of Barrow, he was elected Lucasian Professor of Mathematics, a post he held for 26 years. One of the Professor's duties was to 'lecture and expound' some mathematical discipline of his choice at least once each week at a prearranged time during the seven months of the Cambridge academic year. He was also required to make himself freely available for a two-hour period at stated times (twice a week during term and once a week otherwise, when in residence) to answer 'queries and difficulties' put to him by anyone who came along.

The statutes further ordained that each autumn the Professor should deliver to the Vice-Chancellor for deposit in the university library 'polished copies of no fewer than ten of the lectures given by him during the previous year'. A certain laxity of interpretation was permitted, even in Barrow's time, and in fact Newton is believed to have made only two deposits of lectures during the first 15 years of his occupancy of the Chair. In 1674 he deposited an improved version of his optical lectures of 1670–72; in 1685 he handed over a digest of 97 lectures on algebra that he had delivered between 1673 and 1683. He carried out his minimal lecturing duties, but, as his secretary Humphrey Newton (no relation) records many years later:

so few went to hear him, and fewer that understood him, that ofttimes he did in a manner, for want of hearers, read to the walls . . . he usually stayed about half an hour; when he had no auditors, he commonly returned in a 4th part of that time or less.

Humphrey also provides some engaging personal details:

His carriage then was very meek, sedate, and humble . . . I cannot say I ever saw him laugh but once . . . I never knew him to take any recreation or pastime either in riding out to take the air, walking, bowling, or any other exercise whatever, thinking all hours lost that was [sic] not spent in his studies . . . He very rarely went to dine in the Hall, except on some public days, and then [as] if he has not been minded, would go very carelessly, with shoes down at heels, stockings untied, surplice on, and his head scarcely combed.

This, however, is not the whole story. Newton had a circle of personal friends and did a certain amount of academic entertaining. Humphrey tells us that, 'when invited to a treat, he used to return it very handsomely, freely and with much satisfaction to himself.' His niece, Catherine Conduitt, relates how one of his cronies – a lecturer in chemistry – fell from favour. 'He told a loose story about a nun, and then Sir Isaac left off all confidence with him.'

There were several periods of Newton's life when he largely lost interest in mathematics and physical science: his single-minded search for truth took other directions. During the 1670s his life was dominated by chemical and alchemical studies, both experimental and bibliographical, and he greatly resented being interrupted by mathematical or optical queries. His intense interest in alchemy, which he maintained for more than thirty years, must strike us as odd, even perverse. We must remember, however, that throughout the seventeenth century alchemy remained a living force. Ben Jonson's hostile portrayals in *The Alchemist* were far from typical of the attitude of the general public.

At about this time Newton took up a new field of study – theology. His researches led him to espouse the historical Arian (or Unitarian) position, as opposed to the Trinitarian doctrine of Athanasius which became the orthodoxy of the Church in the fourth century. This could have been extremely serious for Newton's academic future, but he managed to obtain a royal dispensation exempting the Lucasian Professor from taking holy orders. We do not know how this was achieved, but the influence at court of his friend Barrow could well have been decisive. As Westfall remarks:

Now at last, despite gross heresy sufficient to make him a pariah, he had surmounted the final obstacle and found himself secure in his sanctuary. And he had demonstrated a new facet of his genius: he could have his cake and eat it too.

By the mid-1670s Newton had virtually cut himself off from the scientific community. He began a monumental history of the Church in the fourth and fifth centuries, which was to show how events had fulfilled the divine prophecies. He undertook a major revision of his earlier treatise on the Book of Revelation and composed a new statement of his Arian views. He also devoted much effort to writing alchemical treatises and performing chemical experiments. Newton continued his studies in theology, ancient chronology and the interpretation of prophecy until his old age. His voluminous writings on these subjects are now almost unreadable, but they enjoyed considerable popularity for many years after his death. Lord David Cecil, in *The Young Melbourne*, his biography of William Lamb (later Viscount Melbourne and Prime Minister), quotes an entry from the diary of his notorious wife, Lady Caroline Lamb, shortly after their marriage in 1805:

William and I get up about ten or later (if late at night) – have our breakfast – talk a little – read Newton on the Prophecies with the Bible . . .

Remarkable morning reading for a freethinking, newly married couple! The book they were reading was presumably Newton's *Observations on the Prophecies*, which appeared in print, together with *The Chronology of Ancient Kingdoms Amended*, in 1728, the year after his death. It now seems that this presumption was almost certainly mistaken. Some remarkable new evidence on the case is presented in Appendix 7, to which the reader should now turn.

The next landmark in Newton's life was Halley's fateful journey to Cambridge in August 1684, which led to the writing and publication of the *Principia.* We have no contemporary account of what happened during that famous visit; indeed, the most detailed account is contained in a memorandum dated November 1727, more than forty years after the events it describes. It was made by the Huguenot mathematician Abraham de Moivre (1667–1754), and transcribed by John Conduitt, who married Newton's niece and succeeded him at the Mint. This is what we are told:

In 1684 Dr. Halley came to visit him at Cambridge, after they had been some time together, the Dr. asked him what he thought the Curve would be that would be described by the Planets supposing the force of attraction towards the Sun to be reciprocal to the square of their distance from it. Sir Isaac replied immediately that it would be an Ellipsis, the Doctor struck with joy and amazement asked him how he knew it. Why saith he I have calculated it, whereupon Dr. Halley asked him for his calculation without any further delay, Sir Isaac looked among his papers but could not find it, but he promised him to renew it, and then to send it to him . . .

Newton redeemed his promise by writing a short tract, *De motu corporum*, which reached Halley in November 1684. He was delighted with it and, after transcribing it and registering a copy with the Royal Society, he returned to Cambridge and found Newton hard at work on a greatly expanded version. The *De motu* tract is, in fact, the first of a number of papers, ranging from short notes to full-scale treatises on central-force dynamics, which were eventually subsumed in the *Principia* of 1687.

After two years of intense intellectual concentration, Newton's academic life was disturbed, never to be fully re-established, in the very year the *Principia* was going through the press. The upheaval started when King James II instructed the Vice-Chancellor that a Benedictine monk – one Father Francis – be admitted as Master of Arts without taking the required oaths of supremacy and allegiance. Newton was deeply disturbed, not so much by the religious and political implications as by the cynical flouting of University statute law. He was a prominent member of the University delegation whose appeal before the Ecclesiastical Commission was tersely rejected by Judge Jeffreys with the words: 'Go your way and sin no more, lest a worse thing come unto you.'

Two years later, after the 'Glorious Revolution', Newton reaped his reward. In January 1689 he was elected as one of the two University members of the Convention Parliament. (It was a very close contest: Newton 122 votes, the other two candidates 125 and 117.) He remained in London for thirteen months, until the Convention Parliament was dissolved, and was actively involved in the preparation for a new declaration of allegiance to be sworn by members of the universities. During this time he established friendships with John Locke, the philosopher, and Christiaan Huygens, among others, and was painted by the fashionable Godfrey Kneller. Although he returned to Cambridge in 1690, the taste for political power and public prestige had been firmly implanted. It never left him, and within six years he was to leave Cambridge for the metropolis, never to return again.

Newton's final years at Cambridge were, with one exception, externally uneventful. In 1693, at the age of 50, he suffered some kind of mental breakdown: he lost, in his own words, 'the former consistency of his mind'. The medical details are obscure, and will probably always remain so. His illness certainly impaired his zest for scientific pursuits, but he made a good recovery and his intellectual powers remained formidable. There seems no doubt, however, that there was some deterioration of character: he became increasingly arrogant and dictatorial, less sensitive to the feelings of others and more suspicious of their motives. In September 1693 we find him writing a moving letter to his friend Locke:

Being of opinion that you endeavoured to embroil me with women and by other means, I was so much affected with it, as that when one told me you were sickly and would not live, I answered, 'twere better if you were dead. I desire you to forgive me this uncharitableness. For I am now satisfied that what you have done is just, and I beg your pardon for my having hard thoughts of you for it . . .

A month later (15 October 1693), when he had largely recovered, Newton explained matters further to Locke:

The last winter, by sleeping too often by my fire, I got an ill habit of sleeping; and a distemper, which this summer has been epidemical, put me farther out of order, so that when I wrote to you, I had not slept an hour a night for a fortnight together, and for five nights together not a wink.

During his last years at Cambridge Newton was strongly urged to produce a second edition of the *Principia*; this eventually appeared in 1713 (see Illustration 7). He wished to include the results of his further researches on lunar motion, the Solar System and the behaviour of comets. In order to do this he needed the most accurate astronomical data he could get. He naturally turned to Greenwich Observatory, and pressed the luckless John Flamsteed, the first Astronomer Royal, with scant regard for the poor man's feelings. Thus in June 1695 we find Newton abruptly telling Flamsteed: 'I want not your calculations, but your observations only.'

The great divide in Newton's life came in April 1696, on his appointment as Warden of the Royal Mint. This entailed moving to London, where he lived for the remaining thirty years of his life. The post was offered to him by his old Cambridge friend Charles Montague, at that time both Chancellor of the Exchequer and President of the Royal Society. Montague explains that:

The office is the most proper for you. 'Tis the Chief Officer in the Mint. 'Tis worth five or six hundred pounds per An, and has not too much business to require more attendance than you may spare.

In the event, however, Newton entered into the duties of his office with such zeal that his scientific activities almost ceased. The great natural philosopher became a most faithful and conscientious public servant. In 1699 Newton was promoted to the position of Master of the Mint at a salary of £1500 a year; in 1701 he resigned his professorship; in 1703 he was elected President of the Royal Society, a post he held until his death; in 1705 he was knighted by Queen Anne at Trinity College, Cambridge. The retiring scholar had become the complete establishment grandee. He was, in effect, the chief scientific adviser to the Government, and was consulted on many major national issues, among them the long-standing problem of determining longitude at sea. He was, of course, heavily involved in such matters as the reintroduction of copper coinage, the vexed question of bimetallism, and the effects of changes in the relative market values of gold and silver.

With the second edition of the *Principia* safely through the press, Newton's main scientific preoccupation during his old age was the

PHILOSOPHIÆ
NATURALIS
PRINCIPIA
MATHEMATICA.

AUCTORE

ISAACO NEWTONO,
EQUITE AURATO.

EDITIO ULTIMA

Cui accedit ANALYSIS *per Quantitatum* SERIES, FLUXIONES *ac* DIFFEREN-
TIAS *cum enumeratione* LINEARUM TERTII ORDINIS.

AMSTÆLODAMI,
SUMPTIBUS SOCIETATIS.
M. D. CCXXIII.

Illustration 7. Title page of the second Amsterdam reprint (1723) of the
second edition of Newton's *Principia*

distressing priority dispute with Leibniz and his supporters, which we shall be looking at later in this chapter. In 1725 his health began to fail, and he died on 20 March 1727 in his 85th year. He was buried with full national honours in Westminster Abbey.

3. Newton the man

Before looking at Newton's mathematical and scientific achievements, something should be said on the lesser-known aspects of the work and personality of this many-sided genius. When Newton left Cambridge for London in 1696, he packed his papers and some books in a chest which passed, after his death, into the possession of his niece, Catherine Barton, and then to her daughter, the Countess of Portsmouth. The scientific portion of the Portsmouth papers was given to the University of Cambridge in 1888; the rest, consisting of several million words, much in Newton's own handwriting, was retained by the family until 1936 when the papers were dispersed in auction rooms. Fortunately, however, the mathematically educated economist J. M. Keynes was able to reassemble about half the material and return it to Cambridge. What light, we may ask, do these papers throw on the activities and interests of their author? Keynes' eloquent summary of his conclusions (see Reference 5) is worth quoting at some length:

In the eighteenth century and since, Newton came to be thought of as the first and greatest of the modern age of scientists, a rationalist, one who taught us to think on the lines of cold and untinctured reason.

I do not see him in this light. I do not think that anyone who has pored over the contents of that box which he packed up when he finally left Cambridge . . . can see him like that. Newton was not the first of the age of reason. He was the last of the magicians, the last of the Babylonians and Sumerians, the last great mind which looked out on the visible and intellectual world with the same eyes as those who began to build our intellectual inheritance rather less than 10 000 years ago. Isaac Newton, a posthumous child born with no father on Christmas Day, 1642, was the last wonder-child to whom the Magi could do sincere and appropriate homage.

Keynes calls Newton a magician because he regarded the Universe as a secret which could be solved by applying pure thought to certain clues left by God. These clues were to be found partly in the natural world, and partly in certain mystic traditions and documents handed down since Babylonian times. The Portsmouth papers reveal the manic intensity with which Newton pursued his studies over some 25 years in order to read the cosmic riddle. The material has been classified into four groups: theology, chronology, history and alchemy.

Newton's religious views were, as we have seen, Unitarian. He abandoned belief in the Trinity, as Keynes puts it,

not on rational or sceptical grounds, but entirely on the interpretation of ancient authority. He was persuaded that the revealed documents give no support to the Trinitarian doctrines, which were due to late falsifications.

Newton was able to conceal his dreadful heresy throughout his life – and indeed for long after his death. Another large section of the papers is concerned with various apocalyptic writings, 'from which he sought to deduce the secret truths of the Universe'. (Some of these writings were published soon after his death.) There are also hundreds of pages of Church history. Finally, we have a mass of material devoted to alchemy – the search for the philosopher's stone and the elixir of life – to which Newton was undoubtedly addicted. We know, for example, of several periods, each of about six weeks, during which his laboratory fire was never allowed to go out. Here is Keynes' verdict, after looking over about 100 000 words of Newton's alchemical writings:

It is utterly impossible to deny that it is wholly magical and wholly devoid of scientific value; and also impossible not to admit that Newton devoted years to it.

We have seen how, when Newton moved to London, his way of life changed completely. He enjoyed the power and privileges of high public office and the adulation of London society in the Augustan age. In Keynes' words: 'Magic was quite forgotten. He has become the Sage and Monarch of the Age of Reason.' Scholars have yet to evaluate the vast mass of Newtonian material that is now at Cambridge once again. We still have much to learn about the mind and motivations of this astonishing man – a colossus who bestrides the centuries with one foot in the Middle Ages and the other in our own times. Meanwhile, let Lord Keynes have the last word:

As one broods over these queer collections, it seems easier to understand ... this strange spirit who was tempted by the Devil to believe ... that he could reach all the secrets of God and Nature by the pure power of mind – Copernicus and Faustus in one.

4. Newton the natural philosopher

In the rest of this chapter and in the next we shall be looking at some aspects of Newton's mathematical and scientific work. While the *Principia* is the crowning glory of his middle age, most of his major discoveries were made when he was still in his twenties – between 1664 and 1672. His unique genius embraced the whole of seventeenth-century mathematics and natural philosophy, but in this chapter we shall concentrate our attention on six of his major fields of research – four primarily mathematical, and two concerned with the physical world – as indicated by the section headings.

5. Analysis by infinite series

By 1664 the young Newton had absorbed the mathematical discoveries of his predecessors and was ready to strike out on his own. The essence of what he called his 'analytick method' was to combine the concepts and techniques of three hitherto largely separate branches of mathematics: coordinate geometry, the calculus and the expansion of functions as infinite series. Let us look first at the last of these. We have seen that during his student days Newton had made a thorough study of Descartes's *La Géométrie*. From it he took his algebraic symbolism and the use of algebraic equations to describe geometrical curves. However, Descartes's algebra was confined to finite expressions, usually polynomials. By developing a method of dealing with *infinite series*, and manipulating them as if they were polynomials, Newton greatly increased the power of mathematics, particularly when allied to the other two components of his complete analytical method.

The crucial element in Newton's treatment of infinite series was his 'discovery' of the general binomial theorem. The binomial expansion for a positive integral power n had been known for some time. It may be written as

$$(1 + x)^n = \binom{n}{0} + \binom{n}{1}x + \binom{n}{2}x^2 + \ldots + \binom{n}{n}x^n \qquad (8.1)$$

where

$$^nC_r \text{ or } \binom{n}{r} = \frac{n(n-1)(n-2)\ldots(n-r+1)}{1.2.3\ldots r}$$

is the general term in Pascal's triangle (p. 165) and $\binom{n}{0}$ is set equal to 1. Clearly $\binom{n}{r} = 0$ when $r > n$, so the expansion consists of $n + 1$ terms.

In 1665 Newton put forward the proposition (more strictly, the conjecture) that the expansion (8.1) can be generalized for negative and fractional values of n. Now, however, it takes the form of an infinite series, given by

$$(1 + x)^n = \binom{n}{0} + \binom{n}{1}x + \binom{n}{2}x^2 + \ldots + \binom{n}{r}x^r + \ldots$$
$$(8.2)$$

(A meaning can no longer be attached to ^nC_r, regarded as the number of ways of selecting r objects from a set of n objects.)

How did Newton arrive at his generalization? For negative integral values of n he was able to extend Pascal's triangle backwards, as shown in Table 8.1. A new entry is computed as the difference between the entry immediately to the right and the entry immediately above, e.g. $-35 = (-20) - 15$. The coefficients in the expansion of $(1 + x)^n$ are obtained by reading down the column whose second entry is n, for example

$$(1 + x)^{-4} = 1 - 4x + 10x^2 - 20x^3 + 35x^4 - \ldots$$

For fractional values of n, Newton devised a scheme for interpolating in Table 8.1, based on a study of the numerical patterns to be found in the extended Pascal's triangle. Some of his interpolated values at 'half intervals', i.e. for values of $n = $ integer $+ \frac{1}{2}$, are given

Table 8.1. Newton's backward extension of Pascal's triangle

⋯	1	1	1	1	1	1	1	1	1	1	1	1	1	⋯
⋯	−5	−4	−3	−2	−1	0	1	2	3	4	5	6	7	⋯
⋯	15	10	6	3	1	0	0	1	3	6	10	15	21	⋯
⋯	−35	−20	−10	−4	−1	0	0	0	1	4	10	20	35	⋯
⋯	70	35	15	5	1	0	0	0	0	1	5	15	35	⋯
⋯	−126	−56	−21	−6	−1	0	0	0	0	0	1	6	21	⋯
⋯	210	84	28	7	1	0	0	0	0	0	0	1	7	⋯

Table 8.2. Some of Newton's 'half-interval' values interpolated in Table 8.1

1	1	1	1	1	1	1	1	1	1	1
−2	$-\frac{3}{2}$	−1	$-\frac{1}{2}$	0	$\frac{1}{2}$	1	$\frac{3}{2}$	2	$\frac{5}{2}$	3
3	$\frac{15}{8}$	1	$\frac{3}{8}$	0	$-\frac{1}{8}$	0	$\frac{3}{8}$	1	$\frac{15}{8}$	3
−4	$-\frac{35}{16}$	−1	$-\frac{5}{16}$	0	$\frac{1}{16}$	0	$-\frac{1}{16}$	0	$\frac{5}{16}$	1
5	$\frac{315}{128}$	1	$\frac{35}{128}$	0	$-\frac{5}{128}$	0	$\frac{3}{128}$	0	$-\frac{5}{128}$	0
−6	$-\frac{693}{256}$	−1	$-\frac{63}{256}$	0	$\frac{7}{256}$	0	$-\frac{3}{256}$	0	$\frac{3}{256}$	0

in Table 8.2. By generalizing his argument, he was able to formulate a 'rule' for evaluating the general binomial coefficient $\binom{m/n}{r}$, where m, n and r are integers, with n and r positive.

Newton attached considerable importance to his 'method of infinite series', or expansion in *power series* as we would now say. It enabled him to represent a wide variety of curves (both algebraic and transcendental) by equations of the form

$$y = a_0 + a_1 x + a_2 x^2 + \ldots \tag{8.3}$$

The use of such expansions, enabling each term to be treated separately, provided an indispensable tool for effecting the quadrature of curves and the rectification of arcs (p. 140). Newton was especially interested in the quadrature of the circle given by $x^2 + y^2 = a^2$, and the two hyperbolas given by $y^2 - x^2 = a^2$ and $y(a + x) = 1$. He was able to expand the expressions for y, namely $(a^2 - x^2)^{1/2}$, $(a^2 + x^2)^{1/2}$ and $(a + x)^{-1}$, as power series in x and then simply 'integrate' term by term. (Newton was one of the first to use the modern notation for fractional or negative exponents; he had it from Wallis.)

Newton's formulation of the general binomial theorem remained a conjecture, as he was not able to devise a formal proof. So he checked his expansions whenever possible – by direct multiplication, division or root extraction. Thus, for example, putting $n = \frac{1}{2}$ yields

$$(1 - x^2)^{1/2} = 1 - \frac{x^2}{2} - \frac{x^4}{8} - \frac{x^6}{16} - \cdots$$

which he checks by multiplying the series by itself to give $1 - x^2$, 'the remaining terms vanishing by the continuation of the series to infinity', as he puts it. As a double check, he extracted the square root of $1 - x^2$ 'in an arithmetical manner. And the matter turned out well.'

Although Newton discovered the general binomial theorem in 1665, he kept it to himself for more than ten years. The first communication is contained in a famous letter dated 13 June 1676 to Henry Oldenburg, the Secretary of the Royal Society, in response to a request from Leibniz for information on British work on infinite series. Newton states his 'rule' in the form

$$(P + PQ)^{m/n} = P^{m/n} + \frac{m}{n} AQ + \frac{m - n}{2n} BQ + \frac{m - 2n}{3n} CQ + \cdots$$

where each of $A, B, C, \ldots$ denotes the immediately preceding term. In a second letter, dated 24 October 1676, he gives a fuller account of his researches on the subject. Although the *convergence* of infinite series did not become a matter of general concern until the eighteenth century, Newton was well aware of the dangers and frequently mentions the need to ensure that the variable, x, 'shall be small enough'.

6. The fluxional calculus

All the basic ideas and techniques of the Newtonian calculus were formulated during the 'prime years' of 1664–6. Newton gave his first comprehensive treatment of the subject in an unpublished tract dated October 1666. It opens with eight propositions 'for resolving problems by motion', followed by a full table of quadratures (i.e. integrals) of simple algebraic functions; his findings are then applied to 17 problems concerned with tangents, curvature, arc length, quadrature and centres of gravity. Newton's approach to the calculus was dynamic, in the tradition of Galileo and Kepler. He regarded a plane curve ($f(x, y) = 0$, say) as the path traced out by a moving point. During an infinitely small time interval, which he denotes by o, the values of x and y increase by $o.p$ and $o.q$, so p and q may be thought of as the *rates* at which the point (x, y) moves horizontally and vertically. The slope of the tangent to the curve at the point (x, y) is taken to be the ratio of these instantaneous rates of movement of y and x, i.e. q/p. Newton's procedure for evaluating this ratio is essentially the same as that used by Fermat (p. 142).

In 1669 Newton produced a monograph entitled *De analysi per aequationes numero terminorum infinitas* (On Analysis by Equations Unlimited in the Number of Their Terms), which was circulated

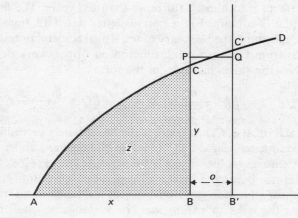

Figure 8.1.

among a few friends. He addressed it to Barrow, almost certainly the only man at Cambridge who would have been able to understand it. (We shall see in Chapter 10 what Barrow did with it.) Newton starts with three rules, each illustrated by numerous examples, for finding the quadrature of (i.e. the area beneath) 'simple curves':

Rule 1 If $ax^{m/n} = y$, then will $na/(m + n).x^{(m+n)/n}$ equal the area ABC.
Rule 2 If the value of y is compounded of several terms the area will also be compounded of the areas which arise separately from each of those terms.
Rule 3 If the value of y or any of its terms be more compounded than the foregoing [i.e. other than of the form $a_i x^i$], it must be reduced to simpler terms, by operating in general variables in the same way as arithmeticians in decimal numbers divide, extract roots or solve affected equations.

The objective of Rule 3 is to reduce an equation of the form $f(x, y) = 0$, say, to a power series expansion, $y = F(x)$. Newton expounds several techniques for doing this, including division, root extraction and successive approximation by iteration (see Section 10).

Newton's 'proof' of Rule 1 is presented by means of a particular example, although the generality of the procedure is apparent. In Figure 8.1, ACD is any curve. We set AB = x and BC = y, and denote the shaded area ABC by z. Now consider the augmented area AB'C', with base AB' = $x + o$. Draw PQ as shown, so that the rectangle BPQB' has the same area as CC'B'B. Setting BP = v, the augmented area AB'C' is equal to $z + ov$. Newton's chosen example is the curve for which $z = \frac{2}{3}x^{3/2}$ or $z^2 = \frac{4}{9}x^3$. He sets $(z + ov)^2 = \frac{4}{9}(x + o)^3$ and obtains, after cancelling through by o,

$$\tfrac{4}{9}(3x^2 + 3xo + o^2) = 2zv + ov^2$$

He continues thus:

If we now suppose BB' to be infinitely small, that is, o to be zero, v and y will be equal and terms multiplied by o will vanish and there will consequently remain $\frac{2}{3}x^2 = zy$, which reduces to $y = x^{1/2}$.

Conversely, he says if $y = x^{1/2}$, then $z = \frac{2}{3}x^{3/2}$. He then generalizes by taking

$$z = \frac{an}{(m + n)} x^{(m + n)/n}$$

and derives, by a similar argument, the equation $y = ax^{m/n}$, 'as was to be proved'. It is apparent that Newton already recognizes differentiation and integration (to use the modern terms) as mutually inverse operations.

Newton was very much aware of the logical difficulties associated with the notion of infinitely small quantities and the limiting values of the ratios of two such quantities. Such methods, he says, are 'shortly explained rather than accurately demonstrated'. In about 1670 he adopted a new approach in which variables are regarded as flowing quantities generated by the continuous motions of points, lines, etc. He calls such a variable quantity (x or y, say) the *fluent*, and its rate of change the *fluxion* of that quantity, which he denotes by $\dot{x}$ or $\dot{y}$ – the Newtonian 'dot' notation. The two fundamental problems of the calculus can now be succinctly stated as: Given a relation between two fluents, find the relation between their fluxions, and vice versa. Newton remarks that it is helpful, but not necessary, to think of these variable quantities as changing with time. He developed these ideas in a tract known as the *Methodus fluxionum et serierum infinitorum* (The Method of Fluxions and Infinite Series), which he wrote in 1671; it was not published until 1736, after his death. As one example he considers the curve whose equation is

$$x^3 - ax^2 + axy - y^3 = 0$$

from which he derives the fluxional equation

$$3x^2\dot{x} - 2ax\dot{x} + a\dot{x}y + ax\dot{y} - 3y^2\dot{y} = 0$$

giving the slope of the tangent at any point as

$$\frac{\dot{y}}{\dot{x}} = \frac{3x^2 - 2ax + ay}{3y^2 - ax}$$

Newton made one further attempt to devise a more logically satisfactory treatment of limiting processes. He sought to eliminate the concept of the infinitely small quantity because, as he says, 'in mathematics the minutest errors are not to be neglected'. The new method is based on the concept of 'prime and ultimate ratios' (from his Latin text), or in English, 'first and last ratios'. The method is expounded in his *Tractatus de quadratura curvarum* (A Treatise on the Quadrature of Curves) which was written in about 1693, but not published until 1704. He had in fact already adopted this approach in the *Principia* (1687), so it will be convenient to defer discussion until the next chapter. Suffice it to say that Newton failed in his attempt to resolve the difficulties posed by the use of limiting processes, which continued to worry his successors throughout the eighteenth century and well into the nineteenth.

7. Optical experiments and speculations

Newton's lifelong interest in optical matters dates from his under-graduate days. He probably read Galileo, Kepler and Descartes on the subject; he certainly attended Barrow's optical lectures at Cambridge – indeed, he helped with their publication in 1669. We know that he was grinding lenses for telescopes as early as 1663, and his epoch-making experiments on the spectrum of sunlight can almost certainly be dated to 1666, which he spent at Woolsthorpe.

Newton's failure to overcome the problem of chromatic aberration (to be explained shortly) led him to construct a reflecting telescope as early as 1668. In 1671 he built a second one and was prevailed upon to send it to the Royal Society, where it aroused great interest. In January 1672 he wrote to Oldenburg that he would send the Society an account of a discovery 'being in my judgement the oddest, if not the most considerable detection, which hath hitherto been made in the operations of nature.' The paper, Newton's first, is dated 19 February 1672 and describes his famous experiments on the decomposition and analysis of white light. The quotation emphasizes the revolutionary nature of these discoveries: they seemed incredible at the time. In 1675 he sent the Royal Society a second paper in which he put forward some conjectures in order to satisfy those who were demanding some kind of theoretical explanation of the astonishing experimental findings. These papers were widely misunderstood and aroused considerable criticism, led by the magisterial Huygens and the cantankerous Robert Hooke. The experience caused Newton acute distress and generated in him a distaste for publication that remained for most of his life. Hooke died in 1703, and Newton was eventually persuaded to publish his second masterpiece, the *Opticks*, in 1704. He was then 61 years of age and had been settled in London for eight years.

The *Opticks, or a Treatise on the Reflections, Refractions, Inflections and Colours of Light* is a delight to read, equally for its language and for its content. In marked contrast to the *Principia*, it can be enjoyed by the non-mathematical reader. Einstein, in his foreword to the 1931 edition, captures the buoyant spirit of the book when he writes:

Fortunate Newton, happy childhood of science! He who has time and tranquility can by reading this book live the wonderful events which the great Newton experienced in his young days. Nature to him was an open book, whose letters he could read without effort . . . In one person he combined the experimenter, the theorist, the mechanic and, not least, the artist in exposition. He stands before us strong, certain and alone; his joy in creation and his minute precision are evident in every word and in every figure.

The *Opticks* is organized as three books. The first deals with the reflection and refraction of light, the production of images and spectra, and the decomposition of white light into its coloured components. It starts with a few definitions and axioms, followed by 33 'experiments' and 18 'propositions'. These beautiful experiments proved that sunlight is a mixture of 'homogenial' lights of different colours ('of rays differently refrangible', as Newton puts it). 'To the same degree of Refrangibility ever belongs the same colour, and to the same colour ever belongs the same degree of Refrangibility.' He divides the spectrum into seven colours by analogy with the musical scale, orange and indigo corresponding to the semitones. He also gives a detailed explanation of the rainbow, discusses chromatic aberration and its implications for telescope design, provides practical advice on lens grinding, and gives a full discussion of how natural colours are produced.

The phenomenon of chromatic aberration arises directly from the fact that the different colours of the spectrum are 'differently refrangible' – or, as we would say now, have different refractive indices (the μ of p. 147) – when passing through a transparent lens or prism. In particular, this means that a converging lens, as used in a refracting telescope, will bring the blue rays to a focus nearer the lens than the red rays, thereby giving a blurred image, with coloured edges, of a distant object such as a star. A reflecting telescope, which uses a curved mirror instead of a lens to focus the image, does not suffer from this trouble.

Book I is prefaced by the forthright statement:

My design in this Book is not to explain Properties of Light by Hypothesis, but to propose and prove them by Reason and Experiments.

Newton is unable to sustain this position in Book II, which deals with what we now call interference phenomena, such matters as the colours and patterns to be observed on thin films and polished plates. Although most of this Book consists of experimental material, Newton does introduce some highly speculative propositions, for example that a ray of light passing through a refracting surface is put at regular intervals into one or other of two states which he calls 'Fits of easy Transmission and Fits of easy Reflection'.

Book III discusses the phenomenon of diffraction, or 'inflection' as Newton calls it. After presenting 11 'Observations' of the diffraction patterns cast by the shadows of various objects, he breaks off abruptly with the remark:

But I was then interrupted [in his experimental work], and cannot now think of taking these things into farther consideration. And since I have not finished this part of my Design, I shall conclude with proposing only some Queries, in order that a farther search be made by others.

These Queries (there are 31 altogether) are, to many readers, the most fascinating part of the whole book. They contain some of Newton's most daring and imaginative speculations and cover an astonishing range – chemistry, the circulation of the blood, gravitation, the Creation, scientific method and animal sensation, among many other topics. Indeed, Newton introduces so many hypotheses that he gives the impression that he is emptying his mind of the conjectures accumulated during a lifetime of intense intellectual activity.

The Queries were formulated and published, not all at once, but over a period of a dozen years or so. The first (1704) edition of the *Opticks* contained sixteen; seven more were added to a Latin edition which appeared in 1706, and a further eight to the second English edition of 1717. The sixteen original Queries deal with such questions as the relation between light, heat and bodily vibrations; the nature of fire; and the action of the eye in distinguishing colours. The first asks: 'Do not Bodies act upon Light at a distance, and by their action bend its Rays?' (We shall return to this topic in Chapter 17.) The 1706 Queries introduce an entirely new topic – double refraction in Iceland spar (or calcite). After presenting the facts in some detail, Newton puts forward a theoretical explanation which, he argues, constitutes an insuperable objection to Huygens' longitudinal-wave theory of light. (Neither Huygens nor Newton considered the possibility of transverse undulations.) Then follows a critical discussion of the nature of the 'aether' (the medium needed to carry the waves), leading to the forthright conclusion: 'To make way for the regular and lasting Motions of the Planets and Comets it is necessary to empty the Heavens of all Matter.' An aetherial fluid of finite density, however thin,

can be of no use for explaining the Phenomena of Nature . . . There is no evidence for its Existence, and therefore it ought to be rejected. And if it is to be rejected, the Hypotheses that Light consists of Pression or Motion, propagated through such a Medium, are rejected with it.

He then comes down unequivocally in favour of a corpuscular hypothesis: 'Are not the Rays of Light very small Bodies emitted from shining substances?' The next Query foreshadows one of the major themes of twentieth-century physics:

Are not gross Bodies and Light convertible into one another, and may not Bodies receive much of their activity from the Particles of Light which enter into their Composition? . . . The changing of Bodies into Light, and Light into Bodies, is very conformable to the Course of Nature, which seems delighted with such Transformations.

He then illustrates his contention with a variety of examples drawn from physics, chemistry and natural history. In the last of the 1706 Queries, Newton really lets himself go. His conclusion, after a wide-ranging survey of the 'Phenomena of Nature', is that

All these things considered, it seems probable to me, that God in the beginning formed Matter in solid, massy, hard, impenetrable, moveable

Particles of such sizes and shapes, and endowed with such properties, as are needed to cause the world to be as it is.

In the 1714 Queries Newton returns to the vexed question of the nature of the aether. After his earlier dogmatic rejection of 'this rare and subtle Medium', he has second thoughts. Indeed, the 'aether problem' continued to exercise him for many years. We shall see in the next chapter that he returned to the subject in the second edition of the *Principia*.

The *Opticks* went into two more editions, in 1721 and 1730, and retained its scientific influence and its popular appeal throughout the rest of the century. At the beginning of the nineteenth century, however, the situation began to change. Dr Thomas Young enunciated the 'principle of interference' and established the wave theory of light for the next hundred years in what was thought to be an unassailable position. Throughout the nineteenth century Newton's corpuscular theory was seen as a noble failure, as an error into which even the deepest of thinkers can sometimes fall. How times have changed!

We shall see in Chapter 17 how in 1905 Einstein resuscitated the corpuscular theory when he explained the photoelectric effect by postulating the existence of discrete *quanta* (now called photons) of light. Einstein's ideas were soon confirmed experimentally, but the body of nineteenth-century experimental evidence for the wave theory of light remained intact. To add to the confusion, Louis de Broglie predicted the existence of matter waves, also confirmed experimentally when a beam of electrons was found to produce a diffraction pattern. By the 1930s physicists were prepared to entertain the incomprehensible: that the corpuscular hypothesis and the wave hypothesis were both, in some sense, true. Now, this blending of the two concepts is exactly what we find in the *Opticks*: Newton postulated that particles of light and particles of matter are able to interact in some way, and his corpuscular theory included a periodic feature to enable the light particles to be put into 'alternating fits of easy reflection and easy transmission'. This, while unacceptable in the nineteenth century, has a clear affinity with the ideas of the twentieth. Once again we can read Newton's second masterpiece not merely as a historical landmark, but for its living scientific interest.

8. Geometrical researches

In the late 1660s Newton embarked on a major study of algebraic curves, using the methods of coordinate geometry. By this time the conic sections had been thoroughly explored analytically (notably by Wallis, as we shall see in Chapter 10), so it was natural for Newton to turn his attention to cubic curves. Using the method of coordinate transformation, he showed that the general cubic equation containing ten terms may, when the coefficient of xy^2 is not zero, be expressed in the simpler form

$$bxy^2 + dx^3 + gx^2 + hy + kx + l = 0 \qquad (8.4)$$

If, however, b is zero, the equation may be reduced to one of the following forms:

$$dx^3 + ey^2 + kx + l = 0 \qquad (8.5)$$

$$dx^3 + fxy + l = 0 \quad (\text{if } e = 0) \qquad (8.6)$$

$$dx^3 + hy = 0 \quad (\text{if } e = f = 0) \qquad (8.7)$$

Equations (8.5)–(8.7), to which Newton gave the names *divergent parabola*, *Cartesian trident* and *Wallisian cubic*, respectively, had been studied by earlier mathematicians. Newton's achievement was to explore in detail the most general form, equation (8.4).

The crucial step in his analysis was the introduction of what he called the 'diametral hyperbola'. Since equation (8.4) is quadratic in y, a general ordinate will meet the cubic at two points, P and Q say. It turns out that the locus of the mid-point of PQ is a hyperbola whose asymptotes are the x- and y-axes. (Its equation is $2bxy + h = 0$.) Newton exploits this result to great effect when investigating the asymptotic properties of cubic curves. By examining various possibilities, he isolates six cases of the general reduced form (8.4), some of which he further subdivides into 'species' according to the signs of various terms. Altogether, including forms (8.5)–(8.7), Newton distinguishes 16 species of the general cubic, some of which he further subdivides into 'grades'. He describes each grade in detail, and draws examples of most of them.

In about 1690 Newton revised and extended his earlier researches and was persuaded to publish his *Enumeratio linearium tertii ordines*

(Enumeration of Lines of the Third Order) in 1704 as one of two mathematical appendices to the *Opticks* (the other was the *Tractatus de quadratura curvarum*). The paper starts with a number of general theorems about cubic curves, their asymptotes and their associated conics. There then follows a detailed description and analysis of no fewer than 72 different types of cubic. (Six more were added later by other writers.) In the *Enumeratio* he derives his types by purely geometrical considerations, thereby superseding his earlier treatment based on changes of coordinate axes. He skilfully exploits one of his most elegant results, which he states casually and without proof: just as all conics are projections of a circle, so any cubic curve can be generated as a projection (or 'optical shadow', as Newton calls it) of a divergent parabola (equation (8.5) above), of which he distinguishes five different types. It was a long time before the eighteenth-century geometers were able to devise a proof of this result.

During the early 1690s Newton was engaged in elaborate geometrical studies, 'all but unknown to modern mathematical historians' until they were brought to light by Whiteside in the 1970s (Reference 10). It seems that Newton was planning a comprehensive three-part treatise, the *Geometria*, but the work was never completed. He shared the prevailing seventeenth-century view that the Greeks could not have made their impressive geometrical discoveries solely by means of the classical methods that have come down to us: their elegant, synthetic proofs must surely, it was argued, have been preceded by some kind of analysis (since lost) which probably made use of techniques not too dissimilar from those of seventeenth-century analytical geometry. Pappus, as we have seen (p. 88), had sought to describe such a method of analysis in his *Mathematical Collection*. After setting out the basic aims of 'geometrical analysis' as expounded by Pappus, Newton applies his general analytic strategy to a variety of specific problems, most requiring for their solution the linked concepts of a set of algebraic equations and the corresponding set of plane geometrical curves, or loci. The problem is essentially solved when the points of intersection of these curves are determined. Newton sums it up in these words: 'Geometry in its entirety is nothing else than the finding of points by the intersections of loci.'

In Chapter 3 we mentioned a lost treatise of Euclid, the *Porisms*. Newton sought to reconstruct the main content of this work, based on guidelines supplied by Pappus. In his several drafts he incorporates a number of concepts of projective geometry, foreshadowing the work of the nineteenth-century geometers by some 150 years. Then, quite suddenly, he seems to lose interest in the Greeks and embarks on a thoroughly contemporary advanced treatise on curves and their properties, which he discusses from the Cartesian, projective and fluxional points of view, giving special emphasis to their asymptotic features. Newton drafted further geometrical papers, but all are incomplete.

9. Early thoughts on mechanics and gravitation

Newton has left few writings on mechanics, terrestrial or celestial, apart from those connected with the composition of the *Principia*. We do know, however, that he began to study the subject as early as the mid-1660s. Here, from his recollections in old age, is the long sentence immediately preceding the quotation on p. 171.

In the same year [1666] I began to think of gravity extending to the Orb of the Moon, and having found out how to estimate the force with which a globe revolving within a sphere presses the surface of the sphere, from Kepler's Rule of the periodical times of the Planets being in a sesquialterate proportion of their distances from the centres of their Orbs, I deduced that the forces which keep the Planets in their Orbs must [be] reciprocally as the squares of their distances from the centres about which they revolve: and thereby compared the force requisite to keep the Moon in her Orb with the force of gravity at the surface of the Earth, and found them [to] answer pretty nearly.

(Probably, though, in 1666 these ideas were not so fully developed.)

Newton left several early notes on the dynamics of uniform circular motion. One, the most elegant, which uses the geometry of the circle, goes like this. Consider a body moving with constant speed v in a circle of radius r around a centre C. In a short time, t, the body moves from A to D, as shown in Figure 8.2. If not constrained, it would move from A to B, where $AB = vt$. In fact, the body is pulled towards C (by what Newton later called 'centripetal

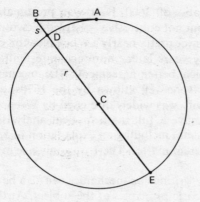

Figure 8.2.

force') through a distance BD = s, say. There is a well-known 'circle' theorem which states that $BA^2 = BD.BE$, or $(vt)^2 = s(s + 2r)$. Since s/r is very small, we may write this as $v^2 t^2 \approx 2sr$. Newton now makes use of Galileo's result (enunciated in the *Dialogues Concerning Two New Sciences* as 'Theorem Two of the third day') that distances traversed from rest in uniformly accelerated motion vary as the squares of the times. We may express this by the well-known equation $s = \frac{1}{2}at^2$, where a is the constant acceleration directed towards C. Eliminating t yields $a = v^2/r$, the familiar formula first stated by Huygens, but in 1666 not yet published. If T is the time taken by the body to complete one revolution, then $vT = 2\pi r$, giving $a = 4\pi^2 r/T^2$, which we can write as $a = (4\pi^2/r^2).(r^3/T^2)$. Now, by Kepler's third law r^3/T^2 is constant for all planets, and the inverse square law of gravitational attraction follows at once.

Newton's next step was to apply this result to the motion of the Moon. Experiments with a pendulum enabled him to measure the gravitational acceleration g at the surface of the Earth, and to compare it with the acceleration necessary to hold the Moon in its orbit (computed from the v^2/r formula). He found a ratio of about 4000 : 1. At this time the Moon's orbital radius was thought to be about 60 times the radius of the Earth, so the inverse square law should yield a gravity ratio of 3600 : 1. It was presumably this comparison – suggestive but not conclusive – that led Newton to say that he found them to 'answer pretty nearly'. When Newton came to write the *Principia* twenty years later, more accurate figures were available which gave much better agreement. The mechanics of uniform circular motion were well known by the 1670s, and the inverse square law of gravity was widely discussed by Wren, Hooke and others. However, to give a full mathematical analysis of the dynamics of the Solar System, including an explanation of Kepler's three laws, was a very different matter. There was only one man who could do that!

Most of Newton's early writings on mechanics can also be seen as an attempt to clarify the basic principles of the subject. At the time there was considerable confusion, much of it deriving from the speculations of Descartes, about such basic concepts as inertia, force, impulse, mass and weight. The laws of motion as Newton understood them in the 1660s were very different from those he enunciated in the *Principia*.

10. Researches in numerical mathematics

In his younger days, Newton was quite prepared to tackle lengthy numerical calculations, for example to evaluate particular values of a function, such as a logarithm, which can be expressed as an infinite series, but where a large number of terms have to be computed to give the required accuracy. At the age of 22 he read a Latin translation of Viète's last work, a tract written in 1600 entitled *De numerose potestatum ad exegesin resolutione* which is concerned with the numerical solution of higher-degree equations. Viète devised a cumbersome 'trial and error' method for obtaining the value of the desired root, digit by digit. By 1665 Newton had developed a much improved iterative method, which he expounded in the *De analysi*. He takes as an example the equation $x^3 - 2x - 5 = 0$. Clearly there is a root near 2, so Newton sets $x = p + 2$ to give

$$p^3 + 6p^2 + 10p - 1 = 0$$

Now, he says, we know that p is small, so we may safely neglect the terms in p^3 and p^2, to give $10p' - 1 = 0$, or $p' = 0 \cdot 1$, where p' is an approximation to p. He then sets $x = q + 2·1$ and obtains an equation in q, which yields a further correction term of $-0·0054$. He repeats the procedure twice more to obtain the final result of $2·094\,551\,47$, which is in error only by one unit in the eighth place. It is interesting to compare Newton's procedure with that of Chu Shih-chieh (p. 94).

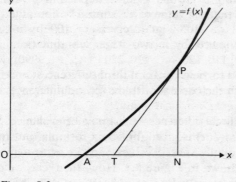

Figure 8.3.

Newton also devised an iterative method, still widely used, for obtaining approximate roots of either algebraic or transcendental equations. Consider the problem of finding a root of the equation $y = f(x) = 0$, as represented in Figure 8.3 by the line OA. If $ON = x_0$ is an approximate root, then a better approximation is given by $OT = x_1$, where $PN = y_0 = f(x_0)$ and PT is the tangent at P. We see that $PN/TN = \dot{y}_0/\dot{x}_0$, where $\dot{x}_0$ and $\dot{y}_0$ are the fluxions of x and y at P, and so $x_1 = x_0 - y_0\dot{x}_0/\dot{y}_0$. The process may be repeated as often as desired to give better approximations, $x_2, x_3, \dots$. The procedure is known as the Newton–Raphson iterative method; it was first published in its final form by Joseph Raphson in 1690. Taking the example treated above, we have

$$y = x^3 - 2x - 5 \quad \text{and} \quad \dot{y} = (3x^2 - 2)\dot{x}$$

so $y = -1$ and $\dot{y} = 10\dot{x}$ when $x = 2$, giving, as before, $x_1 = 2·1$. In fact the two methods are equivalent, but the second one has the great advantage of being applicable to transcendental as well as to algebraic equations.

Newton's most important contribution to numerical mathematics was to develop a range of methods of *interpolation* by means of *finite differences*. His early work on the subject goes back once again to the mid-1660s, but it seems that his interest was reawakened by an inquiry in 1675 from one John Smith, an accountant, who approached Newton for help in a project to compile a table of $n^{1/p}$, with n going from 1 to 10 000 in unit steps, and $p = 2, 3$ or 4. Newton suggested that the value of $n^{1/p}$ should be computed directly for $n = 100, 200, \ldots, 10 000$ at intervals of 100 by an iterative method, to be followed by a two-stage subtabulation: first for $n = 10, 20 \ldots, 90; 110, 120, \ldots; 210, 220, \ldots; \ldots, 9990$, with the differences adjusted to yield constant third differences; and finally at unit intervals within each decade with the second differences adjusted to be constant.

The concept of 'finite differences' must now be explained. Suppose that a function $y = f(x)$ is given, not by a formula, but in a table of y for a set of equally spaced values of x, i.e. $a, a + w, a + 2w, a + 3w, \ldots$, as shown in Figure 8.4. How, then, can we compute the values of y for other values of x, such as ON? The answer is provided by the theory of 'interpolation', which has been described as the science of reading between the lines of a numerical table. The first step is to form a *table of differences*, as in Table 8.3 where, for example,

$$d_1(a) = f(a + w) - f(a)$$

$$d_1(a + w) = f(a + 2w) - f(a + w)$$

$$d_2(a) = d_1(a + w) - d_1(a)$$

$$d_3(a + 2w) \cong d_2(a + 3w) - d_2(a + 2w)$$

Now, if $f(x)$ is a polynomial in x of degree n, the nth differences are constant and so all higher differences are zero. Table 8.4 is part of the difference table for the cubic

$$y = f(x) = x^3 + 2x^2 - 4x + 3 \tag{8.8}$$

Newtonian interpolation is based on the assumption that a function $f(x)$ may be adequately represented – at any rate over the range of

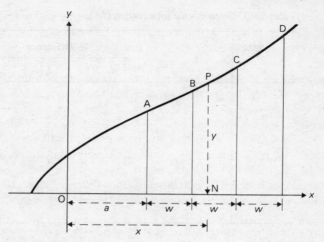

Figure 8.4.

Table 8.3. The principle of finite differences (for six values of x)

		Order of differences			
x	y	1st	2nd	3rd	$\cdots$
a	$f(a)$				
		$d_1(a)$			
$a + w$	$f(a + w)$		$d_2(a)$		
		$d_1(a + w)$		$d_3(a)$	
$a + 2w$	$f(a + 2w)$		$d_2(a + w)$		$\cdots$
		$d_1(a + 2w)$		$d_3(a + w)$	
$a + 3w$	$f(a + 3w)$		$d_2(a + 2w)$		$\cdots$
		$d_1(a + 3w)$		$d_3(a + 2w)$	
$a + 4w$	$f(a + 4w)$		$d_2(a + 3w)$		
		$d_1(a + 4w)$			
$a + 5w$	$f(a + 5w)$				

interest – by a polynomial in x of degree n. Newton's formula for computing $f(x)$ for $x = a + pw$ (where, usually, $0 < p < 1$) is

$$f(a + pw) = f(a) + \binom{p}{1} d_1(a) + \binom{p}{2} d_2(a) + \ldots + \binom{p}{n} d_n(a)$$

$$(8.9)$$

Table 8.4. Difference table for equation (8.8)

x	y	Difference d_1	d_2	d_3
−1	8			
		−5		
0	3		4	
		−1		6
1	2		10	
		9		6
2	11		16	
		25		6
3	36		22	
		47		
4	83			

The coefficients of the difference terms are the same as those in the binomial expansion. Thus, for example, the cubic polynomial (8.8) yields

$$f(1{\cdot}3) = f(1) + 0{\cdot}3d_1(1) + \frac{(0{\cdot}3)(-0{\cdot}7)}{2} d_2(1)$$

$$+ \frac{(0{\cdot}3)(-0{\cdot}7)(-1{\cdot}7)}{6} d_3(1)$$

$$= 2 + (0{\cdot}3) \times 9 - (0{\cdot}21) \times 8 + 0{\cdot}357$$

$$= 3{\cdot}377$$

as may be verified by direct substitution in (8.8).

In most cases of practical interest, the differences of a certain order settle down to nearly constant values, at any rate over a limited range, and higher-order differences can be ignored. While equation (8.9) is the basic interpolation formula, Newton established four others using *central differences* instead of the *forward differences* used in (8.9). The former have the great advantage of being drawn from entries that are on, or as close as possible to, a single horizontal line of the difference table.

Newton's second notable achievement in numerical analysis was to generalize his formulae so that they can be used even if the function is tabulated at unequal intervals – as might happen, for instance, with a series of astronomical observations interrupted by cloud. In fact, during the mid-1670s Newton laid the foundation of the modern theory of 'finite difference interpolation', but he kept back his discoveries for nearly forty years, with the single exception of a short lemma in the *Principia*. In 1710 William Jones, who was preparing a collection of Newton's mathematical writings for publication, persuaded Newton to write up his early notes on interpolation to produce the *Methodus differentialis*, which was published in 1711.

11. Later years: the priority dispute

When Newton left Cambridge for London in 1696 he did not abandon mathematics entirely. In January 1697 he was moved to respond to two challenges issued 'to the sharpest mathematicians in the whole world' by Jean Bernoulli. The first of these was the celebrated 'brachistochrone problem':

There needs to be found out the curved line ADB in which a heavy body shall, under the force of its own weight, most swiftly descend from a given point A to any other given point B.

Newton solved both problems (according to his niece–housekeeper) in the course of a single evening after returning home from a hard day's work at the Mint. On the following morning he sent his solutions to his friend Charles Montague. Newton's solution to the first problem is very brief, at 77 words. He assumes, without proof or supporting argument, that the required curve is an inverted cycloid, and contents himself with giving an elegant geometrical construction for it. His working papers are lost, but we know that he had mastered the essentials of the problem some years earlier.

Newton's long-lasting aversion to publication was notorious. As the London mathematician Augustus de Morgan was later to remark: 'Every discovery of Newton had two aspects. Newton had to make it and then you had to find out that he had done so.' The situation at the turn of the century was that, with one minor exception, the *Principia* was still Newton's only published work. However, his earlier mathematical discoveries were becoming increasingly well known and quoted by others, and he was eventually persuaded to prepare some of his own writings for publication. In 1704, as we have seen, he revised two of his earlier mathematical tracts and published them as appendices to his *Opticks*. In 1707 William Whiston, his successor as Lucasian Professor, published the text of Newton's 'deposited' lectures on algebra (p. 171) as the *Arithmetica universalis*. In 1722, in what was probably his last mathematical effort, Newton supervised the preparation of a corrected second edition of the *Arithmetica*, which was to become his most widely read work.

As the title suggests, Newton regarded algebra as a 'universal arithmetick' which mirrors the structure of common arithmetic, but is rendered universal by the introduction of free variables in place of constants. After stating his general position and laying down some notation, he goes through the basic algebraic operations, modelling each of them on its arithmetical equivalent. His strong interest in practical problem-solving is evident throughout. The techniques that we now call 'model building' are discussed in some detail, especially the transition, as he puts it, from 'the question verbally enunciated' to 'the same algebraically'. He then applies the principles he has expounded to nearly a hundred worked problems – arithmetical, geometrical, mechanical and astronomical. Here are four examples to indicate his range:

(1) If cattle a should eat up a meadow b in time c, and cattle d an equally fine meadow e in time f, and if the grass grows at a uniform rate, how many cattle will eat up a similar meadow g in time h?

(2) To describe a circle which shall pass through a given point and touch two other circles given in position and magnitude. [One of the Apollonian problems.]

(3) Where a stone drops down a well, from the sound of the stone striking the bottom to ascertain the depth of the well.

(4) From four observed positions of a comet crossing the sky with a uniform rectilinear motion, to gather its distance from the Earth and the direction of its motion, supposing the Copernian hypothesis.

Having dealt at length with the formulation of problems as algebraic equations, Newton devotes the rest of the book to a discussion of the properties of the equations themselves and of techniques for 'resolving' them numerically.

In the years between the publication of the second edition of the *Principia* in 1712 and the death of Leibniz four years later, Newton's main scientific preoccupation was the calculus priority dispute with Leibniz and his supporters. The many twists and turns of this most famous – or perhaps infamous – of mathematical quarrels make a fascinating story of human frailty at the highest intellectual level. The issues are, however, no longer of live interest, except perhaps to students of historical pathology. The basic facts are not now in dispute, and the 'verdict of history' may be summarized as follows:

(1) Newton was the first inventor of the infinitesimal calculus. He devised his method of fluxions as early as 1664–66 and soon developed it into a general operational procedure.

(2) The differential and integral calculus, as we have it today, was created by Leibniz between 1675 and 1685 (see Chapter 11).

(3) Newton published none of his mathematical papers until 1704, although from 1669 onwards he imparted some of his results in manuscript tracts or letters to selected friends. Leibniz, on the other hand, published his first – admittedly obscure – account of his differential calculus in a learned journal in 1684 (see Chapter 11).

(4) For many years the two men were on cordial terms of mutual respect. Relations between them were eventually soured by a combination of factors: misinterpretation of documents, acceptance of inaccurate hearsay stories, ignorance of what the other side was doing or had done, and – only too often in the later stages – exaggerated claims, malicious accusations and plain distortions of the facts. The supporters led the hue and cry, with the two principals reluctantly joining in later.

In 1712 the Royal Society appointed a committee to look into the issues and report. The Society accepted the committee's report and ordered its publication, with extracts from the relevant documents. These date mainly from the period between 1669 and 1676 and effectively establish Newton's priority claim. The report, usually known simply as the *Commercium epistolicum*, comes down firmly on Newton's side – which is not surprising as Newton drafted most of it himself. To consolidate his position even more firmly, Newton (writing anonymously) published his own account in the Society's *Philosophical Transactions* of February 1715. He gives a lucid summary of his early discoveries on infinite series and fluxions, assembling his historical arguments with considerable forensic skill, and ably defends his 'philosophy of nature' against the attacks of the Leibnizians, led by Jean Bernoulli.

Even the death of Leibniz in 1716 did not bring the quarrel to an end, although Newton himself took little further part in the exchanges. The dispute eventually petered out in 1722, with the death of Pierre Varignon of Paris, who made several unsuccessful attempts to effect a reconciliation between the contending parties.

Let us end this chapter with a tribute, as generous as it is well deserved, from Newton's great rival – Leibniz, no less: 'Taking mathematics from the beginning of the world to the time of Newton, what he has done is much the better half.'

9   Newton's *Principia*

'The heavens themselves, the planets, and this centre,
Observe degree, priority, and place,
Insisture, course, proportion, season, form,
Office, and custom, in all line of order.'
William Shakespeare, *Troilus and Cressida* (I, 3)

1. The writing and publication of the *Principia*

Newton's master-work, the *Philosophiae naturalis principia mathematica* (The Mathematical Principles of Natural Philosophy) – or *Principia*, for short – is universally acclaimed as one of the greatest scientific books ever written (see Illustration 7). Its creation in less than three years remains an unrivalled feat of sustained intellectual power. Although few people have read (or even dipped into) the *Principia*, its influence in shaping the way we think about the physical world has been enormous.

We have seen how Newton began his researches on celestial mechanics during the 'marvellous years' of the mid-1660s, and how his interest in the subject was gloriously rekindled by Halley's visit to Cambridge in April 1684 – a visit which completely transformed Newton's intellectual life. The problem of understanding the mechanics of the Solar System seized his imagination and gave him the will and strength to carry through the great undertaking to a triumphant conclusion.

In April 1686 the Royal Society received from Newton a manuscript of what was virtually the first Book of the *Principia*. In June, the Society

ordered that Mr. Newton's book be printed, and that Mr. Halley undertake the business of looking after it, and printing it at his own charge; which he engaged to do.

Poor, devoted Halley. He was not a wealthy man! The rest of the *Principia* (Books 2 and 3) was completed in the autumn of 1686. The first edition of 1687 was quite small, almost certainly fewer than 400 copies, and the book soon went out of print. Newton was urged to bring out a new edition, and during the next twenty years he prepared substantial revisions and extensions. The second edition duly appeared in 1713, under the skilled and tactful editorship of Roger Cotes, whom we shall meet in the next chapter. This edition was of 750 copies, bound copies being sold for one guinea. Newton prepared a list of some 70 recipients of presentation copies, starting with 'the Tsar [Peter the Great] 6 for himself & the principal Libraries in Muscovy'.

In his old age Newton sanctioned a third edition and engaged the physician Henry Pemberton FRS as editor. The changes from the second edition are minor: they deal mainly with some new observations on comets and experiments on air resistance. The edition, of 1250 copies, eventually appeared in 1726 when Newton was 83 years of age. All these editions were, of course, in Latin; an English translation by Andrew Motte was published in 1729. A widely read 'explication' of the great book by editor Pemberton was published in 1728; an English translation of Voltaire's popular account appeared ten years later as *The Elements of Sir Isaac Newton's Philosophy*. It was dedicated to his erudite mistress, Madame du Châtelet, and was designed to have a strong appeal to women.

2. The content of the *Principia*
(a) *The mathematical style*
The *Principia* is not – and was not intended to be – an easy book to read. The intrinsic difficulties presented by the material are likely to

be exacerbated for modern readers who have little experience of 'pure' geometry in the Euclidean tradition. When dealing with the geometrical aspects of mechanics Newton uses not the new coordinate geometry of Descartes, but the older synthetic geometry of Apollonius. It is not difficult to see why.

We must bear in mind that the relative standing of the two disciplines in the late seventeenth century was, in one sense, a mirror image of what it is today. To Newton's contemporaries classical geometry was familiar and well understood. By contrast, the newfangled algebraic methods of Descartes had not yet gained general acceptance among the mathematical community; and the analytical treatment of the calculus, in either the Newtonian or the Leibnizian style, was even more recent. Indeed, Newton had not yet published any of his researches in the subject. However, the mathematical edifice of the *Principia* could not have been constructed without recourse to the basic ideas of the calculus, in particular the notion of 'the approach to the limit'. So what Newton did was to treat this concept in a strictly geometrical manner. In fact, as we shall see, he begins Book 1 with a careful but essentially intuitive discussion of limiting processes as applied to curves and their tangents, arcs, chords and quadratures. The spirit is that of Newtonian calculus; the form is that of classical geometry.

(b) *The overall strategy*

The Principia, as we have seen, is organized as three Books. Books 1 and 2 are mainly mathematical, while Book 3 is a prose essay with substantial mathematical insertions. Newton's method is to lay down an axiomatic framework, and then to construct a simplified and idealized mathematical model (as we would now call it) of the physical situation he wished to study – in this case the Solar System. The consequences of the operation of the model are deduced mathematically and compared with experience; the model is then developed to give greater physical verisimilitude, and further deductions are made. This refining process is continued in the first two Books as far as is necessary or practicable. In Book 3 Newton applies his mathematical principles to 'demonstrate the frame of the System of the World'. The essence of the Newtonian method, then, is to use mathematics to explain the workings of nature. For the Solar System his approach was outstandingly successful, but his attempt to apply the same methods to matter on a small scale was a failure; success here had to wait until the nineteenth century. Newton's strategy of maintaining a tight interplay between mathematical analysis and physical experience has been followed by scientists ever since; indeed, in these days of mathematical modelling we take it for granted. In his time it was a revolutionary innovation.

It is perhaps worth remarking that, although Newton the scientist concentrated his investigations on problems of physical interest, Newton the mathematician could not resist the urge to generalize and explore interesting byways. The *Principia* provides many examples of such excursions.

(c) *The axiomatic framework*

The *Principia* is written, as we have said, in classical Euclidean form. It contains nearly 200 propositions, either theorems or problems, and many corollaries. Newton also makes use of lemmas to establish the mathematical results he needs, and scholia for general comments and interpolations in his main argument.

The book opens with eight definitions of such basic concepts as mass, momentum and force (of various kinds). Newton coins the neutral term *centripetal force* for 'that by which bodies are drawn . . . towards a point as to a centre'. After a long and unsatisfactory scholium in which he attempts to distinguish between various kinds of time, position and motion, we reach the famous axioms, or laws of motion. The first law embodies the principle of inertia, the second tells us how to measure the force acting on a moving body, and the third declares 'that to every action, there is always opposed an equal reaction'. The laws are followed by five corollaries on such basic notions as the parallelogram of forces, the conservation of momentum, and the motion of the centre of gravity of a composite system. We must remember that when Newton began his work a science of dynamics just did not exist – he had to create it from first principles. The many revisions and discarded drafts bear witness to the difficulties he had in putting the new subject on a sound basis.

3. Book 1: The motion of bodies in 'empty' space

(a) *The mathematical foundations*

Book 1 is organized in 14 sections, of which Newton recommended that particular attention be paid to the first three. Section 1 is headed 'The method of first and last ratios of quantities, by the help of which we demonstrate the propositions that follow'. This section contains the essence of the Newtonian fluxional calculus. It consists of 14 lemmas and two scholia in which Newton explains how he proposes to handle limiting processes, and establishes the main results he will need. To see what he means by 'first and last ratios', let us take the case of two neighbouring points, P and Q, on a continuous curve, with P fixed and Q movable, and consider the ratio of the arc PQ to the corresponding chord. The 'first ratio' is the one that is obtained when Q is about to move away from P along the curve; the 'last' is when Q is about to coalesce with P. In either case, of course, the 'ultimate ratio will be the ratio of equality', to quote one of Newton's favourite phrases.

Lemma 1 states that:

Quantities and the ratios of quantities, which in any finite time converge continually to equality, and before the end of that time nearer to each other than by any given difference, become ultimately equal.

The proof is by contradiction. The next three lemmas, which may be termed integrating lemmas, compare sets of rectangles inscribed in or around curves with the areas under the curves. The others, differentiating lemmas, deal with such matters as the limiting ratios of arcs, chords and tangents, and the corresponding triangles with vertices at a central point. Lemma 10 states (in modern terminology) that a force of magnitude F acting on a body for a short time t displaces the body by an amount proportional to Ft^2. The point of specifying a short time (or, as Newton puts it, 'in the very beginning of the motion') is, of course, to allow both the magnitude and the direction of the force to be taken as constant during the time interval.

(b) *The explanation of Kepler's laws of planetary motion*

Section 2, containing ten propositions, is entitled 'The determination of centripetal forces'. Proposition 1 proves that Kepler's second law (that equal areas are swept out in equal times) applies to the motion of a body (actually a 'point-mass', a term that will be explained shortly) under the action of any centripetal force whatsoever. Newton phrases it thus:

The areas which revolving bodies describe by radii drawn to an immovable centre of force do lie in the same immovable planes, and are proportional to the times in which they are described.

He treats the force as a sequence of discrete impulses acting at equal small intervals of time.

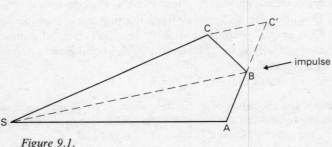

Figure 9.1.

His argument goes like this. In the first time interval the body, by its 'innate force' (S is the centre of force), will move from A to B (Figure 9.1); in the second, if not hindered, it will continue in a straight line to C′, where BC′ = AB. Now, says Newton, 'when the body is arrived at B, suppose that a centripetal force acts at once with a great impulse, and . . . compels the body to continue its motion along the line BC.' At the end of the second time interval the body will be at C, where CC′ is parallel to SB, and BC = AB. Now,

the triangles SAB, SBC′ and SBC all have the same area, so equal
areas are swept out by the 'radius vector' to S in the two equal
times. Clearly, the argument can be extended to further equal time
intervals. He continues:

Let the number of these triangles be augmented, and their breadth diminished
ad infinitum, then ABC . . . will be a curved line . . . [and] . . . the centripetal
force . . . will act continuously.

The demonstration of Kepler's second law follows at once.

We see, then, that in his very first proposition Newton deploys a
limit argument. Although his proof of the area law derives from the
impulsive form of the second law of motion (impulse equals change
of momentum), he moves at once to the more familiar continuous-
motion statement of the law (force equals mass times acceleration).
Proposition 2 proves the converse of Proposition 1; Proposition 4
establishes the 'v^2/r' formula for the centripetal acceleration in
uniform circular motion. Newton notes in a corollary that Kepler's
third law (p. 195) implies an inverse square law of centripetal force
in this particular case.

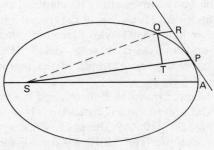

Figure 9.2.

Proposition 6 provides a general measure of centripetal force in
any path. In Figure 9.2, a body (again, actually a 'point-mass') at P,
revolving about the centre of force S, describes the curve APQ. PR
is the tangent at P, and Q is a neighbouring point on the curve (PQ
is 'any arc just then nascent', as Newton puts it). RQ is parallel to,

and QT perpendicular to, SP. Now, the area of the triangle SPQ is $\frac{1}{2}$SP.QT and, by Proposition 1, is proportional to the short time t during which the body moves from P to Q. Newton now makes use of a result that he has established in Section 1, namely that 'in the limit' the force that deflects a body from P to Q is proportional to QR/t^2. This follows from the result that when Q is very near P, the deflection QR may be taken to be equal to the product of half the acceleration and t^2. Since the magnitude of the deflecting force F is proportional to this acceleration, it is also proportional to QR/t^2; or, using Proposition 1, it is inversely proportional to $SP^2.QT^2/QR$, 'taken as that magnitude which it ultimately acquires when the points P and Q coincide'. The centripetal force is thus expressed solely in terms of the geometry of the orbital path – a result which Newton uses to great effect later on.

In Section 3 he moves on to consider 'the motion of bodies in eccentric conic sections'. At this point he asks what law of force will give rise to an elliptic orbit if the force is directed to one of the foci of the ellipse. Now, it follows from Proposition 6 that, in order to establish the inverse square law, we need to prove that QR/QT^2 remains constant as the body moves along its elliptical path. This Newton achieves in Proposition 11. In view of the crucial importance of this theorem and the archaic (to us) style of Newton's demonstration, we will present and analyse it in some detail. His elegant geometrical proof provides a good example of the mathematical methods he uses in the *Principia*.

Figure 9.3, which is adapted from the *Principia* and is a developed version of Figure 9.2, shows Newton's construction. It represents a body moving in an elliptical path APQB, where once again Q is taken to be near P on the ellipse. The centripetal force is directed to S, one of the foci of the ellipse, and has deflected the body at P from proceeding to R by an amount RQ, where RQ is parallel to SP. H is the other focus and C the centre of the ellipse. PCG is the diameter through P; DCK is the conjugate diameter, i.e. the diameter parallel to the tangent RPZ at P (p. 59). Now, in Newton's words, 'Draw SP cutting the diameter DK in E, and the ordinate Qv in x; complete the parallelogram QxPR. Draw QT perpendicular to SP.' (Note that

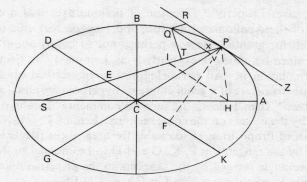

Figure 9.3.

the point x is on SP, and the point v on CP.) The rest of the construction will be apparent.

In the course of his proof, Newton uses four properties of the ellipse; all were proved by Apollonius and would be familiar to Newton's readers. They are (see Figure 9.3):

> *Property 1*   angle SPR = angle HPZ
> *Property 2*   SP + PH = 2CA
> *Property 3*   Gv.Pv/Qv² = CP²/CD²
> *Property 4*   CB.CA = CD.PF

The first two are straightforward and are easily proved. To prove Property 3, it is convenient to use a pair of conjugate diameters (CP and CD in Figure 9.3) as oblique axes of reference. Taking the coordinates of the point Q as $x' = Cv$, $y' = Qv$ (Newton refers explicitly to the 'ordinate Qv'), the equation of the ellipse can be shown to be

$$\frac{x'^2}{a'^2} + \frac{y'^2}{b'^2} = 1$$

where $a' = CP$ and $b' = CD$. Now,

$$\frac{Gv.Pv}{Qv^2} = \frac{(a' + x')(a' - x')}{y'^2} = \frac{a'^2}{b'^2} = \frac{CP^2}{CD^2}$$

which proves Property 3. The use of oblique axes was a central feature of the Apollonian tradition, and Newton was able to take this result for granted. He was perhaps not so happy about Property 4, since he states it explicitly – as Lemma 12 in Section 2. His formulation is: 'All parallelograms circumscribed about any conjugate diameters of a given ellipse or hyperbola are equal among themselves.' Then follows the laconic comment: 'This is demonstrated by the writers on the conic sections.' Lemma 12 is invoked in the proof of Proposition 11 to enable the area of the parallelogram formed by the tangents at P, K, G and D to be equated to the area of the rectangle formed by the tangents to the principal diameters. This gives Property 4, i.e. that CB.CA = CD.PF.

We are now ready to go through Newton's demonstration of Proposition 11. His first step is to prove that EP = CA. Since HI is parallel to EC, and SC = CH, then, by similar triangles, SE = EI. Now it follows from Property 1 that the triangle PIH is isosceles, so PI = PH. Hence EP = $\frac{1}{2}$(SP + PI) = $\frac{1}{2}$(SP + PH) = CA, from Property 2. Newton then establishes four equalities of ratios (he calls them 'proportions') as given below, where L denotes the *latus rectum* (p. 59) of the ellipse:

$$\frac{L.QR}{L.Pv} = \frac{QR}{Pv} = \frac{PE}{PC} = \frac{AC}{PC}$$

$$\frac{L.Pv}{Gv.Pv} = \frac{L}{Gv}$$

$$\frac{Gv.Pv}{Qv^2} = \frac{PC^2}{CD^2}$$

$$\frac{Qx^2}{QT^2} = \frac{EP^2}{PF^2} = \frac{CA^2}{PF^2} = \frac{CD^2}{CB^2}$$

The third of these proportions uses Property 3; the fourth uses Property 4 to obtain the final equality.

Newton now takes a crucial step. He asserts, referring back to his lemma in Section 1, that 'when the points P and Q coincide, $Qv^2 = Qx^2$'. Making this substitution and replacing L by its value

2BC^2/AC, we can multiply together corresponding sides of the four proportions. This gives, after simplifying,

$$\frac{L.QR}{QT^2} = \frac{AC.L.PC^2.CD^2}{PC.Gv.CD^2.CB^2} = \frac{2PC}{Gv}$$

Newton continues: 'But the points Q and P coinciding, 2PC and Gv are equal, and so $L.QR = QT^2$.' We have proved, therefore, that QR/QT^2, being equal to $1/L$, is constant during the motion: it follows from Proposition 6 that 'the centripetal force is inversely as $L.SP^2$, that is inversely as the square of the distance SP'.

Before we leave this famous proposition, let us review what Newton has done. First, then, what equipment does he use? In fact, he draws from three branches of mathematics:

(1) *dynamics*: the centripetal force F acting on a moving body is proportional to the acceleration a, and the distance (QR in Figure 9.3) through which the body is deflected during a short time t may be taken as proportional to at^2;

(2) *geometry*: the four properties of the ellipse listed above;

(3) *the calculus*: the operation of 'proceeding to the limit', as when two points P and Q on a curve move towards coincidence.

Secondly, what has he achieved? He has proved that:

(1) for any orbit governed by a central force, the force F varies inversely as $SP^2.QT^2/QR$ (Figure 9.2);

(2) QT^2/QR is constant for an elliptical orbit with the force directed to one of the foci; from which it follows that

(3) F varies as $1/SP^2$.

One final comment: what Newton has proved in Proposition 11 is that motion in an elliptical orbit implies an inverse square law of force directed to the focus. However, to answer Halley's question (p. 174) requires the converse theorem: that an inverse square law of force entails motion in an elliptical orbit – or, more generally, in one of the conic sections. Newton apparently regarded this as a trivial extension: he relegated it to a corollary. After establishing similar results for a body moving in a hyperbolic or parabolic orbit, he proves Kepler's third law for elliptic orbits, thereby completing the explanation of Kepler's laws for a simplified 'one-body point-mass' model; that is to say, for a body moving in a plane under the influence of a central force (i.e. a force that is always directed to a fixed point) and where the whole mass of the body is assumed to be concentrated at a single point – the 'centre' of the body.

In 1977, the 250th anniversary of his death, Newton was accorded the supreme accolade of replacing Britannia on the reverse of the £1 banknote. The design contained a mathematical diagram (our Figure 9.3). Alas, Newton's fiscal triumph was short-lived: the life of the £1 banknote came to an end in 1986. However, in March 1987 the U.K. Post Office issued a set of four stamps commemorating the *Principia* tercentenary. The 18p stamp portrays the legendary apple in the Woolsthorpe orchard, with a diagram that is the left half of our Figure 9.6; the 22p shows some planets moving in elliptical orbits with a diagram that is our Figure 9.4; the 31p depicts the optical spectrum; and the 34p represents the 'System of the World'.

Sections 4 and 5 of Book 1, while perhaps not in the main line of Newton's argument, provide an impressive demonstration of his skill as a synthetic geometer. We are given a number of elegant constructions for conics which satisfy various conditions. There is, among much else, a set of six propositions for constructing conics to pass through $5 - n$ points and to touch n lines, where $n = 0, 1, 2, 3, 4$ and 5.

Section 6 deals with the problem of finding orbital positions at any given time. In Section 7 Newton investigates the motion of a body moving in a straight line directly towards the centre of force. He first considers an inverse square law of attraction and then proceeds to the general case, 'a centripetal force of any kind'. Section 8 is entitled 'The determination of orbits in which bodies will revolve, being acted on upon by any sort of centripetal force'. Newton has generalized Section 7 from one to two dimensions. The most significant proposition in this section is Proposition 41:

Supposing a centripetal force of any kind, and granting the quadratures of curvilinear figures [i.e. the ability to perform integration]; it is required to find as well the curves in which bodies will move, and the times of their motions in the curves found.

Here Newton achieves a mathematical result of great power and generality. No restriction is placed on the form of the force function,

$f(r)$ say: it is left unspecified. Such a generalization, while unlikely to have a direct application to the physical world, will be appreciated by all mathematicians. It exemplifies yet again the range of Newton's mathematical interests and talents.

(c) *Orbital stability*

In Section 9 Newton introduces an important new feature into his dynamical model: the concept of a movable orbit (call it M) which revolves about the centre of force. This he compares with a fixed orbit of the same size and shape (call it F). In modern notation, if the polar equation of F is $r = f(\theta)$, that of M is $r = f(k\theta)$. In Figure 9.4, P and P' are corresponding positions in the two orbits, A'P'B' is the instantaneous position of the orbit M at time t, with M coinciding with F when $t = 0$. Then PS = P'S = r, $\angle$ ASP = $\angle$ A'SP' = θ and $\angle$ ASP' = $k\theta$. Newton now proves a remarkable result – that an additional centripetal force is needed to produce the orbital rotation and that it is proportional to the inverse cube of r. If F is an ellipse, the force is proportional to $L(k^2 - 1)/r^3$.

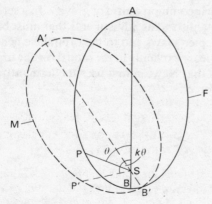

Figure 9.4.

In the next proposition Newton restricts himself to the case of nearly circular orbits and gives a general analysis of the motion of the apsides, and hence of the stability of the orbits. (An *apse*, plural *apsides*, is a point at which the radius from the centre of force meets the orbit at right angles; the radius in question is called an *apse line*.) In Figure 9.4, A and B are the apsides of F; A' and B' the apsides of M. Newton proves the general result that if the centripetal force is proportional to $1/r^n$, the angle between consecutive apse lines remains constant at $\beta = 180°/\sqrt{(3 - n)}$. If $n > 3$, the orbit is unstable. Clearly $\beta = 180°$ when $n = 2$ and $\beta = 90°$ when $n = -1$, the orbits being elliptical in both cases. Virtually all other values of n give rise to rotating orbits. In particular, orbital stability with $\beta = 180°$ implies an inverse square law of force.

Section 10 introduces a further generalization, to orbits whose planes do not pass through the centre of force. This section also contains a very thorough discussion of the cycloidal pendulum, with its remarkable isochronous properties (p. 164).

(d) *From a one-body to a many-body model*

Section 11 breaks more new ground by introducing a second 'point-mass', at the centre of force. We now have two mutually attracting bodies orbiting about their common centre of gravity. In a set of nine propositions Newton establishes the adjustments that must be made to the formulation of Kepler's laws and to the orbital elements. The concept of mass now appears explicitly in the model for the first time. Also, it is in this section that Newton first uses the term 'attraction' instead of 'centripetal force'.

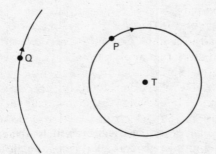

Figure 9.5.

Newton must now face a more serious challenge. The Solar System consists not of two, but of many mutually attracting bodies. The general problem was beyond his powers, so he concentrated on the physically most important case of three bodies – a central body T orbited by two other bodies, P and Q, as in Figure 9.5. He investigates how the attraction of Q will perturb the motion of P. Clearly T, P and Q could represent either the Earth, Moon and Sun, or the Sun and two other planets. In fact, no fewer than 22 corollaries are appended to the main proposition (No. 66); most of them relate to the irregularities of the lunar motion. In the last five, however, P is taken to be, not a single body, but a continuous ring of particles circling the central body T (representing the Earth), with a radius equal to that of the Earth itself. In Corollaries 18 and 19 the ring of particles becomes a fluid annulus, and is used to explain the ebb and flow of the tides. In Corollaries 20–22 the ring 'becomes hard' and attached to the Earth. The attraction of the Sun and Moon on the Earth's equatorial bulge is shown to account for the precession of the Earth's axis.

(e) *From 'point-masses' to finite bodies*

One more major step towards physical realism remains to be taken. The Solar System is not made up of point-masses, but contains heavy bodies of finite size. In Section 12 Newton addresses this question. If the attraction of a body is assumed to be simply the sum of the attractions of its constituent particles, how is this sum to be evaluated? In his famous Moon/apple 'thought experiment' of 1666, Newton sought to compare the centripetal force acting on the Moon to keep it in its orbit with the force of gravity on the surface of the Earth (p. 194). He had to assume that all the particles of the Earth somehow combine to attract an apple with a force that depends on its distance, not from the surface, but from the centre of the Earth. Was such an idea credible? Newton told Halley that he had grave doubts until he succeeded in proving it mathematically in 1685. The crucial proposition is No. 71, which proves that a homogeneous spherical shell, composed of particles each of which attracts according to the inverse square law, attracts an external particle as if the whole mass of the shell were concentrated at its centre.

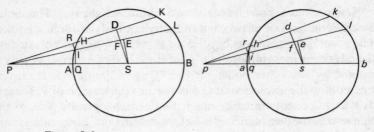

Figure 9.6.

Newton invites us to consider two such particles P and p at different distances from the shell. (For clarity, two separate circles are shown in Figure 9.6.) The two pairs of lines PIL, PHK and pil, phk are drawn such that HK = hk and IL = il; the other lines are drawn as shown. When the circles are rotated about PS and ps, each pair of lines cuts two differential rings (Newton calls them 'circular surfaces') from the shell so formed, two of radii IQ and iq, based on IH and ih, and two others (not shown) based on KL and kl. Newton's strategy is to compare the attraction on P of the ring HI with the corresponding attraction on p of the ring hi. Once again, he uses the concept of ultimate ratios: 'Let the angles DPE, dpe vanish,' he says; then, because DS = ds and ES = es, PE may be taken as equal to PF, pe as equal to pf, and DF as equal to df. Hence we have

$$\frac{PI}{PF} = \frac{RI}{DF} \quad \text{and} \quad \frac{pf}{pi} = \frac{df}{ri}$$

giving

$$\frac{PI.pf}{PF.pi} = \frac{RI}{ri} = \frac{\text{arc HI}}{\text{arc hi}}$$

since DF = df. Also

$$\frac{PI}{PS} = \frac{IQ}{SE} \quad \text{and} \quad \frac{ps}{pi} = \frac{se}{iq}$$

giving

$$\frac{PI.ps}{PS.pi} = \frac{IQ}{iq}$$

since SE = se. Multiplying corresponding sides of these two equations, we get

$$\frac{PI^2.pf.ps}{pi^2.PF.PS} = \frac{arc\,HI.IQ}{arc\,hi.iq} = \frac{area\,of\,ring\,HI}{area\,of\,ring\,hi}$$

Now, every particle of the ring HI is the same distance from P, and the direction of attraction is always at the same angle to PS. It is clear from symmetry that the attraction of the ring HI on P acts along PS, and so must be multiplied by cos IPQ = PF/PS; similarly for the ring hi. Thus we have

$$\frac{attraction\,of\,ring\,HI\,on\,particle\,at\,P}{attraction\,of\,ring\,hi\,on\,particle\,at\,p}$$

$$= \frac{(area\,of\,ring\,HI)(1/PI^2)(PF/PS)}{(area\,of\,ring\,hi)(1/pi^2)(pf/ps)}$$

$$= \frac{PI^2.pf.ps}{pi^2.PF.PS} \cdot \frac{pi^2}{PI^2} \cdot \frac{PF.ps}{PS.pf} = \left(\frac{1}{PS^2}\right)\Big/\left(\frac{1}{ps^2}\right)$$

The same result will obtain for the other pair of rings, and 'indeed for all the rings into which each of the spherical surfaces may be divided by taking sd equal to SD.' By adjusting the angles KPS and kps to maintain this equality, we can simultaneously exhaust the area of the shell from both P and p. This completes the proof that the ratio of the attractions of the two particles equals the ratio of the inverse squares of their distances from the centre of the shell. The result is easily extended to the mutual attraction of two solid spheres, either of uniform density or made up of radially homogeneous shells. After a short general discussion of the attraction of non-spherical bodies, the final section takes an unexpected turn – it deals with optical topics and contains, among other things, an ingenious mathematical demonstration of Snell's law of refraction.

4. Book 2: The motion of bodies in resisting media

The first three sections of Book 2 give a thorough treatment of the motion of bodies in resisting media when the fluid resistance varies as the velocity, as the square of the velocity, or as a combination of the two. Section 2 contains the famous 'fluxional' lemma in which Newton sets out the rules for differentiating expressions of the form $x^p y^q z^r$, where the indices may be fractional or negative. Section 4 deals with various kinds of circular motion in a resisting medium, while Section 5 introduces the new topic of hydrostatics.

Much of the rest of Book 2 is, in essence, a sustained attack on the vortex theory of Descartes, and so is now mainly of historical interest only. Descartes postulated that the Universe is full of tiny, invisible particles that swirl round the Sun in a gigantic whirlpool, dragging the planets along with them. His theory certainly explained why all the planets move very nearly in the same plane and orbit the Sun in the same direction. It also provides an intuitively acceptable collision mechanism for planetary motion, avoiding the unsatisfactory notion of instantaneous 'action at a distance'. However, it could not account for the precise motion of the planets in accordance with Kepler's laws. The vortex theory was severely attacked – and eventually completely discredited – by Newton and his successors, as we shall see shortly.

In this Book Newton also investigates the physical causes of fluid resistance, which he attributes to the inertia of the particles of the fluid that are set in motion by the moving body. This theory implies that the resistance is proportional (very nearly) to the product of three factors: the density of the fluid, the square of the velocity of the moving body, and its cross-sectional area and shape. Section 6 returns to the topic of the cycloidal pendulum, but the oscillations now take place in a resisting medium. 'From these Propositions,' Newton continues, 'we may find the resistance of mediums by pendulums oscillating therein. I found the resistance of the air by the following experiments.' He describes these experiments and analyses their results in considerable detail; his theory is handsomely confirmed. Newton then goes on to compare the resistance of different fluids by pendulum experiments in water and quicksilver.

In Section 7 Newton calculates the resistance offered by different shapes (such as cylinders, spheres and cones) and attempts to analyse the flow of water running out of a leaky container. An interesting mathematical point is that here he gives, without proof, the geometrical conditions a surface of revolution must satisfy so as to offer the least resistance to axial motion through a fluid. This may well be the first problem to be solved by applying the techniques of a branch of mathematical analysis known as the 'calculus of variations' (p. 288). Section 7 concludes with a detailed discussion of various experiments with falling bodies. Thus, for example, we are told how:

From the top of St. Paul's Church [presumably Wren's new cathedral] in London in June 1710, there were let fall together two glass globes, one full of quicksilver, the other of air; and in their fall they described a height of 230 English feet.

Section 8 examines how motion is propagated through fluids, and contains one of the first rigorous analyses of wave motion. A notable, if perhaps incidental, achievement here is the demonstration that the velocity of sound in a fluid is proportional to the square root of the ratio of the 'elastic force' of the fluid to its density. The final section completes the demolition of Descartes by a careful analysis of the circular motion of fluids. Newton is able to show that a vortex motion cannot possibly produce a velocity pattern which accords with Kepler's three laws of planetary motion. His triumphant conclusion is that 'the hypothesis of vortices is utterly irreconcilable with astronomical phenomena, and rather serves to perplex than explain the heavenly motions.' He provides the true explanation in Book 3.

5. Book 3: The System of the World

By the propositions mathematically demonstrated in the former books, in the third I derive from celestial phenomena the force of gravity with which bodies tend to the Sun and the several planets. Then from these forces, by other propositions which are also mathematical, I deduce the motions of the planets, the comets, the Moon and the sea. I wish we could derive the rest of the phenomena of Nature by the same kind of reasoning from mechanical principles.

Thus says Newton in the preface to the first edition. Book 3, which has its own preface, opens with four 'Rules of Reasoning in Philosophy'. Rule 2, for example, asserts that 'to the same natural effects we must . . . assign the same causes'. Since Newton is now investigating 'the world', i.e. the workings of the Solar System, he needs some observational data on which to build. These are set out in the form of six Phenomena, and he is ready to begin his main argument.

Now, although Newton takes Kepler's laws for granted, he makes no mention of elliptical orbits until Proposition 13. Instead, he starts with the simplest possible assumption – that of circular motion. The Phenomena summarize the astronomical evidence that the motions of the six planets, and the satellites of Jupiter and Saturn, are in accord with Kepler's third law – at any rate, very nearly. Hence, he argues, they are subject to an inverse-square force of attraction which he now feels entitled to call 'gravity'. It is worth remarking that, for the Moon and the planets, Newton derives his main argument for the inverse square law, not from Kepler's laws, but from the stability of the orbits, i.e. the very slow motion of the apse lines. In Proposition 4 he carries out a numerical exercise to identify the centripetal force acting on the Moon with terrestrial gravity. He is then able, by appealing both to terrestrial experiments and to Kepler's third law, to establish that the gravitational force exerted by a body is proportional to its mass, and that 'inverse square law distances' are to be measured from the centre of the body.

At this point Newton starts to refine his physical model of the Solar System, just as he has done in Book 1 with his mathematical model. One such refinement, that of elliptical rather than circular orbits, is introduced almost casually in Proposition 13, the main concern of which is to establish that the effect of planetary perturbations is, in most cases, negligible. The next topics to be discussed are the figure of the Earth (i.e. the polar flattening) and the phenomenon of the tides. We then have a set of mathematical propositions in which Newton grapples with the difficult problem of demonstrating how the irregularities of the Moon's motion can be accounted for, in quantitative terms, by the gravitational pull of the Sun. In Proposition 39 Newton achieves one of his most impressive feats – the accurate calculation of the conical motion of the Earth's axis that we know as the precession of the equinoxes. To reach his goal he requires several lemmas, in one of which he uses his 'method of fluxions' to integrate two polynomial expressions of the form $\Sigma_i a_i x^i$. What Newton is doing in the central part of Book 3 is applying the principles of gravitation to account for, not only the overall dynamics of the Solar System, but also the small deviations of the actual motions from the simple pattern laid down by Kepler's three laws.

The rest of Book 3 is concerned with another very difficult problem – the calculation of cometary orbits. The difficulty is that, from observations made from a body moving in an elliptical orbit in one plane, we have to determine the motion of another body moving in a conic-section orbit in a different plane. What Newton does, with the aid of 11 lemmas, is to devise a successive approximation procedure for determining the parabolic orbit of a comet from three evenly spaced observations. If the observations are not evenly spaced, some interpolation will be necessary. To deal with this, Newton inserted a famous lemma, 'To find a curved line . . . which shall pass through any given number of points', which states the two forward difference interpolation formulae (for equal and unequal tabular intervals) that bear his name (pp. 199–201).

To test his procedure Newton chose the great comet of 1680–81, which was observed by Flamsteed and by Newton himself. Newton plotted its orbit and carried out his corrections manually on a large-scale diagram. The calculations were checked and extended by Halley, as we shall see in the next chapter.

6. The General Scholium

The elevation of comets to full membership of the Solar System (they had previously been regarded as atmospheric phenomena) brought the first edition of the *Principia* to an end. The work was widely acclaimed, but was criticized on theological and philosophical grounds. Newton's response was to append a short coda – the famous General Scholium – to the second edition of 1713. The first edition contained no pronouncements on such matters as the nature of the 'aether', the cause of gravity or the role of the Deity. Presumably the General Scholium was intended to clarify his position on these difficult matters – and, if possible, to forestall further controversy.

After opening with a criticism of the vortex theory of Descartes, Newton affirms his belief in the 'dominion of an intelligent and powerful Being'. He develops his ideas as to the nature of such a Being, adopting an austerely deistic position: 'We have ideas of His attributes, but what the real substance of anything is we know not.' The paragraph ends with the cryptic remark: 'And thus much concerning God, to discourse of Whom from the appearance of things, does certainly belong to Natural Philosophy.' We then have the famous penultimate paragraph in which Newton gives his most explicit formulation of the universal law of gravitation:

Hitherto we have explained the phenomena of the heavens and of our sea by the power of gravity, but have not yet assigned the cause of this power. This is certain, that it must proceed from a cause that penetrates to the very centres of the Sun and planets, without suffering the least diminution of its force; that operates . . . according to the quantity of the solid matter which they contain, and propagates its virtue on all sides to immense distances, decreasing always as the inverse square of the distances . . . But hitherto I have not been able to discover the cause of these properties of gravity from phenomena, and I frame no hypotheses. Hypotheses . . . have no place in experimental philosophy . . . to us it is enough that gravity does really exist . . . and abundantly serves to account for all the motions of the celestial bodies, and of our sea.

Newton, then, makes no attempt to provide a physical explanation of gravitation: he is not to be drawn. Indeed, the nature of gravitation remains as mysterious today as it was in the seventeenth century. In his rejection of philosophical speculations, together with his fruitful use of mathematics, Newton is setting the pattern for the great advances in physical science that were to follow. It must be emphasized, however, that his prohibition on the framing of hypotheses is not put forward as a general proposition; rather it relates specifically to the public position he wishes to adopt in the *Principia*. Indeed, he formulated many hypotheses in his other writings – notably, as we have seen, in the Queries in his *Opticks*.

In the last paragraph of the General Scholium Newton does give a hint, no more, of his views on the nature of the aether. 'And now we might add something concerning a certain most subtle Spirit which pervades and lies hid in all gross bodies.' It is, he says, 'the force and action of which Spirit' that determines, not only the nature of the physical world, but also 'all sensation' experienced by 'the members of animal bodies'. His concluding sentence comes as something of an anticlimax:

But these are things that cannot be explained in a few words, nor are we furnished with that sufficiency of experiments which is required to an accurate determination and demonstration of the laws by which this electric and elastic spirit operates.

As we have said, the *Principia* is not an easy book to read. Indeed, Newton himself warned against the detailed study of every proposition, which 'might cost too much time, even to readers of good mathematical learning.' But this is, indeed, part of its attraction. One can return to the *Principia* again and again and always find something new to deepen understanding and stretch the mind. Its treasures are inexhaustible.

10   Newton's circle

'Mathematics – the unshaken Foundation of Science, and the plentiful Fountain of Advantage to human affairs.' Isaac Barrow

1. Introduction

From the time of the publication of the *Principia* and throughout the eighteenth century, the Newtonian world-view was the dominant influence on English intellectual life, especially in the fields of mathematics and natural philosophy. We must therefore be careful not to overlook the contributions of others who stood on lower rungs of the ladder. In this chapter we look briefly at four of these others: two of them older men who helped to start the young scholar from Lincolnshire on his long and fateful journey, and two younger colleagues who were intimately associated with the publication and editing of his greatest work.

2. John Wallis

John Wallis (1616–1703) was the most influential English mathematician of the generation before Newton (see Illustration 8). In the course of his long life he became first a contemporary of Newton's, and then one of his closest friends. Thus, in 1695 we find him gently admonishing the great man: 'Consider, that 'tis now about thirty years since you were master of these notions about Fluxions and Infinite Series; but you have never published aught of it to this day.'

Illustration 8. John Wallis, 1616–1703 (Mansell Collection)

Wallis was born at Ashford in Kent and educated at Felstead School. Looking back in later life on the state of mathematical education during his childhood, he remarked that 'mathematics at that time with us was scarce looked on as academical but rather mechanical – as the business of tradesmen.' He entered Emmanuel College, Cambridge (one of the Puritan colleges), with the intention of becoming a physician. The subject of his 'academic dissertation' was the circulation of the blood; this is believed to be the first occasion on which Harvey's great discovery was presented in public. In the event, however, Wallis's mathematical interests proved too strong; he was elected to a fellowship at Queens' College, Cambridge, and took holy orders. Although his political sympathies in the Civil War lay broadly with the Royalists, he was quite prepared to place his skill in decoding secret dispatches at the service of the Parliamentarians.

In 1649 Wallis was appointed Savilian Professor of Geometry at Oxford. When the monarchy was restored in 1660 he became one of the King's Chaplains. He was also a founder-member of the Royal Society. One of his Oxford colleagues in his enterprise was Christopher Wren, the Savilian Professor of Astronomy (who, but for the Great Fire of London, might have been remembered as a mathematician instead of an architect).

3. Wallis's algebraic investigations

In 1655 Wallis published his *Tractatus de sectionibus conicas*, in which he established the properties of the conic sections using only the methods of plane coordinate geometry. His starting point was the three defining second-degree equations given in Chapter 3 (pp. 58–59). In the same year he brought out his most important book, the *Arithmetica infinitorum*, in which he applied analytical techniques to the 'method of indivisibles' (p. 79). The whole thrust of his work was to replace geometrical with algebraic concepts and procedures wherever possible. The young Newton's study of this book was a major influence in his 'discovery' of the general binomial theorem.

Wallis followed the fashion of his time in thinking of a plane surface as made up of infinitely many lines, but he abandoned the conventional geometrical model once he had associated each such line with a number proportional to its length – or with an algebraic symbol representing the number. Thus, he takes the lengths of

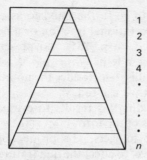

Figure 10.1.

successive lines in the triangle in Figure 10.1 as $1, 2, 3, \ldots$, up to n. Clearly,

$$\frac{0 + 1 + 2 + \ldots + n}{n + n + n + \ldots + n} = \frac{\text{area of triangle}}{\text{area of rectangle}} = \frac{1}{2}$$

He then seeks to evaluate a similar ratio involving sums of squares. He argues that if $n = 1$, then

$$\frac{0^2 + 1^2}{1^2 + 1^2} = \frac{1}{2} = \frac{1}{3} + \frac{1}{6}$$

if $n = 2$, then

$$\frac{0^2 + 1^2 + 2^2}{2^2 + 2^2 + 2^2} = \frac{5}{12} = \frac{1}{3} + \frac{1}{12}$$

and, in general,

$$\frac{0^2 + 1^2 + 2^2 + \ldots + n^2}{n^2 + n^2 + n^2 + \ldots + n^2} = \frac{1}{3} + \frac{1}{6n}$$

The limiting value of this ratio as n gets larger is clearly $\frac{1}{3}$. This is equivalent to saying that the quadrature of (i.e. the area beneath) the curve $y = x^2$ from $x = 0$ to $x = 1$ is $\frac{1}{3}$. Further arithmetical explorations led Wallis to the general result that the corresponding quadrature of $y = x^m$ is $1/(m + 1)$, where m is a positive integer.

The intrepid analyst was prepared to go further, and to assert that this result remains valid when m is a rational fraction or a negative integer (excluding -1, as we noted on p. 145). To preserve the generality of the result it is necessary to write $\sqrt{n} = n^{1/2}$, $\sqrt[3]{n} = n^{1/3}$, $1 = n^0$, $1/n^2 = n^{-2}$, etc. This led Wallis to introduce the notation for negative and fractional indices that we use today.

Wallis had no difficulty in extending his results to evaluate the quadrature of curves with equations of the form $y = x^p(a^n - x^n)^m$, provided p, m and n are positive integers. His prime objective, however, was to 'square the circle', i.e. to effect the quadrature of the curve $y = (1 - x^2)^{1/2}$, by expanding y in a power series in x^2 of the form $a_0 + a_1x^2 + a_2x^4 + \ldots$. In this he was unsuccessful; it was left, as we have seen, to the young Newton to achieve success here. Instead Wallis deployed an elaborate argument, involving much guesswork but guided by sound mathematical intuition, which eventually yielded the well-known infinite product for π:

$$\frac{\pi}{2} = \frac{2 \times 2 \times 4 \times 4 \times 6 \times 6 \times \ldots}{1 \times 3 \times 3 \times 5 \times 5 \times 7 \times \ldots}$$

In 1685 Wallis published his *Algebra* (it was considerably enlarged in 1695) which contains, among much else, what is believed to be the first systematic use of algebraic formulae in the modern manner. For example, in the well-known formula $s = vt$ for the uniform motion of a particle in a straight line, he regards s as the *number* which measures the ratio of the distance travelled to a prescribed unit of length, and similarly for v and t. Earlier writers would have written $s_1 : s_2 = v_1 t_1 : v_2 t_2$.

Wallis was a prolific expositor and editor, and contributions from other writers are sometimes to be found in his books. It was, for example, in the 1695 edition of his *Algebra* that Newton's two letters of 1676 to Leibniz (p. 183) were published, by permission, for the first time.

4. Isaac Barrow

Isaac Barrow (1630–77) – Newton's teacher, colleague and friend – was born in London and educated at Charterhouse, Felstead and

Illustration 9. Isaac Barrow, 1630–77 (Trinity College, Cambridge)

Cambridge (see Illustration 9). At Charterhouse he was a most troublesome pupil, much given to fighting and to promoting fighting among the other boys. Indeed, we are told that his father was heard to pray that if it should please God to take any of his children, he could best spare Isaac. Fortunately the boy was not taken and, when he was moved to Felstead School, soon proved to be an excellent all-round student. His brilliance and versatility were maintained throughout his life; his interests embraced not only mathematics but also the classics, theology, poetry and science. He was a prolific writer in both Latin and English – writing, we are told, 'with a sustained and somewhat stately eloquence' – and was distinguished in several fields of scholarship.

In 1644 Barrow entered Trinity College, Cambridge, graduating in 1648; in the following year he was elected a Fellow. In 1655 he applied for the post of Professor of Greek at Cambridge, a position for which he was eminently qualified. He did not get it, apparently because of religious difficulties. At Cambridge at this time the influence of the Independents (non-conformists) was very strong, whereas Barrow was suspected of Anglican, even High Church, sympathies. Deeply disappointed at this failure, the young scholar sold his books and left England to spend the next four years travelling, mainly in Eastern Europe.

Barrow was a small, lean man of unusual physical strength and courage. He needed both in some of his encounters with Mediterranean pirates! He was untidy of dress and a heavy smoker. His learning, charm and lively companionship won him many friends, not least among the ambassadors and other English residents of the countries he visited.

In 1659 Barrow returned to England, took holy orders and was successful in obtaining, at the second time of asking, the Greek Chair. In 1662 he was elected Professor of Geometry at Gresham College, London. Things happened quickly in those days, and in 1663 Barrow became the first holder of the Chair of Mathematics at Cambridge that had been established by Henry Lucas (1610–67), one of the University's MPs. In 1669 occurred the event for which he is perhaps best known today: he resigned his Lucasian Chair – ostensibly in order to be able to give more time to theological studies – and was succeeded by his friend and former pupil, the young Isaac Newton.

In the following year Barrow, like Wallis before him, became Chaplain to Charles II. The King was an admirer of Barrow, as scholar, conversationalist and wit, and was anxious to secure his presence at court. This royal appointment does not seem to have impaired Barrow's academic career; in 1672 he was appointed Master of Trinity, and three years later he was chosen to be Vice-Chancellor of the University, a post he held until his sudden death of a fever in May 1677. He was buried in Westminster Abbey. The final sentence of Aubrey's 'brief life' of Barrow is worth quoting: 'As he laye unravelling in the agonie of death, the Standers-by could hear him say softly, *I have seen the glories of the world.*'

Barrow made substantial contributions to mathematics and natural philosophy, particularly to optics, and was one of the first to be elected to membership of the Royal Society. As a mathematician, he was strongly conservative and disliked the analytical methods favoured by his Oxford rival, Wallis. He greatly admired the work of the classical masters, and his knowledge of Greek and Arabic enabled him to produce carefully edited and widely read translations (into both Latin and English) of the writings of Euclid, Apollonius, Archimedes and others.

Most of Barrow's own contributions are contained in three published collections of his professorial lectures. The *Lectiones opticae* appeared in 1669, the *Lectiones geometricae* the following year; both were supervised for publication by Newton. The third work, the *Lectiones mathematicae*, was not published until 1683, after the author's death. Of the three, the *Lectiones geometricae* is the most important as it contains the essential foundations of the calculus in geometrical form, as we shall see in Section 6.

5. Barrow and Newton

Although much has been written on the subject, the precise relationship between Barrow and Newton remains something of a mystery. While Barrow was never Newton's tutor in any formal sense, a strong working association based on mutual respect and friendship developed over the years between the young scholar and his college senior. At that time no mathematical lectures were given to undergraduates, but we know that in 1664–65 Newton attended Barrow's lectures on space, time and motion. Newton's heavily annotated copy of Barrow's 1655 edition of Euclid's *Elements* (a standard textbook of the time) is preserved at Cambridge.

In this connection it is amusing to read, in the somewhat unreliable account of Newton's life and work that was written after his death by John Conduitt, that:

soon after he stood to be a Scholar of the House [Trinity College] and Dr. Barrow examined him in Euclid, which he knew so little of that Dr. Barrow conceived a very indifferent opinion of him. The Dr. never asked him about Descartes's *Geometry*, not imagining that any one could be master of that book without first reading Euclid, and Sir Isaac was too modest to mention it himself, so that he was not made Scholar of the House till the year following. Upon this Sir Isaac read Euclid over again, and began to change his opinion of him.

Newton's annotations leave no doubt as to the depth of this second reading!

Barrow helped his young friend in a variety of ways, not least by providing him with much-needed encouragement to pursue his early studies in the calculus. In July 1669 Barrow sent John Collins a copy of Newton's *De analysi* (p. 184) for limited private circulation. He described the work as that of 'a friend of mine here, that hath a very excellent genius to these things'. Three weeks later Barrow wrote to Collins again and disclosed the identity of his friend, whom he introduced to the scholarly world outside Cambridge as 'a fellow of our College . . . very young . . . but of an extraordinary genius and proficiency'. John Collins (1625–83) operated an unofficial London centre for the exchange of mathematical information; he married the Queen's laundress.

Barrow had an excellent mathematical library which he generously placed at Newton's disposal, but probably his greatest service – both to Newton himself and to posterity – was to place his young colleague in a secure academic position. The reasons why Barrow decided to relinquish the Lucasian Chair are no longer accurately known; the story that he resigned because he recognized Newton's superior fitness for the post is now dismissed by most scholars as a sentimental gloss on the scanty facts that are still available. What we know of Barrow's forceful and ambitious character suggests that we should look elsewhere for his motives.

The Royal Statutes governing the Lucasian Chair were expressly designed to prevent the Professor from taking an active part in university administration or politics; he was not allowed to hold any university or college position except a fellowship. It is therefore reasonable to regard Barrow's resignation as a necessary preliminary to his being considered for the Mastership of the College – a post that he did indeed obtain before very long. Be this as it may, once Barrow had decided to resign he was in no doubt as to who should succeed him.

The power of appointment was vested in the surviving executors of Henry Lucas's will. These two old men – a London lawyer and the University Printer – were, after a little judicious flattery, only too willing to endorse Barrow's recommendation and Newton was elected on 29 October 1669. The new professor paid a well-deserved tribute to his predecessor when he devoted his first course of lectures (in 1670) to the subject of optics, initially as a continuation of Barrow's *Lectiones opticae*, which had just appeared in book form.

Although the mathematical association between the two men declined from this time onwards, the mutual personal regard remained. Newton concluded a letter to Collins in 1672 thus: 'We are very glad here that we shall enjoy Dr. Barrow again especially in the circumstances of Master, nor doth any rejoice at it more than I.'

6. Barrow's geometrical lectures

The thirteen lectures published in 1670 as the *Lectiones geometricae* contain a varied collection of theorems. Most of them relate to the dominant problems of the time – the finding of tangents, maxima and minima, quadrature and rectification of curves. The treatment is almost entirely geometrical, which does not make for easy reading today. Nevertheless, the work had considerable influence, not only on other British mathematicians, but also on Continental mathematicians, Leibniz in particular.

In his preface Barrow says that he did not find the presentation very satisfactory, but instead of editing the lectures he decided to send them forth 'in Nature's garb', which presumably means verbatim, as they were delivered in the public schools of Cambridge in 1668. Although Newton saw the work through the press for his friend, he made – or was allowed to make – few changes.

The 1660s were a time of much mathematical activity. Barrow wished to present the current 'state of the art', and he included a full account of such recent discoveries as were known to him. The first theorem of Lecture X may serve to illustrate both the ingenuity and the complexity of Barrow's geometrical approach. Figure 10.2 shows the construction. From any point E on the curve AEG a straight line EF is drawn in a prescribed direction (here taken parallel to the *y*-axis), such that EF is equal to the arc AE. The curve AFI is the locus of F. The line TE is the tangent to the curve AEG at E, and TE is made equal to EF. Barrow's theorem asserts that TF touches the curve AFI at F, which he proves by showing that the extended line, TF, always lies on the same side of the curve AFI.

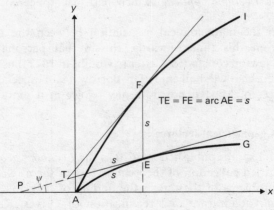

TE = FE = arc AE = *s*

Figure 10.2.

On the face of it the theorem is no more than an unnecessarily complicated method of constructing tangents, but closer inspection shows that it provides a geometrical demonstration of the basic formula for the arc length *s* of a curve, which may be expressed in fluxional notation as $\dot{s}^2 = \dot{x}^2 + \dot{y}^2$. To see this, we consider the construction with respect to x- and y-axes as shown, set $s = $ arc AE = EF = ET, and introduce the angle EPA = ψ. The coordinates of E and F are (x, y) and $(x, y + s)$ respectively, so the slope of TF is $(\dot{y} + \dot{s})/\dot{x}$. It is also equal to $(s + s \sin \psi)/s \cos \psi = \sec \psi + \tan \psi$. Now, since $\tan \psi = \dot{y}/\dot{x}$, we get $\dot{s}/\dot{x} = \sec \psi = (1 + \tan^2 \psi)^{1/2}$, or $\dot{s}/\dot{x} = (1 + \dot{y}^2/\dot{x}^2)^{1/2}$, from which the desired result, $\dot{s}^2 = \dot{x}^2 + \dot{y}^2$, follows at once.

Barrow then gives several related theorems, followed by some applications to special curves. He concludes this part of his exposition with the remark:

I think these are sufficient to indicate the method by which, without the labour of calculation, one can find tangents to curves and at the same time prove the constructions. Nevertheless, I add one or two theorems, which it will be seen are of great generality, and not lightly to be passed over.

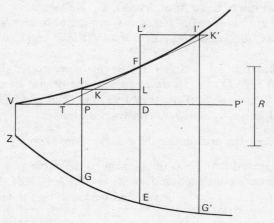

Figure 10.3.

We shall not pass them over either – they are the most important theorems in the book! Here is Barrow's statement of the first of these general theorems:

Let ZGE [Figure 10.3] be any curve of which the axis is VD; and let ordinates applied to this axis, VZ, PG, DE, continually increase from the initial ordinate VZ; also let VIF be a curve such that, if any straight line EDF is drawn perpendicular to VD, cutting the curves in the points E and F, and VD in D, the rectangle contained by DF and a given length R is equal to the intercepted space VDEZ; also let DE : DF = R : DT, and join TF. Then TF will touch the curve VIF.

To prove the theorem, Barrow considers any point I on VIF ('first on the side of F towards V'), and draws the lines IPG and IKL as shown. This gives FL/LK = DF/DT = DE/R, and hence R.FL = LK.DE. Now, since the construction applies to any point on VIF, we have R.IP = area VPGZ, as well as R.FD = area VDEZ. By subtraction, we get R.FL = area PDEG, and hence LK.DE = area PDEG < DP.DE. It follows that LK < DP = LI. Taking a point (call it I′) on the curve on the other side of F, and making the same construction, we can show similarly that L′K′ > L′I′; 'from which it is clear,' says Barrow, 'that the whole of the line TKFK′ lies within or below the curve VIFI′.' He then completes the argument by examining what happens when the ordinates VZ, PG, DE continually decrease and 'the curve VIFI′ is concave to the axis VD'.

To see what Barrow is doing, let us introduce some algebraic notation. Taking VD as the x-axis and V as the origin, let us write the curve VIFI′ as $y = f(x)$ and the curve ZGEG′ as $z = g(x)$, with DF = y and DE = z. Barrow's construction makes

$$R.y = \text{the area under the } z\text{-curve from O to } x \qquad \text{(A)}$$

He has also proved that TF is the tangent to the curve $y = f(x)$ at F. Now, Barrow's construction also defines the length of the subtangent DT (= t, say) as $t = R.y/z$, and we know that the slope of the y-curve = y/t. Hence

$$R \times \text{slope of the } y\text{-curve} = R.y/t = z \qquad \text{(B)}$$

Barrow's theorem can thus be expressed as

statement (A) implies statement (B)

This is one half of the complete theorem, now known as the 'fundamental theorem of the calculus', expressing the inverse relationship between tangents and quadratures (or, as we now say, between differentiation and integration).

Barrow does not reach the other half of the fundamental theorem until Theorem 19 of Lecture XI. There he defines the ordinate we have called z to be $R.y/t$ and proves that the quadrature of $z = g(x)$ from O to x is equal to $R.y$. That is to say, he proves the converse of the previous theorem, namely:

statement (B) implies statement (A)

This completes Barrow's double proof of the fundamental theorem. It can be stated concisely, in the modern notation that we shall be presenting in the next chapter, as

$$\text{If } \int y \, dx = A, \text{ then } dA/dx = y, \text{ and vice versa}$$

Here A is an area function (we have used z above) and y is an ordinate.

We must now return to Lecture X. After establishing a set of 14 theorems by his preferred geometrical methods, he continues thus:

We have now finished in some fashion the first part, as we declared, of our subject. Supplementary to this we add, in the form of appendices, a method for finding tangents by calculation frequently used by us. Although I hardly know, after so many well-known and well-worn methods of the kind above, whether there is any advantage in doing so. Yet I do so on the advice of a friend; and all the more willingly because it seems to be more profitable and general than those which I have discussed.

We know that the friend was Newton because he says in a letter:

A paper of mine [the *De analysi*] gave occasion to Dr. Barrow to shew me his method of Tangents before he inserted it into his 10th Geometrical Lecture. For I am that friend which he there mentions.

Barrow acknowledges Newton's assistance in his preface to the book.

In these appendices Barrow abruptly changes his form of presentation and gives an analytical treatment of the tangent problem which is very similar to that of Fermat (whose work was not known to him) and of the young Newton (whose work certainly was). Barrow illustrates his new approach by five examples, two of them being the folium of Descartes (p. 134) and the quadratrix (p. 27).

We see here that Barrow – apparently with some reluctance, and presumably under Newton's influence – departs from his usual Euclidean approach and presents, in effect, an algebraic algorithm for calculating the slope of a curve of the form $f(x, y) = 0$. His treatment is entirely modern. Indeed, in his appendices Barrow takes a giant stride from Antiquity to the modern world. It must be said, however, that his conservative approach in the rest of his writings made it difficult for him to make effective use of the relationship between the inverse tangent and quadrature (indeed, he probably did not realize the full significance of his own discoveries) and for his contemporaries to understand his books. However, he made ample amends for any obscurity by his encouragement of, and timely assistance to, his young colleague.

7. Edmond Halley

Edmond Halley (1656–1742) is widely remembered for two things: his fateful journey to Cambridge in 1684 (p. 173), and his comet. John Aubrey tells us that:

He was born in Shoreditch parish, at a place called Haggerston, the backside of Hogsdon . . . the eldest son of a Soape-boyler, a wealthy Citizen of the City of London.

Edmond went to St Paul's School and then, at the age of 16, to Queen's College, Oxford, where he soon made his mark as an observational and a theoretical astronomer. In 1676, while still a student, he was given leave of absence for a year to visit the remote island of St Helena in order to map the stars of the southern hemisphere, using the improved telescopic equipment that had recently become available. In the same year he published his first paper, in the *Philosophical Transactions of the Royal Society*, on an improved computation of the orbital elements of Jupiter and Saturn. He gave both a geometrical and an algebraic treatment of the problem. Three years later he was sent to Danzig by the Royal Society to negotiate over a dispute with the famous astronomer Johannes Hevelius on the use of the new telescopic sights. Halley was able to convince the much older Hevelius of their value and to restore amicable relations.

During the 1680s Halley consolidated his position as one of the leading mathematical astronomers of the day. He was prominent in London scientific society and served as Assistant Secretary of the Royal Society from 1685 to 1693. During these years, while encouraging Newton to write the *Principia* and then seeing it through the press, Halley found time to produce a series of mathematical papers on such topics as the solution of higher-degree equations, logarithms, infinite series and – inevitably – the cycloid and its generalizations. He was also an accomplished classicist. He restored, from hints by Pappus and others, the lost eighth book of the *Conics* of Apollonius (p. 57) and brought out a scholarly edition of the complete work. He also translated into Latin an Arabic version of one of Apollonius' minor works, the *Cutting of a Section* (having, it is said, learnt Arabic for the purpose).

In 1691 Halley was refused the Savilian Professorship of Astronomy at Oxford, on grounds of suspected atheism; perhaps it was some consolation to him to be appointed Deputy Controller of the Royal Mint at Ipswich in 1695. Two years later he was given command of the sloop *Paramore* with orders to chart the magnetic variation in southern latitudes. During the two-year voyage he penetrated the Antarctic and explored the South Atlantic in considerable detail. His next maritime task was nearer home – to make a thorough survey of the tides and currents in the English Channel. In 1703 his outstanding talents could be denied no longer: he succeeded Wallis as Professor of Geometry at Oxford. In 1713 he was elected Secretary of the Royal Society and in 1721 became the second Astronomer Royal, succeeding the luckless John Flamsteed. Halley had in fact been doing pioneer work in astronomy for many years before his official appointment. His investigations included a study of variable stars, establishing the 'proper motion' of stars by comparing contemporary observations with those of Hipparchus, measuring the precession of the equinoxes, and, above all, calculating cometary orbits. At the age of 61 he embarked on a long-term project – carried through to completion – of observing the Moon's motion throughout its complete nodal cycle of 19 years.

Halley's physical researches also covered a wide range. He investigated the variation of atmospheric pressure with height, worked out a theory of the circulation of trade winds and monsoons, and improved diving equipment. Also, he was no mean mathematician. For example, in a paper of 1691 he demonstrated that not all infinite quantities are equal. This may be said to foreshadow – albeit faintly – the work of Cantor, some 200 years later, which we shall be looking at in Chapter 16. His fame spread throughout Europe: he received favours from the Emperor in Vienna and the friendship of Peter the Great of Russia. In 1729 he was elected a foreign member of the Académie des Sciences in Paris. Halley was indeed a man of parts: he combined outstanding ability, wide culture and unbounded energy; during his long life his interests ranged from archaeology to diplomacy, from surveying to Classical Greek, from the motion of the Moon to the integrity of the currency.

8. Halley's cometary studies

It was in 1681, while in Paris, that Halley started to calculate cometary orbits. In 1695 we find him writing to Newton: 'I intend as far as I can to limit the Orbs of all the comets that have been hitherto observed.' His point of departure was Newton's geometrical method of computing parabolic orbits from three fairly equally spaced observations (Proposition 41 of Book 3 of the *Principia*). Halley was able to improve on this procedure, as Newton handsomely acknowledged:

But afterwards Dr. Halley did determine the orbit [of the great comet of 1680–81] to a greater accuracy by an arithmetical calculus than could be done by graphic operations.

Halley's more refined calculations, which took account of the gravitational influence of Jupiter and Saturn, led him to suspect that the motion of some comets was periodic: that they moved in highly elongated elliptical orbits rather than parabolic orbits. Newton and Halley exchanged a number of letters on the subject. Halley's scientific studies were then interrupted by official duties, and his major work on comets did not appear until 1705. Its centrepiece is a table of the orbital elements of 24 comets that had been seen between 1337 and 1698 and had been sufficiently well observed for their orbits to be computed. The table, he tells us, 'was the result of a prodigious deal of calculation'.

Halley became increasingly convinced that the comets of 1531, 1607 and 1682 'were one and the same comet'. He predicted that it would return in 1758, as indeed it did, and has done three times since. This famous object, universally known as Halley's Comet, is now fully accepted as a permanent member of the Solar System with a period of about 76 years. As to the other comets in the table, both Newton and Halley thought that the comet of 1680–81 was also periodic. Newton's statement in the second edition of the *Principia* is unequivocal: 'Moreover, Dr. Halley, observing that a remarkable comet had appeared four times at equal intervals of 575 years' – in September 44 BC, in 531, February 1106 and at the end of 1680 – 'set himself to find out an elliptic orbit . . . in which a comet might revolve in 575 years.' Newton then gives the computed orbital elements, and sums up thus:

The observations of this comet from the beginning to the end agree as perfectly with the motion of the comet in the orbit just now described as the motions of the planets do with the theories from whence they are calculated, and by this agreement plainly evince that it was one and the same comet that appeared all the time, and also that the orbit of that comet is here rightly defined.

It seems, however, that on this occasion the two great men were wrong: recent studies indicate that the 1680 comet has a period of some 9000 years! Be that as it may, Halley's place in history – as the godfather of the most famous of all scientific books, and with the most famous of all comets as his eponymous memorial – is secure.

9. Roger Cotes

Roger Cotes (1682–1716) is probably best known as the subject of Newton's remark: 'If he had lived, we might have known something.' Only now are we beginning to appreciate how much this brilliant young man was able to accomplish during his tragically short life. Cotes was born on 10 July 1682, the son of a Leicestershire rector. His mathematical abilities soon attracted attention and, at the age of 11, he was moved from Leicester School to be coached by his uncle, John Smith, a Lincolnshire rector. (John's son, Robert, was to become one of Cotes's closest professional colleagues; he also founded the Smith Prize at Cambridge.) In due course Roger went to St Paul's School and then, in 1699, to Trinity College. He was soon made a Fellow, and formed a close friendship with William Whiston, who succeeded Newton as Lucasian Professor in 1702.

We must now mention two events which proved of great significance for Cotes's career. Dr Thomas Plume, Archdeacon of Rochester, died in 1704 and bequeathed some £2000 to found a Chair of Astronomy and Experimental Philosophy at Cambridge, and to build, equip and maintain an observatory (and, if necessary, a professorial residence as well). In 1700 Dr Richard Bentley, the formidable classical scholar, was appointed Master of Trinity College. He was determined to raise the academic standard of the College, and in particular to foster the study of physical science. His schemes to this end included three major undertakings concerning Cotes, all of which were successfully accomplished. The first was the election of Cotes as the first Plumian Professor, the second was the construction at Trinity College of an observatory financed by Plume's bequest, and the third was the appointment of Cotes – an ardent Newtonian – as editor of the second edition of the *Principia*. Cotes was duly elected to the new Chair in 1707 and took holy orders in 1713. He died suddenly on 5 June 1716 at the age of 34.

The construction of an astronomical observatory at Cambridge was started, under Cotes's energetic direction, soon after his appointment, extra cash being raised by subscription. It was built over the Great Gate of Trinity College. Something of an architectural eyesore, it fell into disuse over the years and was taken down in 1797. Although little practical astronomy seems to have been done there, a few records survive – for example, of Cotes's observations of a solar eclipse in April 1715. He was interested in the more accurate determination of time and devised several instruments for this purpose. We also have some correspondence relating to the delivery of a special clock which was presented to the observatory by Newton. (It now stands in the Master's lodge.)

10. Editor of the *Principia*

Bentley was in no doubt that Cotes was the man for the editorial job, and managed to persuade Newton to accept him. We find Bentley writing to Newton in October 1709, thus:

You need not be so shy of giving Mr. Cotes too much trouble: he has more esteem for you and more obligation to you, than to think that trouble too grievous . . . We will take care that no little slip in a Calculation should pass this fine Edition.

Never was a promise better kept. From 1709 to 1713 Cotes was heavily occupied with his meticulous and skilful editing; he left the *Principia* essentially in the form in which we know it today. Bentley's accounts have survived and show that 750 copies of the second edition were printed, of which 575 were sold for the sum of £315 19s 0d. Bentley made a profit of nearly £200; Cotes, who did all the work, received just 12 copies of the book.

The printing of the first half of the new edition went smoothly enough, Cotes needing only to make 'little alterations' to the text. As the work proceeded, however, he found it necessary to draw Newton's attention to more and more problems in the copy, some of which took a long time to resolve; and also to draft revisions of his own. It was nearly two years before work could begin on Book 3. The most intractable problem proved to be the theory of lunar motion. Although Newton knew and took account of seven 'inequalities' of the Moon's motion, he was unable to devise a satisfactory general theory. Cotes was far from satisfied with some of Newton's arguments, and the work dragged slowly on through most of 1712.

Then a thunderbolt struck! Newton to Cotes, 14 October 1712:

There is an error in the tenth proposition of the second book, which will require the reprinting of about a sheet and a half . . . I was told of it since I wrote to you and I am correcting it.

This proposition is concerned with the motion of a body in a resisting medium under the influence of gravity. A formula given in the first edition contained a numerical value of $\frac{1}{2}$; the correct value is $\frac{3}{4}$. The error had been found by Jean Bernoulli some years earlier when he was reworking one of Newton's examples, but he could not find any mistake in Newton's general argument, from which the particular result was deduced. Indeed, it took Newton himself many days to find the error – a very subtle one. He eventually succeeded and formulated no fewer than five different proofs – all correct – one of which was duly inserted in the new edition.

Newton was extremely annoyed at the turn of events. Cotes took it more calmly, merely remarking: 'I observe that you have increased the Resistance in the proportion of 3 to 2, which is the only change in your conclusions.' Clearly, he did not appreciate the wider implications of the affair. It is sad to record that from this time the relationship between the two men began to deteriorate. To cite but one example, Newton had drafted a preface to the second edition in which he paid Cotes the sort of tribute he had a right to expect. But it was never printed. Why? The answer must be sought in the fact that the priority quarrel between Newton and Leibniz over the calculus was passing through a critical and acrimonious phase. It was most unfortunate that it was Jean Bernoulli, a leading Leibnizian, who had discovered the mistake. Newton feared that the 'Continental mathematicians' would contend that a man who could make such an error was not likely to be the independent creator of the calculus. This is more or less what did happen: all Europe was told how the great man had been saved from a blunder by the Leibnizians. Newton could never bring himself to thank Bernoulli for drawing his attention to the error, so presumably he felt unable to thank Cotes either. After all, Bernoulli had found what Cotes had missed.

In the event, it was Cotes who provided an excellent preface of his own to the second edition. He starts by explaining the 'manner of philosophy' used in the *Principia* – broadly speaking, what we now think of as the *scientific method*. He then summarizes the arguments needed to deduce

the Principle of Gravity from the Phenomena of Nature in a popular way that it may be understood by ordinary readers and may serve at the same time as a specimen to them to the Method of the whole Book.

Cotes rounds off his essay with a powerful attack on the vortex theory of Descartes (pp. 129 and 222). This curious theory retained a strong hold in England until well into the eighteenth century, and even longer on the Continent. Voltaire, visiting England in 1723, remarked:

A Frenchman arriving in London finds things very different in natural philosophy, as in everything else. He has left the world full, he finds it empty. At Paris they see the Universe as composed of vortices of subtle matter, in London they see nothing of the kind.

11. Cotes's mathematical researches

Cotes published only one paper during his lifetime. It was entitled *Logometria* and appeared in the *Philosophical Transactions of the Royal Society* for March 1714. He first develops the theory of logarithms, which he defines as 'measures of ratios', and then applies his ideas to the integration (or, as he puts it, 'finding the fluents of fluxional expressions') of a variety of rational and irrational algebraic forms, yielding a number of special functions, many of which were not clearly recognized at the time. He also employs the method of continued fractions (p. 51) to obtain rational approximations to irrational numbers, and investigates a number of physical problems, most of them suggested by his study of the *Principia*. The paper ends with a long *Scholium generale* in which he gives solutions (but not proofs) for 13 problems of current interest. These include the rectification of a number of curves and the evaluation of the surface areas and volumes of the associated solids of revolution. The *Principia* influence is again apparent in his choice of problems.

Most of Cotes's papers were published posthumously in 1722 by his cousin, Robert Smith, who followed Cotes as Plumian Professor. Smith entitled his collection *Harmonia mensurarum* and issued it in four parts. The first is essentially the paper already mentioned; the second consists of extensive tables of 18 forms of fluents (i.e. integrals) embodying Cotes's new techniques; the third is a set of 12 illustrative problems, with full solutions. Cotes's creation of these tables was an impressive achievement; they contain many of the so-called 'standard' forms and much else besides. As de Morgan put it in 1857: 'They represent the first substantial advance in the development of integration techniques applied to logarithmic functions and trigonometrical expressions.'

In 1716 Cotes discovered a general method of factorizing expressions of the form $a^n \pm x^n$ into linear and quadratic factors. The geometrical statement of the theorem requires a circle to be divided into $2n$ equal arcs. His discoveries embodying what are known as 'Cotes's properties of the circle' greatly enlarged the range of expressions that could be integrated. Smith edited and extended this work, presenting the results in no fewer than 94 tables of integrals to form the fourth part of *Harmonia mensurarum*.

Cotes's interest in practical astronomy led him to investigate the theory of errors of observation. His results were set out in a tract entitled *Aestimatio errorum in mixte mathesi* (The Estimation of Errors in Practical Mathematics). The processing of astronomical observations usually entails 'solving' either plane or spherical triangles. Cotes was primarily concerned with the differential formulae relating the small variations in two of the six 'elements' of the triangle (the three sides and the three angles) when two other elements are held constant. He gave 28 such formulae – 10 for plane and 18 for spherical triangles. His work on numerical methods derives, once again, from the *Principia*. In his paper 'De methodo differentiali Newtoniana', he proves Newton's interpolation formulae (p. 199) and discusses several other topics, including the approximate quadrature of curves. Cotes's reading of Newton's *Methodus differentialis* in 1711 led to further work on interpolation and subtabulation using central differences (p. 200). He set out his results in a paper called *Canonotechnia*, which Smith included in *Harmonia mensurarum*.

Finally, we must mention Cotes's 'theorem of the harmonic mean'. It was formulated in an unpublished tract, *On the Nature of Curves*, which eventually appeared in 1748 in a treatise entitled *Algebra* by the Edinburgh mathematician Colin Maclaurin, with due acknowledgement to Cotes. The theorem states that: 'If, through a fixed point O, a variable straight line is drawn cutting an algebraic curve in points $P_1, P_2, P_3, \ldots, P_n$, and if a point H is taken on the line such that OH is the harmonic mean of $OP_1, OP_2, \ldots, OP_n$, then the locus of H is a straight line.' The harmonic mean h of n numbers (or line segments) $a_1, a_2, \ldots, a_n$ is defined as

$$\frac{1}{h} = \frac{1}{n}\left(\frac{1}{a_1} + \frac{1}{a_2} + \ldots + \frac{1}{a_n}\right) \quad (10.1)$$

This rather surprising theorem follows directly from the relationships between the roots and the coefficients of a polynomial equation. Let us, as an illustration, consider the cubic equation

$$a_3 x^3 + a_2 x^2 + a_1 x + a_0 = 0 \quad (10.2)$$

If the roots are x_1, x_2 and x_3, we can express equation (10.2) as

$$a_3(x - x_1)(x - x_2)(x - x_3) = 0 \quad (10.3)$$

Expanding this equation and comparing coefficients of powers of x in equations (10.2) and (10.3) yields

$$a_0 = -a_3 x_1 x_2 x_3$$

$$a_1 = a_3(x_2 x_3 + x_3 x_1 + x_1 x_2)$$

$$= a_3 x_1 x_2 x_3 \left(\frac{1}{x_1} + \frac{1}{x_2} + \frac{1}{x_3}\right)$$

So we have $a_1/a_0 = -3/h$, where h is the harmonic mean of the x_i. The result is readily generalized for a polynomial equation of the nth degree, $y = f(x) = 0$ say, to give

$$a_1/a_0 = -n/h \quad (10.4)$$

To prove Cotes's theorem, we let $y = mx$ be the equation of a line through the origin O which intersects the curve $y = f(x)$ in the points $P_1, P_2, P_3, \ldots, P_n$, as illustrated in Figure 10.4. Then the x-coordinates of these points are given by the roots of the equation $f(x) = mx$, which may be written as

$$a_n x^n + a_{n-1} x^{n-1} + \ldots + a_2 x^2 + (a_1 - m)x + a_0 = 0 \quad (10.5)$$

If OH is the harmonic mean of $OP_1, OP_2, \ldots, OP_n$, and H has coordinates $(\bar{x}, \bar{y})$, then, by analogy with equation (10.4), we have

$$(a_1 - m)/a_0 = -n/\bar{x} \quad (10.6)$$

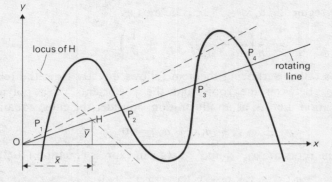

Figure 10.4.

Since $\bar{y} = m\bar{x}$, we can eliminate m to give

$$a_1\bar{x} - \bar{y} + na_0 = 0 \tag{10.7}$$

which is the equation of the locus of H as the line OH rotates about O. Since equation (10.7) is a linear equation in $\bar{x}$ and $\bar{y}$, the locus is a straight line, as was to be proved.

If the roots of $f(x) = mx$ are not all real, the geometrical statement of the theorem is no longer valid. However, the harmonic mean, as defined above, is still real and the algebraic situation is essentially unchanged. The proof is readily adapted to cover the more general case when the equation of the curve is $f(x, y) = 0$ and f consists of the sum of terms of the form $\pm Ax^p y^q$. A detailed study of Cotes's mathematical achievements is to be found in Reference 32.

We must now leave the world of Newton and his friends and cross the English Channel. Most of the more exciting new mathematics during the century that followed the death of Newton in 1727 was created on the Continent of Europe. British mathematicians suffered severely from insularity, the main cause of which was the excessive reverence accorded to the great Sir Isaac and a too rigid adherence to his notations and procedures. A related cause was perhaps the feeling of enmity in some quarters towards Continental works – a legacy of the priority dispute.

11 Leibniz

'Leibniz was one of the supreme intellects of all time . . . his greatness is more apparent now than it was at any earlier time.' Bertrand Russell

'It would be difficult to name a man more remarkable for the greatness and the universality of his intellectual powers than Leibniz.' John Stuart Mill

1. Introduction

Gottfried Wilhelm Leibniz (see Illustration 10) achieved great distinction in two major fields: philosophy and mathematics. However, it is the immense range of his achievements that inspired the two quotations above: he made notable contributions to law, politics, theology, history, philology, logic, geology and several branches of physics – not to mention his construction of calculating machines. The term 'universal genius' has often been applied to this extremely talented polymath. In this respect he may be compared with Descartes; indeed, it was the dominant influence of these two philosopher–mathematicians that created the tradition of philosophic rationalism which, broadly speaking, dominated Continental thought until the present century. In Britain it was Newton, Locke and (later) David Hume who between them created and sustained the contrasting empirical tradition in philosophy and science.

While Leibniz's philosophical status has undoubtedly declined during the last 250 years, his reputation as a mathematician has certainly increased. Many of his philosophical ideas are now generally regarded as 'fantastic' (Russell's word). Leibniz believed that, in spite of appearances, matter is really a collection of rudimentary minds, or souls, which he called 'monads'. His doctrine that 'everything is for the best in this, the best of all possible worlds' has been immortalized by Voltaire in his brilliant satire *Candide*; Leibniz himself is caricatured as the good Doctor Pangloss. On the other hand, as a mathematician Leibniz was – and still is – widely acclaimed as one of the founding fathers of the calculus. In contrast, his pioneering work on combinatorial analysis and symbolic reasoning was largely ignored until the beginning of this century; it is now much admired. He stands out as one of the few first-class mathematicians who is equally at home with the 'continuous' and the 'discrete'.

2. The universal genius

Leibniz was born in Leipzig on 1 July 1646, and died in Hanover on 14 November 1716. His father, who died in 1652, was a Professor of Moral Philosophy. The young Leibniz, although he had good schooling, was largely self-taught by intensive reading in his father's library. At the age of 15 he entered the University of Leipzig as a law student, but also found time to study philosophy and to develop a keen interest in mathematics and science. In 1663, at the age of 17, he received a Bachelor of Philosophy degree. By 1666 he was ready for his doctorate with a thesis, *De arte combinatoria*, on a universal method of reasoning. The Leipzig professoriate, 'bilious with jealousy', refused Leibniz a degree, ostensibly on account of his youth. He thereupon moved to Nuremberg, and during the journey there is said to have composed an essay on a new method, based on the historical approach, of teaching law. The associated University of Altdorf not only awarded Leibniz his doctoral degree at once, but invited him to take up the University's Chair of Law. This he refused, saying that he had 'very different things in view'. In fact he entered the service of the Archbishop of Mainz, and soon became a skilful diplomat and a valued political adviser. In 1670 he suggested a French invasion of Egypt as a means of deflecting Louis XIV's

CODEFROI GUILLAUME
LEIBNITZ,
Né le 3 Juillet 1646 mort le 14 Novembre 1716.

Illustration 10. Gottfried Leibniz, 1646–1716 (Mansell Collection)

aggressions in Europe. Two years later the French Secretary of State asked Leibniz to come to Paris for discussions. The young diplomat was delighted to visit the cultural centre of Europe, and he managed to stay for four years. He was, however, never able to obtain a personal interview with the king and the Egyptian scheme disappeared from practical politics until the time of Napoleon. Leibniz's papers on the subject remained buried in the Hanoverian archives. When Napoleon occupied Hanover in 1803 he was somewhat put out to learn that the idea of a French conquest of Egypt had been put forward by a German philosopher more than a century earlier. Another of Leibniz's grandiose projects, equally unsuccessful, was to unite the Catholic and Protestant Churches.

The years (1672–76) that Leibniz spent in Paris were the happiest of his life. They provided the intellectual stimulus he needed to enable him to realize his great potentialities. He arrived in Paris as a highly educated, urbane young diplomat of great charm and wit, with an engaging zest for life and an infinite capacity for hard work. While his interest in mathematics was intense, his knowledge of the subject was limited. He left the city a mature and well-equipped mathematician, already in possession of the leading ideas of the infinitesimal calculus. He rapidly absorbed the main scientific ideas of the day, and soon realized how much mathematics he needed to learn. He met Huygens, the leading natural philosopher of the 1670s, who gave the young diplomat some lessons in mathematics, advised him on his reading, and set him some problems.

In January 1673 Leibniz's political duties took him to London where, as we saw in Chapter 8, he made contact with, among others, Henry Oldenburg, and may have seen a copy of Newton's *De analysi*. He was elected a Fellow of the Royal Society in April of that year. Leibniz visited London again in 1676, bringing with him his calculating machine. He had made a prototype in 1672, but it took two more years to develop a working instrument. One of his visitors to see the machine in Paris was Étienne Périer, a nephew of Blaise Pascal, who had constructed a mechanical adding and subtracting machine thirty years earlier (p. 152). Leibniz took the next logical step and mechanized multiplication. To do this he used an ingenious device known as the 'Leibniz stepped wheel', a cylindrical drum containing nine longitudinal teeth of different lengths (in steps from 1 to 9), so arranged that a smaller pinion wheel engaged a varying number of these teeth, depending on its position. This elegant device, essentially in the form Leibniz left it, was still being used in some of the last mechanical calculators ever made, in the 1940s. The mechanical ingenuity of this invention shows yet another facet of Leibniz's 'universal genius'. He maintained a keen interest in industrial crafts and processes throughout his life.

The Archbishop of Mainz died in 1673, while Leibniz was in London negotiating on his employer's behalf. He was then offered the post of Counsellor to the Duke of Brunswick-Lüneberg, the Elector of Hanover. Leibniz accepted with some reluctance, and only on condition that he could stay on in Paris. In 1676, however, he was finally called back to Germany and was put in charge of the ducal library at Hanover. The remaining 40 years of his life were spent, often unhappily, in the service of successive Electors of Hanover. Much of his time was occupied in trivial tasks such as genealogical researches into the tortuous early history of the Brunswick family. (When Leibniz died he had got as far back as the year 1005.) During these years he made innumerable journeys all over Europe. Like Mozart in music, Leibniz was able to draft political reports, compose philosophical essays or solve mathematical problems, while being rattled along the atrocious roads (many no more than cow-tracks) of seventeenth-century Europe.

In 1700 Leibniz was invited to Berlin to tutor the young Electress of Brandenberg. While there, he organized the Berlin Academy of Sciences, and became its first Life President. Most of his major philosophical works were written between 1690 and 1716, after his most creative mathematical period had come to an end. In 1714 the Elector of Hanover crossed the water to become King George I of England, but he refused to allow Leibniz to accompany him; his services were no longer needed and he was discarded. It is only fair to say, however, that the hostility to Leibniz in England occasioned by the priority quarrel with Newton over the invention of the calculus may have been a factor in the King's decision. So Leibniz was condemned to remain in his provincial backwater. He died, largely forgotten, two years later; his secretary, we are told, was his only mourner.

3. The *characteristica generalis*

Throughout his life Leibniz was guided by a few general philosophical ideas which he sought to apply in many diverse fields. One of these ideas was what he called the *characteristica generalis*, a general symbolic language into which all processes of reasoning could be translated. The genesis of the idea is to be found in his thesis of 1666, *De arte combinatoria*. The language would have its syntax and grammar in the form of a set of rules which, if followed, would ensure that the argument was correct. Such a language would, he believed, also serve in the process of invention. Needless to say, Leibniz did not succeed in constructing such a language, but his preoccupation with the subject undoubtedly guided his early studies in combinatorics, and later his invention and development of the calculus. In particular it stimulated his interest in the wise choice of symbols and notations, and the search for general rules (he called them *algorithms*) for the manipulation of such symbols.

4. The origins of the Leibnizian calculus

We have seen how Leibniz started his serious study of mathematics under the tutorship of Huygens in Paris. In 1672 Huygens set Leibniz the problem of summing the series

$$S_n = \sum_{r=1}^{n} 2/r(r + 1)$$

(The terms are the reciprocals of the triangular numbers, $1, 3, 6, 10, \ldots$.) Now, Leibniz knew that a sequence b_r may be summed if each term can be expressed as a difference, $a_r - a_{r+1}$. Since

$$\frac{1}{r(r + 1)} = \frac{1}{r} - \frac{1}{r + 1}$$

it follows at once that

$$S_n = 2 - \frac{2}{n + 1}$$

Table 11.1. Leibniz's 'harmonic triangle'

$\frac{1}{1}$	$\frac{1}{2}$	$\frac{1}{3}$	$\frac{1}{4}$	$\frac{1}{5}$	$\frac{1}{6}$	$\cdots$
	$\frac{1}{2}$	$\frac{1}{6}$	$\frac{1}{12}$	$\frac{1}{20}$	$\frac{1}{30}$	$\cdots$
		$\frac{1}{3}$	$\frac{1}{12}$	$\frac{1}{30}$	$\frac{1}{60}$	$\cdots$
			$\frac{1}{4}$	$\frac{1}{20}$	$\frac{1}{60}$	$\cdots$
				$\frac{1}{5}$	$\frac{1}{30}$	$\cdots$
					$\frac{1}{6}$	
						$\cdots$

and the 'sum to infinity' is 2. Leibniz derived many results of this kind, making extensive use of the 'harmonic triangle' (Table 11.1). The analogy with Pascal's 'arithmetic triangle' (Table 7.1, p. 165), and also the method of construction by differencing (e.g. $1/6 = 1/2 - 1/3$ or $1/30 = 1/12 - 1/20$), will be apparent. The sum of the numbers in any row is equal to the first number in the row above. The number in the rth position in the nth row (counting from 1) is the reciprocal of

$$ n \binom{n + r - 1}{n} $$

Although these results were not new, they gave Leibniz the insight that 'the operations of summing sequences and of taking their difference sequences are in a sense inverse to each other.' He saw the problem of quadrature as a summation of equidistant ordinates, and that the difference of successive ordinates gives an approximation to the local slope of the tangent. If the intervals between successive ordinates were chosen to be infinitely small, the approximation would become exact. The logical problem was, of course, to explain precisely what is meant by 'infinitely small'.

Leibniz's next basic idea was that of the *characteristic triangle* (his term). Here again, the concept was not new. He probably got it from Pascal; it had also been used by Fermat and by Barrow. In

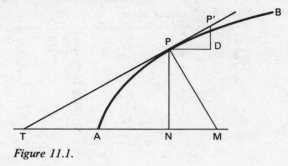

Figure 11.1.

Figure 11.1 APB is an arbitrary curve, TPP′ is the tangent and PM the normal at P. If PP′ is sufficiently small, P′ may be taken to be on the curve – this is the crucial assertion. We see also that the small triangle PP′D (the characteristic triangle) is similar to the two large triangles PMN and TPN. Leibniz used these similarity relations to derive a variety of quadratures, one of which we shall discuss in Section 7.

5. A case study of the process of mathematical invention

By 1675 Leibniz was in possession of all the main ideas that formed the foundation of his calculus. In a period of less than three weeks he brought these ideas together in a series of manuscript notes which together contain all the essential ingredients of his great invention. The notes, which are dated 25, 26, 29 October and 1, 11 November 1675, form a most valuable record of how mathematical inventions – or discoveries, if you prefer – are actually made. To quote Leibniz himself:

It is an extremely useful thing to have knowledge of the true origins of memorable discoveries, especially those that have been found not by accident but by dint of meditation . . . The art of making discoveries should be extended by considering noteworthy examples of it.

In these notes Leibniz considers a series of equally spaced ordinates of a curve, as shown in Figure 11.2. The differences of successive ordinates are denoted by w, an infinitely small 'unit interval' between them being implied. The area OBA is taken to be the sum of the y's (which Leibniz writes as omn.y); the complementary area OBC is

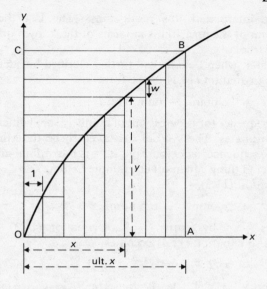

Figure 11.2.

put equal to the sum of the 'thin' rectangles whose representative area is xw (written as omn.xw). The sum of the two areas is the rectangle OABC, whose sides Leibniz denotes by ult.x and omn.w. He expresses the relation between these three areas as

$$\text{omn}.xw = \text{ult}.x, \text{omn}.w - \text{omn}.\text{omn}.w \qquad (11.1)$$

where y has been replaced by omn.w. The last term is to be interpreted as the sum of the y terms, each of which is the sum of a sequence of w terms – in modern notation,

$$\text{omn}.\text{omn}.w = w_1 + (w_1 + w_2) + (w_1 + w_2 + w_3) + \dots$$

Leibniz is now able, by making various substitutions in equation (11.1), to derive other quadrature relations. Thus, for example, setting $w = a/x$, where a is a constant, yields

$$\text{omn}.a = \text{ult}.x, \text{omn}.a/x - \text{omn}.\text{omn}.a/x \qquad (11.2)$$

Now, $y = a/x$ is the equation of a rectangular hyperbola, and so omn.a/x is the quadrature of this curve. Leibniz knew that this quadrature is a logarithm, and so omn.omn.a/x must be a sum of logarithms. He interpreted this result thus: 'The last theorem expresses the sum of the logarithms in terms of the known quadrature of the hyperbola.'

A few days later, when Leibniz had further clarified his ideas, we find him writing equation (11.1) in the form

$$\text{omn.}xl = x, \text{omn.}l - \text{omn.omn.}l \qquad (11.3)$$

He envisages a sequence (or progression) of infinitely small differences of successive ordinates. The symbol l is taken to be the xth such difference in the sequence: as Leibniz put it, 'x is the ordinal number and l is the ordered thing'. In another example Leibniz sets $l = x$ to give, from equation (11.3),

$$\text{omn.}x^2 = x, \text{omn.}x - \text{omn.omn.}x \qquad (11.4)$$

Now he knows – e.g. by setting $l = 1$ in equation (11.3) – that omn.$x = x^2/2$, so equation (11.4) becomes

$$\text{omn.}x^2 = x^3/2 - \text{omn.}x^2/2$$

from which omn.$x^2 = x^3/3$. Clearly, by setting $l = x^n$, we can use the method of mathematical induction (p. 166) to deduce the general result that omn.$x^n = x^{n+1}/(n + 1)$.

In his notes of 29 October we find Leibniz exploring the operational rules for the symbol omn. At this point he introduces an important change of notation: 'It will be useful to write $\int$ for omn, so that $\int l = \text{omn.}l$, or the sum of the l's.' ($\int$, the long script S, stands for *summa*.) He also writes $\int\int$ for omn.omn., and so can express equation (11.3) as

$$\int xl = x \int l - \int\int l$$

which will be recognized as the modern formula for the 'integration by parts' of xl.

After giving various rules for '$\int$', such as $\int (y + z) = \int y + \int z$, Leibniz is ready to tackle the general problem of quadrature. He now introduces his second important piece of notation – the symbol d to denote differentiation, or inverse quadrature. He writes:

Given l, and its relation to x, to find $\int l$. This is to be obtained from the contrary calculus, that is to say, suppose that $\int l = ya$. Let $l = ya/d$; then just as $\int$ will increase, so d will decrease the dimensions. But $\int$ means a sum, and d a difference. From the given y, we can always find y/d or l, that is, the difference of the y's.

Why, we may ask, does Leibniz write y/d for the differences of the y's? He regards variables such as x, y and l, and constants such as a, as one-dimensional line segments. Now, since $\int l$ has the dimensions of an area, he writes $\int l = ya$ to preserve dimensionality. Its difference, namely, l, is a line segment, so if d is to have the same dimensional property as $\int$, we must place it in the denominator; thus 'differences of $ya = l = ya/d$'. In fact, Leibniz did not stick to this dimensional interpretation for very long. In the manuscript of 11 November he is writing dy instead of y/d for the difference of the y's. However, he is still using $\int y$ for the area under the curve $y(x)$; it was some years before he consistently wrote it as $\int y.dx$. In this manuscript Leibniz continues his investigation of the rules for the symbol d; for example, the form of $d(uv)$ and of $d(u/v)$. He then proceeds to solve some specific problems. One of these is to find the curve whose subnormal, p, is inversely proportional to the ordinate y. (The subnormal is NM in Figure 11.1) Now, $dy = p/y.dx$ and, if $p = b/y$, then $dx = y^2/b.dy$. He then integrates (as we would say) to get $x = y^3/3b$.

Leibniz was a master-builder of notation, and the symbols $\int$ and d were carefully chosen to emphasize the analogy between the calculus of sums and differences and the calculus of quadratures and tangents. We see, then, that all the essential ideas of the infinitesimal calculus are to be found in these working notes. At this stage, however, much was still unclear, and it took Leibniz several more years to straighten things out.

6. The modern notation of the calculus

It will be convenient at this stage to summarize the calculus notation that we use today. In Figure 11.3, let the equation of the curve APQR be $y = f(x)$, and let P′ be a point on the curve near P. The 'infinitely small' quantities* PN and P′N are usually denoted by δx

---

* Throughout this book we have followed the practice of past writers in using the description 'infinitely small quantity'. This phrase may, however, strike some modern eyes as self-contradictory, since 'infinite' means 'immeasurably great'; strictly, we should refer to an 'infinitesimal quantity', or simply an 'infinitesimal'. For our purposes we may regard 'infinitely small' as synonymous with 'infinitesimal'.

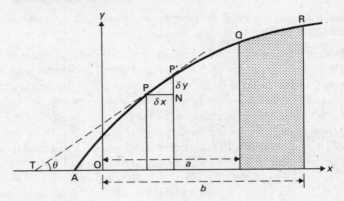

Figure 11.3.

and δy (and the chord, or arc, PP' by δs), where PP'N is the 'characteristic triangle'. The limiting value of the ratio $\delta y : \delta x$ as δx tends to zero is written as dy/dx (or sometimes as Dy or $D_x y$) and is called the *differential coefficient* of the function y at the point P. (Geometrically, $dy/dx = \tan \theta$, the slope of the tangent to the curve at P.) Now, dy/dx, like y, is a function of x; it is sometimes called the *derivative* (or *derived function*) of y and written as $f'(x)$. The derivative of a derivative, i.e. $d/dx \, (dy/dx)$, is written as d^2y/dx^2 or $f''(x)$, and is termed the 'second derivative'. The process can be continued to give the 'nth derivative', written as $d^n y/dx^2$ or $f^{(n)}(x)$.

An equation containing differential coefficients is called a *differential equation*. A well-known example is

$$\frac{d^2y}{dx^2} + A \frac{dy}{dx} + By = f(x)$$

the equation governing the forced oscillations of a damped spring.

Turning now to integration, the value of the shaded area in Figure 11.3 is written as $\int_a^b y \, dx$, and is called a *definite integral* taken between the *limits* $x = a$ and $x = b$. The expression $\int y \, dx$ (with no limits specified) is called an *indefinite integral*, and is undetermined in that the process of integration produces an arbitrary additive constant. For example,

$$\int x^3 \, dx = x^4/4 + C$$

In a definite integral, though, this constant cancels out:

$$\int_2^3 x^3\,\mathrm{d}x \;=\; (3^4/4 + C) - (2^4/4 + C) \;=\; 16\tfrac{1}{4}$$

It must be said, at this point, that the notation presented here is far from satisfactory. In recent years some purists have sought to replace it by something more logically defensible. However, the weight of opinion of applied mathematicians, engineers and scientists has ensured the general retention of the traditional 'Leibnizian' usage. It must be emphasized that '$\mathrm{d}y/\mathrm{d}x$' does *not* mean that a number $\mathrm{d}y$ is to be divided by another number $\mathrm{d}x$ (that is why the d's are printed in roman, not italic). Rather, it denotes the result of a certain operation, $\mathrm{d}/\mathrm{d}x$ (or D_x), applied to $y = f(x)$. The reason for this notation is apparent: $\mathrm{d}y/\mathrm{d}x$ is the limit of the fraction $\delta y/\delta x$. The notation for integration is even more peculiar. The symbols $\int$ and $\mathrm{d}x$ have no meanings when taken by themselves, while $\int \ldots \mathrm{d}x$ (referred to as 'integration with respect to x') is to be regarded purely as an operational symbol for performing the inverse of differentiation.

7. The transmutation rule and a famous series

In 1673 Leibniz discovered what he called the 'transmutation rule'. The basic idea is to evaluate the area under a curve in two different ways: as the sum of small rectangles and as the sum of small triangles. In Figure 11.4, if we make the obvious approximations that result from P′ being near P, we see that the area of OPAB is given by the sum of all the small triangles OPP′ (for a succession of pairs of points P and P′) plus the triangle OAB. We now draw the tangent PT and construct the lines OS, OF and SQQ′, as shown. The property of the characteristic triangle yields PP′/PD = OS/OF, so the area of the triangle OPP′, shown shaded in Figure 11.4, is

$$\tfrac{1}{2}\mathrm{PP'.OF} \;=\; \tfrac{1}{2}\mathrm{PD.OS} \;=\; \tfrac{1}{2} \text{ shaded area RQQ'R'}$$

As the point P on the original curve moves from O to A, the corresponding point Q, constructed as above, moves from O to C, thereby tracing out a new, or transmuted, curve. Finally, we

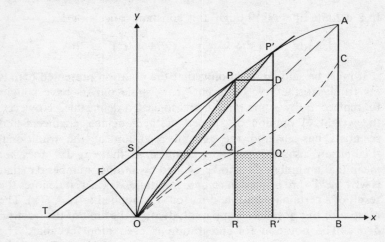

Figure 11.4.

have

$$\text{area OPAB} = \text{(sum of triangles OPP')} + \text{triangle OAB}$$
$$= \tfrac{1}{2}\text{(sum of areas RQQ'R')} + \text{triangle OAB}$$
$$= \tfrac{1}{2}\text{area OQCB} + \text{triangle OAB} \qquad (11.5)$$

This is Leibniz's transmutation rule: it reduces the quadrature of a given curve to that of another curve which is derived from the first by a tangent construction in the manner described.

To see what the rule amounts to, let us introduce some modern notation. In Figure 11.4, let $P = (x, y)$, $A = (x_0, y_0)$ and $Q = (x, z)$. Then, from the method of constructing Q, we have

$$z = y - x\,dy/dx \qquad (11.6)$$

and the transmutation rule may be written as

$$\int_0^{x_0} y\,dx = \tfrac{1}{2}\int_0^{x_0} z\,dx + \tfrac{1}{2}x_0 y_0 \qquad (11.7)$$

By using equation (11.6), this can be reduced to

$$\int_0^{x_0} y\,dx = x_0 y_0 - \int_0^{x_0} x(dy/dx)\,dx \qquad (11.8)$$

which will be recognized as the formula for 'integration by parts' once again.

The rule can be applied either when the quadrature of the new curve is known, or when there is a known relation between the quadratures of the two curves. As an example of the second case, we take the quadrature of the 'higher parabolas', $(y/b)^q = (x/a)^p$, where $p, q > 0$, as studied by Fermat (p. 143). Differentiating, we get $dy/dx = (p/q)(y/x)$, so from equation (11.6) we have

$$z = (q - p)y/q$$

With equation (11.7), this yields

$$\int_0^{x_0} y \, dx = q x_0 y_0/(p + q) \tag{11.9}$$

The quadrature of the 'higher hyperbolas', $(x/a)^p (y/b)^q = 1$, is achieved in a similar manner.

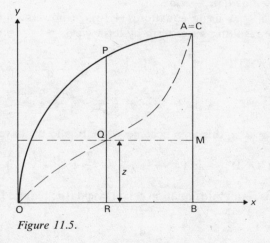

Figure 11.5.

Leibniz also used the transmutation rule to evaluate the quadratures of a segment of a cycloid and a segment of a circle. His solution of the latter problem led him to the famous series for $\pi/4$. Let us see how he did it. In Figure 11.5 the curve OPA is a quadrant of a circle of radius a; its equation is $y^2 = 2ax - x^2$. The curve OQC is the transmuted curve, constructed as before. (Note that the points C and A of Figure 11.4 now coincide.) Applying equation (11.6), we find that

$$\frac{z}{a} = \frac{x}{y} = \frac{y}{2a - x} = \sqrt{\left(\frac{x}{2a - x}\right)}$$

so that the equation of the curve OQC is

$$z^2(2a - x) = a^2x, \quad \text{or} \quad x = 2az^2/(a^2 + z^2) \quad (11.10)$$

By the transmutation rule (11.7), we have

$$\text{area OPAB} = \tfrac{1}{4}\pi a^2 = \tfrac{1}{2}\,\text{area OQCB} + \tfrac{1}{2}a^2 \quad (11.11)$$

To compute the area OQCB $= \int_0^a z\,\mathrm{d}z$, Leibniz uses a clever 'trick' – he changes axes from (x, y) to (z, x) and, as we would say, treats x as a function of z. Since RQ $= z$ and RB $= a - x$, the area OQMB (which is not the area required by the transmutation rule) can be expressed as $\int_0^z (a - x)\mathrm{d}z$.

Expanding $a - x$, using equation (11.10), in powers of z^2 and integrating the resulting series term by term yields

$$\text{area OQMB} = \int_0^z \left\{ a - 2a\left(\frac{z^2}{a^2} - \frac{z^4}{a^4} + \frac{z^6}{a^6} - \dots\right)\right\}\mathrm{d}z$$

$$= az - \frac{2}{3}\frac{z^3}{a} + \frac{2}{5}\frac{z^5}{a^3} - \frac{2}{7}\frac{z^7}{a^5} + \dots$$

When we put $z = a$, this area becomes OQCB, and we have

$$\text{area OQCB} = a^2\left(1 - \frac{2}{3} + \frac{2}{5} - \frac{2}{7} + \frac{2}{9} - \dots\right)$$

Inserting this result into equation (11.11) yields the required series:

$$\frac{\pi}{4} = 1 - \frac{1}{3} + \frac{1}{5} - \frac{1}{7} + \frac{1}{9} - \dots$$

or, alternatively,

$$\frac{\pi}{8} = \frac{1}{1 \times 3} + \frac{1}{5 \times 7} + \frac{1}{9 \times 11} + \dots$$

Leibniz went on to consider several related series, among them

$$\frac{1}{2 \times 4} + \frac{1}{6 \times 8} + \frac{1}{10 \times 12} + \cdots$$

which he knew to be equal to $\frac{1}{4}\log 2$, a remarkable result which can be obtained from the quadrature of the hyperbola. Leibniz was fascinated by this result, and began to suspect a deeper connection between the two basic transcendental problems, the quadrature of the circle and the quadrature of the hyperbola.

8. The publication of the Leibnizian calculus

Although Leibniz was in possession of the basic concepts and procedures of his calculus when he left Paris in 1676, he did not publish until eight years later. His first paper on the subject appeared (in Latin) in the October 1684 issue of the *Acta eruditorum Lipsienium* (Proceedings of the Scholars of Leipzig). The *Acta* had been founded as a monthly journal only two years earlier, largely on Leibniz's initiative; it was the first scientific journal to be produced in Germany.

A likely reason for his delay in publishing was that he realized only too clearly how vulnerable were his infinitesimal concepts to attack on logical grounds. He attempted to disarm his critics by making a drastic change in the definition of the differential. Previously dx, dy, and so on had been infinitely small quantities, as illustrated in Figure 11.1. In the 1684 paper, however, Leibniz introduces a fixed line segment of finite length, which he calls dx, as indicated in Figure 11.6. He then defines the differential dy as the finite line segment given by d$y = y/t \, dx$, where t is the subtangent. The ratios between these finite quantities are the same as those between the old infinitesimal ones, so the operational rules are unchanged. Of course this artifice did not resolve the logical difficulty, and in fact Leibniz continued to write about infinitely small differentials in his later papers.

Let us now attempt to summarize the 1684 paper; it is quite short, only six pages. Here, first, is the English translation of the title in full: 'A new method for maxima and minima as well as tangents, which

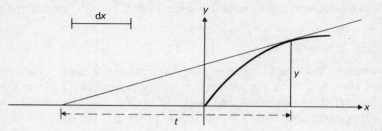

Figure 11.6.

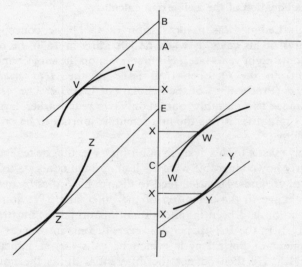

Figure 11.7.

is impeded by neither fractional nor irrational quantities, and a remarkable type of calculus for them'. The text begins with the definition of differentials as line segments (Figure 11.7 is a simplified version of Leibniz's first figure):

Let an axis AX and several curves such as VV, WW, YY, ZZ be given, of which the ordinates VX, WX, YX, ZX, perpendicular to the axis, are called *v, w, y, z,* respectively. Let the tangents be VB, WC, YD, ZE, intersecting the axis at B, C, D, E, respectively. Now, some straight line selected arbitrarily is called dx, and the line which is to dx as v (or w, or y, or z) is to XB (or XC, or XD, or XE) is called dv (or dw, or dy, or dz), or the difference of these v (or w, or y, or z) [i.e. dv : dx = v : x, etc.]. Under these assumptions we have the following rules of the calculus.

Leibniz then gives the rules for differentiating sums, products, powers, roots and so on, but he does not offer any proofs. He goes on to formulate the condition 'dv = 0' for a maximum or minimum, and to discuss the conditions for convexity, concavity and a point of inflection. This leads him to second-order differentials, and he writes: 'Hence a point of inflection occurs when d dv = 0 while neither v nor dv = 0.'

In the next paragraph Leibniz sets out the scope of his invention:

Knowing thus the Algorithm (as I may say) of this calculus, which I call differential calculus, all other differential equations can be solved by a common method. We can find maxima and minima as well as tangents without the necessity of removing fractions, irrationals, and other restrictions, as had to be done according to the methods that have been published hitherto.

He then deploys an argument to enable him to make the transition from the finite line segments defined in his first paragraph to the infinitely small differentials that he calls 'the momentary differences'. He points out that this method can deal with transcendental as well as algebraic curves 'and thus holds in a most general way without any particular and not always satisfied assumptions.' This is believed to be the first time that the term 'transcendental', in the sense of 'non-algebraic', occurs in print.

Leibniz is now in a position to discuss some specific examples. He shows how to determine the tangent at any point of a curve given by an algebraic equation, however complicated, of the form $f(x, y) = 0$. He then discusses the Fermat problem of optical refraction (p. 147), and has no difficulty in deriving Snell's 'law of sines'. He concludes with a flourish:

And this is only the beginning of much more sublime Geometry, pertaining to even the most difficult and most beautiful problems of applied mathematics, which without our differential calculus or something similar no-one could attack with any such ease.

The 1684 paper is undoubtedly a *tour de force*, but some of it is far from clear. Indeed, the Bernoulli brothers, who did so much to popularize and clarify the Leibnizian calculus, called the paper 'an enigma rather than an explication'.

Leibniz's first publication on integral calculus appeared two years later, also in the *Acta*, with the curious title: 'On recondite geometry and the analysis of indivisibles and infinities'. In this paper he tackles the inverse tangent problem, introduces the integral sign, $\int$, and again stresses the fact that his methods can deal with transcendental as well as algebraic quantities. As an example of a transcendental curve he takes the cycloid, giving its equation as

$$y = u + \int du/u, \quad \text{where } u^2 = 2x - x^2$$

From this equation, he tells us that

all properties of the cycloid can be demonstrated. The analytic calculus is thus extended to those curves that hitherto have been excluded for no better reason than that they were thought to be unsuited to it.

The inverse relationship between differentiation and integration is stated explicitly in this paper thus: 'Like powers and roots in ordinary calculations, so here sums and differences, or $\int$ and d, are each other's converses.' The terms 'integral' and 'integral calculus' are not used in this paper. The former first appears in a paper by Jacques Bernoulli in 1690; the latter was introduced by Leibniz and Jean Bernoulli in 1698. Leibniz wrote many more papers on the calculus. For example, in an *Acta* article in 1693 he expresses the inverse relationship by means of a geometrical construction.

Among his many other contributions to mathematics we may mention his treatment of complex numbers and their role in equation-solving, his advocacy of the binary system of numeration and, above all, his many happy inventions in notation and terminology.

9. Summing up: Newton and Leibniz

We conclude with a brief comparison between the sum and difference calculus of Leibniz and the method of fluxions of Newton.

The status of these two men as independent inventors of the infinitesimal calculus rests on three considerations: they both created a coherent system and a set of algorithmic rules for solving problems about curves in general; they both understood the inverse relation between differentiation and integration; and they both invented effective symbolic notations so that the new methods could be applied analytically by manipulating formulae instead of having to use geometrical arguments. What, then, are the main differences between them? We mention four. Newton considered variables as flowing quantities changing with time; Leibniz thought in terms of sequences of infinitesimally close values. Newton's dynamical approach led him to make the rate of change of a quantity, the 'fluxion', a fundamental concept; Leibniz gave this position to the 'differential', the difference between successive values in a sequence. To Newton, the fundamental theorem of the calculus is implied in his definition of integration (given the 'fluxion', to find the 'fluent'); Leibniz conceived of integration as summation, so the fundamental theorem is not implied in the definition of integration – it has to be proved. Finally, Leibniz's symbolism was far more convenient and flexible than Newton's.

Although Leibniz's earlier papers were not easy to understand, his work soon became widely known and accepted on the Continent. This was due, in large measure, to the activities of the Bernoulli brothers, Jacques (or Jakob) (1654–1705) and Jean (or Johann) (1667–1748). An article by Jacques in the *Acta* of 1690 was the first of a series of papers in which Leibnizian methods were applied to a variety of outstanding problems. The first 'textbook' on the differential calculus, l'Hôpital's *Analyse des infiniment petits*, appeared in 1696. His teacher was the young Jean Bernoulli, whose lectures on the integral calculus, although well known by the end of the century, were not published until 1742. In England and Scotland, on the other hand, the situation was very different. Throughout the eighteenth century Britain lived in the shadow of the towering figure of Newton. Indeed, it was not until the mid-nineteenth century that 'the principles of pure D-ism, as opposed to the Dot-age of the University', as Charles Babbage (1792–1871) put it, gained general acceptance.

Finally, we may remind ourselves that both Leibniz and Newton had serious difficulties with the theoretical foundations of the calculus. They both made free use of infinitely small quantities, although well aware of the logical inconsistencies thereby entailed. These were not finally resolved until the concept of 'limit' was clarified in the nineteenth century (see Chapter 16), and the notion of an infinitely small quantity was banished from the mathematical scene.

12 Euler

'Read Euler, read Euler, he is the master of us all.' P.-S. de Laplace

1. Introduction

The esteem in which Leonhard Euler (see Illustration 11) was held by his contemporaries has certainly not diminished in the two centuries since his death in 1783. He was the leading mathematician and theoretical physicist of the eighteenth century: worthy to be ranked just below Archimedes, Newton and Gauss; one of the last, and one of the greatest, of mathematical universalists. He left few branches of mathematics untouched, and – as Dr Johnson said of Oliver Goldsmith – 'he touched nothing that he did not adorn'. Euler was, perhaps, fortunate in the time of his birth. Analytical geometry, the calculus, Newtonian mechanics and universal gravitation had been around for some decades, and a large number of specific problems had been solved by ingenious use of the new techniques. The time was ripe for a period of consolidation and unification, for a systematic attack on the whole body of pure and applied mathematics. Euler was superbly equipped to take up the challenge.

Euler's output was immense. He published more than 500 books and papers during his lifetime, while a further 400 appeared posthumously. It has been computed that his publications during his working life averaged about 800 pages a year. (The Swiss edition of his complete works, now nearing completion, will consist of 74 volumes.) He was also a most effective popularizer of science. While in Berlin he gave lessons in astronomy and physics to the niece of

Frederick the Great, the Princess of Anhalt-Dessau. Euler wrote up his lessons in book form, and his *Letters to a German Princess* was translated into seven languages and became immensely popular all over Europe. All we can do here is to give an impresssion of the range and depth of Euler's contributions, and to put some flesh on the bones, as it were, by selecting a few topics for more detailed discussion.

2. 'Analysis incarnate'

Euler was born in Basle on 15 April 1707, but the family soon moved to a nearby village when his father – himself a gifted amateur mathematician – became the Calvinist pastor there. He wanted his son to succeed him in the village church, but from a very early age Leonhard knew exactly what he wanted to do. However, he dutifully complied with his father's wishes, and entered the University of Basle to study theology and Hebrew. Fortunately for posterity his mathematical abilities attracted the attention of Jean Bernoulli, who gave him a private lesson once a week, and the pupil soon became close friends with his tutor's two sons, Daniel and Nicolas. Leonhard duly took his Master's degree at the age of 17, thereby strengthening the father's theological ambitions for his gifted son. However, Pastor Euler was persuaded – albeit most reluctantly – to abandon these ambitions when told by Jean Bernoulli that Leonhard was destined to be a great mathematician. Even so, Euler's early training and parental influences had struck a deep chord: he remained a devout Calvinist all his life.

We first hear of Euler's mathematical researches in 1727. The Paris Academy had proposed the masting of ships as the prize problem for that year, and Euler submitted a memoir which failed to win the prize but received an honourable mention. (He subsequently won the prize, which was awarded biennially, no less than twelve times.) His next step was to apply for the Chair of Mathematics at Basle. He failed to get the job, but his hopes were kept alive by the prospect of joining Daniel and Nicolas Bernoulli in St Petersburg. The brothers had promised to find a place for their friend at the Academy. Tragically, Nicholas was drowned in 1726,

LEONARD EULER.

London, Published as the Act directs, Oct.13th 1804, by J. Wilkes.

Illustration 11. Leonhard Euler, 1707–83 (Mansell Collection)

but Daniel kept the promise. Euler received his call to St Petersburg in 1727, but the post was officially in the medical section of the Academy. Fate then dealt him a severe blow. The liberal Empress Catherine I (the widow of Peter the Great) died on the very day that Euler set foot on Russian soil. The heir was a minor, and the ruling faction inaugurated a repressive regime in which the Academy was looked upon as an unnecessary luxury. In the resulting confusion, Euler was able to slip quietly into the mathematical section; the medical post was forgotten. By maintaining a low profile and saying nothing, he was able to get down to solid work. Government spies were everywhere and he dared not allow himself a normal social life. As he said some years later, when reproached by the Queen of Prussia for his silence: 'Madame, I come from a country where, if you speak, you are hanged.'

By 1733 Daniel had had enough of Holy Russia and returned to free Switzerland and the Basle Chair. Euler, at the age of 26, was able to step into the leading mathematical position in the St Petersburg Academy. He decided to make the best of a difficult situation, and to marry and settle down. This he certainly did – his wife presented him with thirteen children, of whom five survived to adulthood. It was fortunate that Euler was able to work with children playing around him – even, we are told, with a baby on his lap. Conditions in Russia improved slightly, but only temporarily, when the boy-tsar died in 1730, to be succeeded by the Empress Anna. Euler's output during the next ten difficult years was prodigious, even though he suffered an illness, probably brought on by overwork, which left him without the use of his right eye. During this period he completed the first of his many treatises, the *Mechanica*, which was published in 1736. He also wrote elementary textbooks for the Russian shools, and advised the Government on a variety of matters whenever called upon. In 1741 he was at last able to escape: he was delighted to accept the invitation of Frederick the Great to join the Berlin Academy, although he continued to receive a pension from St Petersburg.

Euler remained at Berlin for 25 years and did most of his greatest work there. It was not, however, a very happy time as Frederick would have preferred a more polished courtier and a more sparkling wit for his top mathematician. Although Frederick appreciated

Euler's talents and engaged him on a variety of practical problems, the two men never became close friends. Indeed, relations grew ever more strained and in 1766, at the age of 59, Euler decided to move back to a very different St Petersburg at the cordial invitation of Catherine the Great. (She was a German princess – another tsar's widow – and a true daughter of the European Enlightenment.) Euler was royally received and was provided with every material comfort. At this time, however, he began to lose the sight of his remaining eye, and soon became totally blind. Undaunted, he accustomed himself to writing his formulae on a large slate and, with his two sons acting as amanuenses, was able to maintain – indeed even to increase – his productivity. He had a phenomenal memory, prodigious powers of mental calculation, and the ability to store long sequences of mathematical arguments in his memory for later dictation. His most impressive feat during his sightless years was his comprehensive attack on that most intractable of eighteenth-century problems, the theory of the Moon's motion. All the complicated analysis was done entirely in his head.

In 1771 Euler suffered another disaster: his house was burnt down in the great fire of that year. The blind old man narrowly escaped with his life; his library was destroyed but most of his manuscripts were saved. Catherine promptly made good the loss and Euler was soon back at work, but in 1776 he suffered an even greater loss in the death of his beloved wife. He married again, to a half-sister of his first wife. The third tragedy was the failure, after some initial success, of an operation to restore the sight of his left eye. Once again he was condemned to total darkness. He died, active to the end, on 18 September 1783, at the age of 76. He did some mathematics on his slate during the afternoon, then dined with some friends and outlined the calculations needed to compute the orbit of the newly discovered planet Uranus. Later in the evening, while playing with his grandson, he suffered a stroke and died immediately.

3. General analysis

The first of Euler's great textbook–treatises, *Introductio in analysin infinitorum* (see Illustrations 12 and 13), was published in 1748 in two parts. The first part is mainly concerned with the infinite processes of analysis: the expansion of functions in infinite series, infinite products or continued fractions; and the summation of a great variety of algebraic and trigonometric series. The publication of the *Introductio* established the concept of a *function* at the heart of analysis. Euler defined it very broadly thus: 'A function of a variable quantity is an analytical expression composed in whatever way of that variable quantity and of numbers and constant quantities.' He also defined a function of several variables.

Euler gave no explicit definition of an analytical expression, and in later years laid less stress on the need for any particular kind of functional form. Like Leibniz, he distinguished between algebraic and transcendental functions, the defining characteristic of the latter being, in his view, that the necessary algebraic operations on the independent variable must be repeated an infinite number of times. It was therefore reasonable to represent such functions by infinite series. Euler believed that any function could be expressed as a power series, either finite or infinite. His robust pragmatism is illustrated by his remark that 'If anyone doubts that every function can be so expressed, then the doubt will be set aside by actually expanding functions.' He also distinguished between single-valued and multivalued functions, between explicit and implicit functions, and between continuous and discontinuous functions. (This last distinction was not quite as we would make it today; the concept of continuity was not clarified until the nineteenth century.)

Euler was a great notation builder: he introduced or established in use many notations that we now take for granted. Here are some examples: π, e, Σ, e^x, $\log x$, $\sin x$, $\cos x$; $f(x)$ for a function of x; A, B, C and a, b, c for the angles and sides of a triangle. In the *Introductio* Euler defined the exponential and logarithmic functions in a new way, as limits, thus:

$$e^x = \text{ limit as } n \text{ tends to infinity of } (1 + x/n)^n \quad (12.1)$$

$$\log x = \text{ limit as } n \text{ tends to infinity of } n(x^{1/n} - 1) \quad (12.2)$$

(He actually wrote $e^x = (1 + x/i)^i$ where i is an 'infinite number'.) From these definitions he obtained the well-known expansions in infinite series and developed the main properties of the two functions,

Illustration 12. Frontispiece of Euler's *Introductio in analysin infinitorum* (1748)

including the inverse relation between them. Later in his life he clarified the meanings to be attached to the logarithm of a negative number and the logarithm of a complex number, thereby bringing a long mathematical controversy to an end.

The expansion of e^x as an infinite series follows at once from equation (12.1) as

$$e^x = 1 + x + \frac{x^2}{2!} + \frac{x^3}{3!} + \ldots + \frac{x^n}{n!} + \ldots \qquad (12.3)$$

Setting $x = 1$ gives the 'exponential constant', e – a transcendental number of value $2 \cdot 7183 \ldots$. The function defined by equation (12.2) is called the *natural* (or Napierian) *logarithm* of x. It is often written as $\ln x$ to indicate that the logarithm is to be taken to the 'natural' *base* of e. There is a simple inverse relationship between the two functions defined by equations (12.1) and (12.2): if $y = e^x$, then $x = \ln y$, and vice versa. *Common logarithms*, widely used for calculations until the advent of the electronic computer and calculator, have base 10. Other bases, such as 2, are also used.

Euler introduced the use of the letter 'i' for $\sqrt{-1}$, so that any complex number (p. 116) could be written as $a + ib$, where a and b are real numbers. If we replace x by ix or $-ix$ in equation (12.3), we get

$$e^{\pm ix} = (1 - x^2/2! + x^4/4! - \ldots) \pm i(x - x^3/3! + x^5/5! - \ldots)$$
$$(12.4)$$

In Chapter 8 of the *Introductio* Euler introduces the trigonometrical functions and develops their first comprehensive analytical treatment. It had been known for some time (to Newton and Leibniz, among others) that the two bracketed expressions in equation (12.4) are the power-series expansions of $\cos x$ and $\sin x$, respectively, so equation (12.4) can be written as

$$e^{\pm ix} = \cos x \pm i \sin x \qquad (12.5)$$

The 'Euler identities' of 1743, namely

$$\cos x = \tfrac{1}{2}(e^{ix} + e^{-ix}) \quad \text{and} \quad \sin x = \tfrac{1}{2}(e^{ix} - e^{-ix}) \qquad (12.6)$$

INTRODUCTION

A

L'ANALYSE INFINITÉSIMALE,

PAR LÉONARD EULER;

Traduite du latin en français, avec des Notes & des Éclaircissements,

PAR J. B. LABEY,

Professeur de Mathématiques aux Écoles Centrales du Département de la Seine.

TOME SECOND.

―――――――

A PARIS,

Chez BARROIS, aîné, Libraire, rue de Savoye, n°. 23.

―――――

L'An Cinquième de la République Française (1 7 9 7).

Illustration 13. Title page of a French translation (1797) of Volume 2 of Euler's *Introductio in analysin infinitorum*

follow at once. In all these formulae the angle x must of course be measured in radians (p. 159).

In the *Introductio* Euler also establishes the famous formula

$$(\cos x + \mathrm{i} \sin x)^n = \cos nx + \mathrm{i} \sin nx \qquad (12.7)$$

and generalized it for all real values of n. The formula is always associated with Abraham de Moivre, but in fact he never stated it explicitly; it follows at once from the equality of $(e^{ix})^n$ and $e^{i(nx)}$. In 1746 Euler astonished his contemporaries by proving that an imaginary power of an imaginary number can be a real number: for example, $i^i = e^{-\pi/2}$. He later showed that there are infinitely many values for i^i.

4. Infinite series

Euler's work on infinite series started as early as 1730, and aroused tremendous interest. At that time the concept of convergence of series was not understood, and Euler himself was certainly confused at times. He did point out the risks of working with divergent or alternating series, but did not always heed his own warning. For example, he argued that

$$1 - 1 + 1 - 1 + \ldots = \tfrac{1}{2}$$

a result he obtained by setting $x = -1$ in the expansion of $1/(1 - x)$. Many of his proofs would not pass muster today.

Here is another early example of Euler's carefree approach. (He gave another derivation in the *Introductio*, to meet criticism.) An nth-degree polynomial equation $f(x) = 0$ whose roots are $x_1, x_2, \ldots, x_n$ can be written as

$$a(x - x_1)(x - x_2) \ldots (x - x_n) = 0$$

Euler knew that there are an infinite number of roots of the equation $\sin x = 0$, namely $0, \pm \pi, \pm 2\pi, \pm 3\pi, \ldots, \pm n\pi, \ldots$, so he went ahead and 'factorized' $\sin x$ as an infinite product:

$$\sin x = Ax \left(1 - \frac{x^2}{\pi^2}\right)\left(1 - \frac{x^2}{2^2\pi^2}\right)\left(1 - \frac{x^2}{3^2\pi^2}\right) \ldots \qquad (12.8)$$

He also knew that the power-series expansion for $\sin x$ is

$$\sin x = x - \frac{x^3}{3!} + \frac{x^5}{5!} - \frac{x^7}{7!} + \dots \tag{12.9}$$

To evaluate A in equation (12.8), Euler used the fact – obvious from equation (12.9) – that $\sin x$ tends to x as x tends to 0, giving $A = 1$. Equating the coefficients of x^3 and x^5 in equations (12.8) and (12.9) gives

$$\sum_{n=1}^{\infty} (1/n^2) = \pi^2/6, \qquad \sum_{n=1}^{\infty} (1/n^4) = \pi^4/90$$

Further results of the same form are obtained by equating coefficients of higher powers of x.

The general series of this type, sometimes written as

$$\zeta(s) = \sum_{n=1}^{\infty} (1/n^s)$$

has many remarkable properties, especially when s is allowed to be a complex number. It is now called the 'Riemann zeta function', after the great nineteenth-century geometer Bernhard Riemann (1826–66). Euler was able to derive a general expression for $\zeta(s)$ when s is an even integer. He was not able to do so when s is odd, but he did evaluate $\zeta(s)$ for the first few odd values of s.

The function $\zeta(s)$ is convergent for real $s > 1$. When $s = 1$ we have the harmonic series given by

$$S = 1 + \frac{1}{2} + \frac{1}{3} + \frac{1}{4} + \dots \tag{12.10}$$

Although this series diverges to infinity, it does so extremely slowly. (The sum to 1000 terms, for example, is less than 8.) That it does diverge is easily demonstrated. By grouping the terms in equation (12.10) as

$$S = 1 + \frac{1}{2} + \left(\frac{1}{3} + \frac{1}{4}\right) + \left(\frac{1}{5} + \frac{1}{6} + \frac{1}{7} + \frac{1}{8}\right)$$

$$+ \left(\frac{1}{9} + \dots + \frac{1}{16}\right) + \dots$$

we see that $S > T$ where

$$T = 1 + \frac{1}{2} + \left(\frac{1}{4} + \frac{1}{4}\right) + \left(\frac{1}{8} + \frac{1}{8} + \frac{1}{8} + \frac{1}{8}\right)$$

$$+ \left(\frac{1}{16} + \ldots - \frac{1}{16}\right) + \ldots$$

$$= 1 + \frac{1}{2} + \frac{1}{2} + \frac{1}{2} + \frac{1}{2} + \ldots$$

Since T clearly diverges, so also does S.

Euler's study of the harmonic series led him to two remarkable results. The first was the discovery of a new mathematical constant, now always called 'Euler's constant' and denoted by γ. He was able to show that the harmonic series and the logarithmic function diverge together, in the sense that the function

$$F(n) = 1 + \frac{1}{2} + \frac{1}{3} + \ldots + \frac{1}{n} - \ln n$$

approaches ever closer to a constant limit, γ, as n (a positive integer) gets larger. To this day we do not know whether the real number $\gamma \ (= 0.5772\ldots)$ is rational or irrational. The second result reveals an unexpected link between analysis and number theory. Euler was able to demonstrate, by a fairly simple argument, that the divergence of the harmonic series implies that the number of primes is infinite, and vice versa. Some years later, by generalizing his argument, Euler discovered a remarkable identity which expresses the zeta function as an infinite product extended over prime numbers only. The theorem states that, if $s > 1$, then

$$\zeta(s) = \prod_{\text{all } p} \frac{1}{1 - p^{-s}}$$

where the product Π is taken over all prime numbers, p. The zeta function is of crucial importance in the theory of numbers; indeed, this theorem of Euler's may be regarded as an analytical expression of the 'fundamental theorem of arithmetic' (Reference 35), which states that a positive integer can be expressed as a product of prime factors in one way only (i.e. prime factorization is unique).

In 1737 Euler took up the subject of continued fractions, which we met in Chapter 3. By showing (in Chapter 18 of the *Introductio*) how to go from a series to its continued-fraction representation, and vice versa, he may be said to have laid the foundations of this branch of analysis. He proved that every rational number can be expressed as a finite continued fraction, and gave many continued-fraction expansions of both rational and irrational numbers. For example:

$$\frac{e + 1}{e - 1} = 2 + \cfrac{1}{6 + \cfrac{1}{10 + \cfrac{1}{14 + \dots}}}$$

Later he proved that both e and e^2 are irrational.

5. Calculus and differential equations

The *Introductio* was followed by two more great treatises: the *Institutiones calculi differentialis* in 1755, and the *Institutiones calculi integralis* (in three volumes) in 1768–70. These volumes not only present a comprehensive analytical treatment of the calculus as Euler inherited it, but also contain a mass of new discoveries of his own. He developed the techniques of integration in several directions and had much to say on *elliptic integrals*. Such integrals, in their most general form as developed by Euler's successors, are expressed as

$$\int R(x, \sqrt{P(x)}) \, dx$$

where $R(x, y)$ is any rational function of x and y, and $P(x)$ is the general polynomial function of the third or fourth degree. A variety of particular cases were studied in detail.

Euler also investigated the properties of a number of 'special' functions, some of his own creation, and made important contributions to the theory of, and techniques for solving, differential equations (p. 264), a flourishing and rapidly expanding eighteenth-century study. Many of Euler's papers on differential equations deal with physical problems, for example the oscillation of a loaded spring or the transmission of sound. Differential equations in more than one independent variable and where differentiation is always performed with respect to one such variable at a time (the values of the others being kept constant) are called *partial differential equations*. The study of such equations really began as part – almost a by-product – of the sustained eighteenth-century attack on the problem of the vibrating string. Euler's most important paper on the subject, 'On the vibration of strings', appeared in 1749. He extended his earlier concept of a function to include discontinuous curves. (In modern terms, the curves he had in mind were continuous but had discontinuous derivatives at a finite number of points.) He established the basic solution of the 'wave equation', and engaged in a lively debate on the problem of the vibrating string with the leading Continental mathematicians of the day.

6. The calculus of variations

It was Jean Bernoulli in 1725 who started Euler on his researches into what we now call the *calculus of variations*. Up to that time a number of isolated problems involving the maximum or minimum values of functions had been solved by *ad hoc* ingenuity. We have seen (p. 223) that one such problem – to determine the shape of a solid of revolution that would offer the least resistance to uniform axial motion through a fluid – was solved by Newton in the *Principia*. Another was the 'brachistochrone problem', which was solved by Newton in his old age (p. 201). However, until Euler took up the subject, no general methods were available. The characteristic features of a problem in the calculus of variations are that:

(1)   we have an integral in which the unknown function, $y(x)$ say, appears in the expression to be integrated;
(2)   we wish to determine the value of this integral which will make it either a maximum or a minimum, subject to specified conditions.

A discussion of Euler's treatment is, regrettably, beyond the scope of this book; suffice it to say that his extensive researches in this field were published in 1744 in a book entitled *Methodus inveniendi lineas curvas maximi minimive proprietate gaudentes*, which effectively created a new branch of mathematics.

The development of the calculus of variations received a strong boost from physics, from the adoption by the eighteenth-century scientists of the 'principle of least action' as a guiding principle in nature. In an appendix to his book, Euler formulated the principle as an exact dynamical theorem for the case of a particle moving along any kind of plane curve. The principle attracted strong theological support, and Euler himself seemed to have believed that all natural phenomena operate in such a way that some combination of physical variables is either maximized or minimized. He expressed himself thus:

For since the fabric of the Universe is most perfect and the work of a most wise Creator, nothing at all takes place in the Universe in which some rule of maximum or minimum does not appear.

7. Analytical and differential geometry

The second part of the *Introductio* of 1748 deals mainly with coordinate geometry. Euler divides curves into algebraic and transcendental, and establishes a number of general theorems about the former. He then investigates the general second-degree equation in two dimensions, and shows that it represents the various conic sections. He also discusses the classification of cubic and quartic curves, and introduces the parametric representation of curves whereby x and y are expressed in terms of a third variable, t say. (For example, a parametric representation of the parabola $y^2 = 4ax$ might be $x = at^2$, $y = 2at$.) He then turns his attention to three-dimensional coordinate geometry, studying the general (x, y, z) equation of the second degree and laying down the rules for a change of axes. By analysing the basic forms of the general equation, Euler was able to identify seven distinct cases: cone, cylinder, ellipsoid, hyperboloids of one and of two sheets, parabolic cylinder and hyperbolic paraboloid (the last of these being his own discovery).

Another branch of mathematics which was virtually founded by Euler is *differential geometry*: the study of those properties of curves and surfaces that vary from point to point (e.g. over the surface of an egg) and so need the techniques of the calculus for their investigation. In his *Recherches sur la courbure des surfaces*, which appeared in 1760, he establishes the theory of surfaces – which he represents by $z = f(x, y)$ – as a branch of geometry:

I begin by determining the radius of curvature of any plane section of a surface; then I apply this solution to sections which are perpendicular to the surface at any given point; and finally I compare the radii of curvature of these sections with respect to their mutual inclination, which puts us in a position to establish a proper idea of the curvature of surfaces.

Euler then defines the two principal normal sections of a surface and the principal curvatures, κ_1 and κ_2. One of his results, known simply as Euler's theorem, gives the curvature κ of any other normal section as

$$\kappa = \kappa_1 \cos^2 \alpha + \kappa_2 \sin^2 \alpha$$

where α is the angle between the specified normal section and the first principal normal section. Euler was also the first to consider the subject of developable surfaces (i.e. surfaces that can be flattened out on a plane). Towards the end of his life he started to write on space curves, which he represented by parametric equations with (x, y, z) expressed as functions of the arc length, s. In this field, as in so many others, we see Euler laying the foundations on which his successors were able to build to such good effect.

8. Topology

We now turn to yet another branch of mathematics of which Euler can claim to be the founding father: *topology*, which the eighteenth century knew as *geometria situs*. Euler's classic paper on the branch of topology we now call 'graph theory' was written in 1736 to solve a particular puzzle – that of the seven bridges of Königsberg (now Kaliningrad) spanning the River Pregel where tributaries meet at an island, as shown in Figure 12.1. The problem was to devise a route around the city which would cross each of the seven bridges once only. To demonstrate Euler's approach and to illustrate the grace and fluency of his writing, we shall quote extensively from an English translation of this famous paper given in Reference 24. (Other translations are to be found in References 5 and 9.) It begins thus:

In addition to that branch of geometry which is concerned with magnitudes, and which has always received the greatest attention, there is another

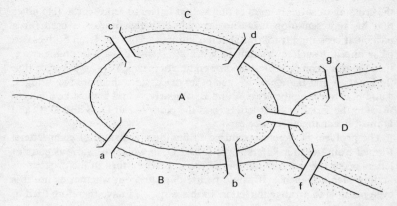

Figure 12.1.

branch, previously almost unknown, which Leibniz first mentioned, calling it the *geometry of position*. This branch is concerned only with the determination of position and its properties; it does not involve measurements, nor calculations made with them. It has not yet been satisfactorily determined what kind of problems are relevant to this geometry of position, or what methods should be used in solving them.

He notes that the Königsberg bridge problem seems a suitable candidate and continues: 'I have therefore decided to give here the method which I have found for solving this kind of problem, as an example of the geometry of position.' Euler then describes the problem and sets out his notation, as given in Figure 12.1. He continues:

From this, I have formulated the general problem: whatever be the arrangement and division of the river into branches, and however many bridges there be, can one find whether or not it is possible to cross each bridge exactly once? . . . My whole method relies on the particularly convenient way in which the crossing of a bridge can be represented. For this I use the capital letters A, B, C, D for each of the land areas separated by the river. If a traveller goes from A to B over bridge a or b, I write this as AB.

If he then goes from B to D over bridge f, Euler writes this as ABD; and similarly for longer journeys. He develops his argument thus:

In general, however many bridges the traveller crosses, his journey is denoted by a number of letters one greater than the number of bridges. Thus the crossing of seven bridges requires eight letters to represent it. [He notes that his representation takes no account of which bridge is used for a particular crossing.] It follows that if a journey across the seven bridges can be arranged in such a way that each bridge is crossed once, but none twice, then the route can be represented by eight letters which are arranged so that the letters A and B are next to each other twice, since there are two bridges, a and b, connecting the areas A and B; similarly, A and C must be adjacent twice in the series of eight letters, and the pairs A and D, B and D, C and D must occur together once each.

The problem is therefore reduced to finding a sequence of eight letters, formed from the four letters A, B, C and D, in which the various pairs of letters occur the required number of times. Before I turn to the problem of finding such a sequence, it would be useful to find out whether or not it is even possible to arrange the letters in this way . . . I have therefore tried to find a rule . . . for determining whether or not such an arrangement can exist.

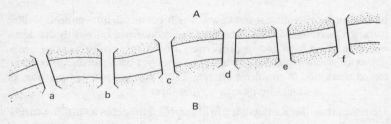

Figure 12.2.

To this end, Euler first considers the situation represented in Figure 12.2. We have a single area A, into which there lead any number of bridges a, b, c, d, etc. For example: 'If three bridges lead to A, and if the traveller crosses all three, then in the representation of his journey the letter A will occur twice, whether he starts his journey from A or not.' In general, if the number of bridges is any odd number, say $2n + 1$, then the number of occurrences of A in the route representation is $n + 1$. In the Königsberg case, therefore, there must be three occurrences of A in the route representation, and two each of B, C and D. In a sequence of eight letters, representing the seven bridges, we must have three A's, two B's, two C's and two D's. This cannot happen, so 'it follows that such a journey cannot be undertaken across the seven bridges of Königsberg'.

As we would expect, Euler then generalizes his findings. We omit his discussion and merely quote his conclusions, which he embodies in three rules:

If there are more than two areas to which an odd number of bridges lead, then such a journey is impossible.

If, however, the number of bridges is odd for exactly two areas, then the journey is possible if it starts in either of these areas.

If, finally, there are no areas to which an odd number of bridges lead, then the required journey can be accomplished starting from any area.

A more concise statement of the position is that a route of the kind specified (known as an 'Eulerian path') is possible if and only if N, the number of areas served by an odd number of bridges, is either 0 or 2. (It is easily shown that N must be even.) Königsberg has four areas served by five, three, three and three bridges, so no Eulerian path is possible. Euler concludes his paper by showing how, after it has been established that a solution does exist, the actual route can be found.

One of the best known of Euler's topological discoveries relates the number of vertices, edges and faces of a polyhedron. We can express his result, with an obvious notation, by the formula $F + V = E + 2$. The basic theorem can be generalized in a variety of directions. Once again, we see Euler opening up new territory for his successors to colonize.

9. The theory of numbers

Euler was one of those mathematicians who were equally at home with the discrete and with the continuous; Leibniz was another. Euler's contributions to the theory of numbers – the most important of which are contained in his *Anleitung zur Algebra* of 1770 – are legion, and we can mention only a few. He derived much of the inspiration for his researches in number theory from his study of Fermat. In 1736 he proved Fermat's theorem, namely that if p is a prime and a is prime to p, then $a^p - a$ is divisible by p (p. 148). In 1760 he generalized this theorem by introducing 'Euler's phi function', $\phi(n)$, which he defined as the number of integers less than and relatively prime to n. (This function has recently found a new application in cryptography.)

Not surprisingly, we find him addressing the problem presented by Fermat's famous 'last theorem' (p. 148): that the equation $x^n + y^n = z^n$ has no integer solutions when $n > 2$. Euler proved its correctness for $n = 3$ and $n = 4$; his 'descent' proof for the latter case has been given in Chapter 7. He also proved several of Fermat's other assertions, for example that every prime number of the form $4n + 1$ can be uniquely expressed as the sum of two squares.

Our next example provides a good illustration of Euler's ingenuity in devising special methods to deal with particular problems. Fermat had conjectured that all numbers of the form $F_n = 2^{2^n} + 1$ are prime. The numbers F_0 to F_4 clearly are, but in 1732 Euler proved that F_5 is composite (i.e. not prime). His proof is based on the observation that the prime number 641 can be expressed in two ways: as $5^4 + 2^4$ and as $5 \times 2^7 + 1$. Hence we have

$$F_5 = 2^{32} + 1 = 2^{28}(5^4 + 2^4) - (5 \times 2^7)^4 + 1$$
$$= 2^{28} \times 641 - (641 - 1)^4 + 1 = K \times 641$$

since the expansion of $(641 - 1)^4$ yields $1 +$ four terms, each divisible by 641. Many of the higher F_n have since been proved to be composite; indeed, no prime F_n has been found beyond F_4.

Euler also considered a number of classic Diophantine problems, and stated several important theorems that were proved rigorously later on. He established such basic concepts as congruence, linear and quadratic residues, and formulated the law of quadratic reciprocity, all of which played crucial roles in the later development of number theory. We must remember that in Euler's time the theory of numbers consisted of little more than a collection of isolated results. Modern number theory may be said to have begun with the publication of Gauss' *Disquisitiones arithmeticae* in 1801 (see Chapter 14). Once again we find Euler laying foundations on which his successors were able to build.

10. Physics, mechanics and astronomy

Like most of his contemporaries, Euler made no distinction between pure and applied mathematics. He drew his subject matter from many branches of physics, astronomy, navigation, cartography and commerce. We have already pointed out how the vibrating string problem provided the main motivation for his work on partial differential equations. As another example, we may mention his extensive optical researches, which were brought together in his three-volume treatise *Dioptrica* published in 1771.

Euler's contributions to analytical dynamics were outstanding. While his early work dealt mainly with particle dynamics (his first treatise, the *Mechanica*, came out, as we have seen, in 1736), his most important researches were concerned with the dynamics of a rigid body. He derived the general equations of motion both for a free body and for a rigid body rotating about a fixed point. By locating the fixed point at the origin, and choosing moving axes aligned along the principal axes of inertia of the body, he obtained the 'Eulerian equations of motion', familiar to applied mathematicians. The term 'moment of inertia' is due to him; in 1750 he discovered the existence of principal axes and moments of inertia of a rigid body. In 1776 he further clarified the foundations of the subject when he proved the basic theorem that the rotation of a rigid body about a point is always equivalent to a rotation about a line through that point.

Euler also laid the foundations of yet another subject, hydrodynamics, by establishing the general equations of flow of an inviscid (i.e. frictionless) fluid, first for incompressible flow and then, three years later, for compressible flow. His equations of fluid flow take the form of a set of non-linear partial differential equations in four independent variables (three of space and one of time) which are very difficult to handle. As Euler put it in his 1755 paper:

If it is not permitted to us to penetrate to a complete knowledge concerning the motion of fluids, it is not to mechanics, or to the insufficiency of the known principles of motion, that we must attribute the cause. It is analysis itself which abandons us here, since all the theory of the motion of fluids has just been reduced to the solution of analytical formulae.

Even so, Euler continued to work on the subject, and indeed was engaged in writing a treatise on hydromechanics at the time of his death.

Euler pursued his astronomical researches throughout his working life. He seems to have been the first to express Newton's second law of motion as a system of differential equations, namely $F_x = m\,\mathrm{d}^2x/\mathrm{d}t^2$ (and two similar equations for y and z). For a fixed body of mass M at the origin and a moving body of mass m at (x, y, z), the axial components of the gravitational force are given by $F_x = GMmx/r^3$ (and two similar equations for y and z). The extension to the situation when both bodies move under their mutual attraction is straightforward. In his *Theoria motuum planetarum et cometarum*, published in 1774, he gave a complete analytical treatment of the two-body (Sun and planet) problem. (Newton had treated it geometrically in the *Principia*.) Euler collected his extensive researches on lunar motion – that major preoccupation of so many eighteenth- and nineteenth-century mathematicians – in two further volumes: the *Theoria motus lunaris* (1753) and the *Theoria motuum lunae* (1772). The latter treatise, as we have mentioned earlier, was composed when he was totally blind. Although he was not able to devise a convenient numerical method for computing the Moon's motion, his results provided the data for Mayer's widely used lunar tables. For this work Mayer's widow was voted a sum of £5000 by the British Parliament. Euler's services were also recognized: he received an honorarium of £300 from the same source. Finally, we must mention his work on the theory of the tides and the sailing of ships. His researches in this area are presented in two remarkable books: *Scientia navalis* (1749) and *Théorie complète de la construction et de la manoeuvre des vaisseaux* (1773).

In this chapter we have sought to give some indication of the range and depth of Euler's achievements. His influence on the science and art of mathematics and theoretical physics has been, and still is, immense. Much of his work, written with his characteristic charm and lucidity, remains as fresh and as stimulating today as when it was first penned more than two centuries ago.

13 D'Alembert and his contemporaries

'The art of numbering and measuring exactly a Thing whose Existence cannot be conceived.' Voltaire, on the status of the calculus in the mid-eighteenth century

1. Introduction

The seventeenth century, which saw the completion of the transition from the medieval to the modern world, was also a time of rapid growth in mathematical activity and in the size of the mathematical community. The expansion has continued, and at an accelerating pace, ever since. During the eighteenth century mathematical concepts, theorems and techniques became ever more specialized and esoteric. An unfortunate consequence of this burgeoning activity has been to put much of the 'new' mathematics of the last 250 years or so beyond the reach of the general reader, and therefore beyond the scope of this book. So our choice of material in the remaining chapters is severely restricted: we must abandon any attempt to present a balanced overall picture of the subject. In this spirit we have selected one man to 'represent' the mathematicians of the late eighteenth and early nineteenth centuries. First, however, we must mention some of the other distinguished practitioners who laboured between 1750 and 1850 in the shadows of the two giants, Euler and Gauss. Pride of place must be accorded to the modest Joseph-Louis Lagrange (1736–1813), second only to Euler in genius, Napoleon's 'lofty pyramid of the mathematical sciences' and author of the influential *Mécanique analytique*. Next we have the magisterial Pierre-Simon de Laplace (1749–1827), the peasant's son who became a marquis, one of Napoleon's chief scientific advisers, and author of the mighty *Traité de mécanique céleste* – the *Almagest* of the modern world. We then have the versatile analyst Adrien-Marie Legendre (1752–1830) – the third of the three Ls – with his distinguished contributions to the calculus (particularly to function theory, differential equations and elliptic integrals), number theory and the theory of errors of observation. Our fourth figure is the engineer–officer Jean Victor Poncelet (1788–1867), left for dead on the battlefield during Napoleon's retreat from Moscow, and held for two years as a prisoner of war in Saratov before returning to France to elevate the study of projective geometry – virtually dormant since the days of Desargues and Pascal – to new heights of elegance and generality. A little later the centre of the stage is occupied by the devout Catholic and Royalist Augustin-Louis Cauchy (1789–1857), apostle of rigour in mathematical analysis and virtual creator of group theory and the theory of functions of a complex variable. Finally, there are two of the most tragic figures in the mathematical pantheon: the gentle Norwegian Niels Henrik Abel (1802–29), the genius of analysis who died of poverty and neglect at the age of 26, two days before the news came through of his appointment as Professor of Mathematics at the University of Berlin; and the dashing, foolhardy, brilliant Évariste Galois (1811–32), killed in a duel at the age of 20 after sitting up most of the previous night dashing off his mathematical last will and testament. Let us now focus on our chosen standard-bearer (a very personal choice) and thus the subject of this chapter: Jean d'Alembert.

2. Jean d'Alembert

Jean le Rond d'Alembert (see Illustration 14) was born in Paris on 16 November 1717. He was abandoned as an infant by his mother – later discovered to be the Marquise de Tencin, a gifted writer and the sister of a Cardinal – on the steps of the Church of St Jean le Rond, near Notre Dame. The child was taken to the Foundling Hospital and put out to nurse in the country for six weeks. His father, the Chevalier Destouches, who had been away

Illustration 14. Jean d'Alembert, 1717–83 (Mary Evans Picture Library)

from Paris when his son was born, then placed the child with a foster-mother, a Madame Rousseau, the wife of a glazier. He also settled an income of 1200 livres on his son and saw that he had a good education. In later years, when d'Alembert had climbed to the top of the greasy pole of Parisian society, he spurned the overtures of his aristocratic mother, and preferred to be recognized as the son of his humble foster-parents; he ensured that they had a comfortable old age. He adopted the surname of d'Alembert when he was a young man; how or why we do not know.

The young Jean attended a boarding school for eight years until, at the age of 12, he entered the famous Collège Mazarin. This was one of the few colleges in Paris to have a professor of mathematics. Indeed, the Chair had been made famous by Pierre Varignon (1654–1722), an early worker on the analytical development of the calculus (p. 204). When he left the college d'Alembert began to study law, while retaining his keen interest in mathematics. Speaking of himself in the third person, he wrote that:

Having no teachers, hardly any books and not even a friend he could ask for help when he was in difficulty, he used to go to public libraries, where he would read quickly in an attempt to grasp some general ideas and then would return home to try to work out proofs and solutions for himself. On the whole he succeeded, and often discovered important propositions which he thought were new, only to be disappointed (and somewhat satisfied) when he came upon them later in books he had not read.

After completing his law studies, d'Alembert decided to turn to medicine, keeping mathematics as a recreation. His friend the Marquis de Condorcet (1743–94) tells us how d'Alembert took his mathematical books to the house of a friend, and vowed not to look at them until he had qualified in medicine. However, his mathematical mind remained so active that he took back one book after another to look something up. Eventually he yielded to his inclinations and committed himself to mathematics and poverty. 'The years which followed this decision,' says Condorcet, 'were the happiest of his life.'

D'Alembert did not have to wait too long for recognition. In 1739, the Académie des Sciences showed interest in one of his memoirs on the integral calculus, and on 29 May 1741, at the age of 23, he joined the Académie as its astronomical correspondent. Once he had his foot on the first rung of the ladder of fame, his ascent was rapid. In 1754 he became the *secrétaire perpetuel* of the Académie, and was widely acknowledged as France's leading mathematician and natural philosopher. As a prominent philosopher of the eighteenth-century Enlightenment, d'Alembert operated on a broad intellectual front. He was well educated, highly intelligent, witty and urbane, at once sceptical and idealistic. He had many friends – Voltaire among them – and not a few enemies.

D'Alembert was joint editor with Denis Diderot (1713–84) of the famous *Encyclopédie* – or, to give it its full title, *Encyclopédie, ou Dictionnaire raisonné des sciences, des arts, et des métiers* – published in 28 folio volumes between 1751 and 1772. D'Alembert withdrew from the joint editorship in 1758, but continued to write most of the scientific articles, and also contributed the much-admired *Discours préliminaire*. The aim of the work was, as he put it, to show 'the order and concatenation of human knowledge'. The *Encyclopédie* came under heavy attack from the Jesuits for its secularist tendencies. D'Alembert was tireless in its defence, and earned himself the sobriquet of 'the fox of the Encyclopedia'. Another of his interests was the improvement of the teaching of mathematics in schools and colleges. He believed passionately in the power of reason, not only to increase man's control over his physical environment, but also to alleviate his social and moral difficulties. For him, man's life in society was the prime source of morality:

The science of morals is perhaps the most complete of all the sciences, when we consider the truths of which it is composed. It all rests on one simple and incontrovertible fact – the need which men have for one another, and the reciprocal obligations which that need imposes. All the moral laws follow from this. The interests of the individual and the group are never incompatible.

In 1746 d'Alembert was invited by Frederick the Great to join the Berlin Academy. He declined the offer, although he eventually accepted a pension from that monarch. (He sent many of his scientific papers to Berlin.) D'Alembert also declined an invitation from Catherine the Great of Russia to join her Court as tutor to her son. He preferred to remain in his beloved Paris, where he died on 29 October 1783.

3. D'Alembert's correspondents

D'Alembert's post at the Académie placed him at the centre of French – and hence of European – scientific activity. He maintained an extensive correspondence with many of the most eminent scholars of the time. Some of his encounters with the three most distinguished of his mathematical contemporaries, Euler, Lagrange and Laplace, are worth recalling.

Euler in Berlin and d'Alembert in Paris maintained an extensive and cordial mathematical correspondence over many years. They did not aways see eye to eye, though, and in 1757 a disagreement about the vibrating string problem led to an estrangement. In 1759 the young Lagrange joined in the argument with a controversial paper which was criticized by both Euler and d'Alembert. It seems that Lagrange accepted most of Euler's points, because in October 1759 we find Euler writing to Lagrange in the following terms:

I am delighted to learn that you approve my solution . . . which d'Alembert has tried to undermine by various cavils, and that for the sole reason that he did not get it himself. He has threatened to publish a weighty refutation; whether he really will I do not know. He thinks he can deceive the semi-learned by his eloquence. I doubt whether he is serious, unless perhaps he is thoroughly blinded by self-love.

In 1761 Lagrange, seeking to meet the criticisms of d'Alembert and others, gave a different treatment of the vibrating string problem. The debate continued for another 20 years. From 1768 d'Alembert brought out a series of booklets in which he attacked Euler's views; in 1779 Laplace entered the fray and sided with d'Alembert. In fact all the participants advanced incorrect arguments, and the outcome of the debate was inconclusive. The issues in dispute were not resolved until Joseph Fourier (1763–1820) took up the subject in the next century.

By the early 1760s Frederick had become anxious to replace the ageing Euler as his top mathematician, and invited d'Alembert to Berlin to look over the situation. He told an angry Frederick in no uncertain terms that it would be an outrage to appoint any other mathematician above the great Euler. D'Alembert was not the man to allow a personal difference or a professional disagreement to cloud his judgement.

The harshness and lack of generosity displayed in Euler's 1759 letter to Lagrange is most uncharacteristic. In fairness to Euler, we must mention an earlier incident. In 1755, Lagrange – then only 19 – sent Euler an account of his general analytical method for handling problems in the calculus of variations. Euler at once recognized its merit and its superiority to his own, semi-geometric methods. He generously held back some of his own papers on the subject so that Lagrange could publish his method first, 'so as not to deprive you of any part of the glory which is your due.' Not content with that, Euler used his influence with Frederick to get Lagrange elected a foreign member of the Berlin Academy at the early age of 23.

The brilliant career of Lagrange had begun some four years earlier when he was appointed Professor of Mathematics at the Royal Artillery School at Turin, his birthplace. His outstanding talents soon attracted d'Alembert's admiration, and Lagrange became d'Alembert's favourite protégé and lifelong friend. He encouraged his young correspondent not to be too modest, to attack difficult and important problems, to take better care of his health and to avoid overworking. When Lagrange was 45 he became seriously ill; he suffered from attacks of melancholia and psychosomatic disorders. In a poignant last letter, written in September 1783 only a month before his death, d'Alembert reverses some of his earlier advice:

In God's name do not renounce work, for you the strongest of all distractions. Good-bye, perhaps for the last time. Keep some memory of the man who of all in the world cherishes and honours you the most.

Let us now go back 20 years to the time shortly after d'Alembert's visit to Berlin to advise Frederick on his mathematical appointments. D'Alembert and Euler are now reconciled, and we find them directing their joint efforts to the delicate task of getting Lagrange installed as court mathematician at Berlin. After lengthy negotiations they succeeded, and on 6 November 1766 Lagrange was officially welcomed by Frederick, who said he would be honoured to have at his court 'the greatest mathematician in Europe'. Lagrange became director of the physico-mathematical division of the Academy, a post he held with great distinction for 20 years. Frederick was delighted with his new acquisition. Lagrange's appointment paved the way for Euler to leave Berlin for St Petersburg after a stay of 25 years. Soon after settling in Berlin,

Lagrange married a young lady from Turin. D'Alembert wrote to his friend in jocular mood:

> I understand that you have taken what we philosophers call the fatal plunge . . . A great mathematician should know above all things how to calculate his happiness. I do not doubt then that after having performed this calculation you found the solution in marriage.

Lagrange replied: 'I don't know whether I calculated ill or well, or rather, I don't believe I calculated at all.'

Pierre-Simon de Laplace (1749–1827), the future marquis, was born on a Normandy farm. After five years at the University of Caen, he set out for Paris at the age of 20 to conquer the mathematical world. On arrival, he called on d'Alembert and sent in his recommendations. He was not received. The young Laplace sensed what was needed: he returned to his lodgings and wrote a long letter to d'Alembert on the principles of mechanics. In his reply inviting Laplace to call, d'Alembert wrote:

> Sir, you see that I paid little enough attention to your recommendations: you don't need any. You have introduced yourself better. That is enough for me; my support is your due.

A few days later, thanks to d'Alembert, Laplace was appointed Professor of Mathematics at the École Militaire in Paris.

4. The foundations of the calculus

The major efforts of eighteenth-century mathematicians were directed to extending the powers of the calculus, and applying the new techniques to solving practical problems. The very success of this pragmatic approach led to the widespread adoption of a cavalier attitude to the logical foundations of the subject. As d'Alembert himself put it in 1743:

> Up to the present . . . more concern has been given to enlarging the building than to illuminating the entrance, to raising it higher than to giving proper strength to the foundations.

This lack of rigour was recognized, regretted, but largely ignored. Thus we find d'Alembert advising his students 'to continue with their studies of the subject; faith would eventually come to them.' There were, however, some powerful critics of this unsatisfactory state of affairs. The most telling attack was launched in 1734 by Bishop Berkeley (1685–1753) in a tract entitled *The Analyst; Or a Discourse Addressed to an Infidel Mathematician* (generally believed to be Edmond Halley). The Bishop set out his objectives in the subtitle of his tract:

Wherein it is examined whether the Object, Principles, and Inferences of the modern Analysis are more distinctly conceived, or more evidently deduced, than Religious Mysteries and Points of Faith

Berkeley argued that the so-called truths of science are no more firmly based than those of theology; he used the calculus as a prime example of his thesis. His conclusion was that: 'He who can digest a second or third Fluxion, a second or third Difference, need not, methinks, be squeamish about any Point in Divinity.'

There was a major logical difficulty with the status of infinitely small quantities. Did they exist; were they zero or non-zero; or were they, as Berkeley called them, 'ghosts of departed quantities'? D'Alembert took an uncompromising position when he insisted that:

a quantity is something or nothing: if it is something, it has not yet vanished; if it is nothing, it has literally vanished. The supposition that there is an intermediate state between these two is a chimera.

D'Alembert's contribution to the debate – and it was a most important one – was to resurrect the concept of 'limit' as a possible solution of the problem. (The idea had been put forward a few years earlier in England by Benjamin Robbins, and possibly others.) He explains his ideas in a 1765 *Encyclopédie* article, on *Limite*:

One magnitude is said to be the limit of another magnitude when the second may approach the first within any given magnitude, however small, though the first magnitude may never exceed the magnitude it approaches; so that the difference of such a quantity to its limit is absolutely unassignable.

Later in the article he makes the point that, not only can the magnitude never exceed its limit, it cannot actually attain it either. In another *Encyclopédie* article, on *Différentiel* (1754), d'Alembert uses his limit concept to explain and justify the rules of the differential calculus. He draws the familiar chord and tangent diagram and states that: 'The differentiation of equations consists simply in finding the limits of the ratio of the finite differences of two variables included in the equation'. This is really no more than a reformulation of Newton's prime and ultimate ratios of the *Principia*. Indeed, d'Alembert was probably the first Continental mathematician to see that Newton's notion of the derivative was essentially correct. It was the nineteenth-century refinement of the concept of limit that eventually resolved this basic problem. Without any clear idea of a function, d'Alembert was not able to take the final steps necessary to clear things up. For another 50 years his approach remained no more than one of several competing ideas.

In 1768 d'Alembert wrote a short popular exposition of his ideas, entitled *Sur les principes métaphysiques du calcul infinitésimal*. It is a model of clarity, cogent argument and graceful writing. Its purpose, to quote the closing sentence, was to 'provide a sufficient introduction to the subject for those who merely wish to have a general, but correct, idea of its principles'.

5. Differential equations

By 1744 d'Alembert's dynamical researches had thrown up some partial differential equations that he was unable to solve. It was three years later, in a paper delivered to the Berlin Academy, that he had his first success in this field. The context, not surprisingly, was the problem of the vibrating violin string. By assuming that the amplitudes of vibration are small, d'Alembert was able to generalize previous results and show that there are infinitely many curves other than the sine curve that can be modes of vibration. His general solution of the 'wave equation' is given by

$$y = \phi(x + vt) + \psi(x - vt)$$

where y is the displacement of the string at a distance x along its length and at time t, and v is the speed at which displacements are propagated along the string. For any particular problem there are initial and boundary conditions to be satisfied (e.g. $y = f(x)$ when $t = 0$, the 'initial curve', and $y = 0$ when $x = 0$ and when $x = L$

for all t). It is these conditions that determine the form of the functions ϕ and ψ.

Most of d'Alembert's results were in fact given by Euler in his 1748 paper, 'On the vibration of strings', which was written after he had read d'Alembert's earlier papers on the subject. While the two men agreed on the form of their equations, they had very different ideas as to what functions would qualify as initial curves and so as solutions of partial differential equations. Thus, for example, Euler would allow a corner in the initial curve at the point at which the string is plucked. D'Alembert disagreed, maintaining that the second derivative does not exist at such a corner point and so the wave equation does not apply there. It was this disagreement that seems to have led to the coolness between the two men that we noted earlier. The controversy rumbled on inconclusively for many years. D'Alembert made many other significant contributions to the rapidly expanding study of differential equations, but by this time the subject was becoming too technical for us to be able to pursue it here.

6. General analysis

The name of d'Alembert is probably best known in the context of the 'd'Alembert ratio test' for examining the convergence of infinite series. The test states that the series $u_0 + u_1 + u_2 + \ldots$ is convergent if $|u_{n+1}/u_n| \leqslant r$, where $r < 1$, for all sufficiently large values of n. This is the modern formulation, but we must remember that d'Alembert, like his contemporaries, had a very imperfect understanding of the nature of convergence. Thus, for example, in a paper of 1768 we find him applying his test to the series obtained by expanding $(1 + x)^n$ in powers of x, and reaching the striking conclusion that when $n = -2$ and $x = 99/100$, 'the series is divergent up to the 99th term but convergent thereafter'.

D'Alembert spent much time and effort trying to prove what we now know as 'the fundamental theorem of algebra' – namely that every polynomial equation $f(x) = 0$, of degree $n \geqslant 1$ and having real or complex coefficients, has at least one root, real or complex. Although he did not succeed in constructing a complete proof, he worked at it so hard that in France the theorem is widely known as d'Alembert's theorem. What, in essence, he tried to show was that any algebraic operation on a complex number will yield another complex number. We shall return to this topic in the next chapter.

7. Mechanics and astronomy

Most eighteenth-century mathematical practitioners were inspired directly by physical problems, mechanics and dynamical astronomy providing the main challenges. Indeed, d'Alembert wrote in the *Encyclopédie* of a transition from the seventeenth-century age of mathematics to the eighteenth-century age of mechanics. In his view – and that of most of his contemporaries – the main use of mathematics was to serve physics.

In 1743 d'Alembert published his first major work, the *Traité de dynamique*, in which he developed Newtonian mechanics – which until this time had remained much as Newton had left it in the *Principia* – by exploiting the techniques of the calculus. Today d'Alembert is best known in this field for the principle that bears his name. In the 1740s the generally accepted mathematical model envisaged a dynamical system as an assemblage of discrete particles ('point-masses'). The problem was to extend this model to take account of continuous distributions of matter, either solid or fluid. In either case, what we may call 'the effective force on a particle' is a directed quantity equal to the product of the mass of the particle and its acceleration. This force must be equal to the resultant of all the forces acting on the particle. If the particle forms part of a material assemblage, such forces may be divided into two classes: 'external forces', acting from outside the assemblage, and 'internal forces', the reactions due to the remaining particles. D'Alembert's assumption was that the internal forces form, by themselves, a system in equilibrium, from which it follows that the system of effective forces is equivalent to the system of external forces. This is d'Alembert's principle.

The principle may be expressed analytically. Consider a representative particle of the assemblage at the point (x, y, z) relative to fixed rectangular axes; its mass is m and it is acted on by external forces given by (X, Y, Z). The components of the effective force on the particle are $m\ddot{x}$, $m\ddot{y}$, $m\ddot{z}$ (using the 'dot' notation of Newton to denote differentiation with respect to time). Resolving parallel to, and taking moments about, the z-axis, we get

$$\sum(m\ddot{z}) = \sum(Z) \quad \text{and} \quad \sum(x.m\ddot{y} - y.m\ddot{x}) = \sum(xY - yX)$$

where the summation embraces all particles in the assemblage. These summations are, of course, replaced by integrations for a continuous distribution of matter. We can write these two equations as

$$\frac{\mathrm{d}}{\mathrm{d}t}\sum(m\dot{z}) = \sum(Z) \quad \text{and} \quad \frac{\mathrm{d}}{\mathrm{d}t}\sum(x.m\dot{y} - y.m\dot{x}) = \sum(xY - yX)$$

and there are two further pairs of similar equations. Since the axes may be in any position, the equations express what we may call the fundamental properties of linear and angular momentum.

If the body is assumed to be rigid these equations are sufficient, but for fluids or elastic solids additional physical assumptions are needed. D'Alembert's principle can be expressed in many ways; it is common to say that the system of external forces is in equilibrium with that of the effective forces reversed. This brings dynamical problems within the realm of statics – a convenient but unnatural situation. A familiar example of a reversed effective force is the fictitious 'centrifugal force' acting on a stone that is whirled round on the end of a string. Historically, the great value of d'Alembert's principle was that it provided a *general* method of treating dynamical problems. Previously, problems of rigid dynamics had been dealt with individually on the basis of special assumptions, sometimes of dubious validity.

In 1744 d'Alembert published his *Traité de l'équilibre et du mouvement des fluides*, in which he applied his principle to the motion of fluids. It was this work that gave rise to partial differential equations that he was not able to solve. His much admired *Théorie générale des vents*, which dealt with the motion of air, appeared the following year.

D'Alembert also made a number of notable contributions to dynamical astronomy. He was the first to discuss thoroughly the precession of the equinoxes and the variations in the obliquity of the ecliptic. His astronomical researches were collected in the monumental *Système du monde*, which was published in three volumes in 1754. Finally, we should mention his extensive writings on such social topics as the problem of life expectancy, lotteries and the value of annuities, and his – sometimes unorthodox – studies of probability theory.

Here we must take our leave of the foundling who not only became a great mathematician, but also achieved a commanding position in the intellectual life of mid-eighteenth-century Europe; and also of the many other 'makers of mathematics' who were to be found in the Parisian salons, cafés, government offices, lecture theatres and military academies during the turbulent times of the Enlightenment, the French Revolution, the First Empire and the restoration of the traditional French monarchy.

8. The triumph of analysis

We have now reached the end of the eighteenth century, and so may pause briefly to take stock. The seventeenth century has in mathematics been called the century of genius; the eighteenth the century of ingenuity. The great seventeenth-century discoveries – coordinate geometry, projective geometry (on a lesser scale) and, above all, the calculus – did not seriously threaten the primacy of geometry, the treasured legacy of the Greeks. The preference for geometrical methods and concepts is clear enough in the work of Descartes, Pascal, Newton and many others (Wallis is an exception); we even find Leibniz saying 'that the whole matter is reduced to pure geometry, which is the one aim of physics and mechanics.'

By contrast, the most striking feature of eighteenth-century mathematics is the dominance of *analysis* – in the sense of the word as used by Euler and d'Alembert, rather than by Pappus or Viète. Mathematical objects were, increasingly, no longer represented by figures, but were described by algebraic formulae – by finite and infinite algebraic expressions which could be manipulated in accordance with formal rules. The prime reason for such a shift of emphasis was the greater power and versatility of the calculus, as developed in the eighteenth century under the leadership of Euler into the vast subject we now know as 'mathematical analysis'. Another feature of eighteenth-century mathematics is its practical motivation, much of the best work being directly inspired by physical problems. The influence of the *Principia* in this direction was immense. The views of d'Alembert on this matter, as we have noted, may have been somewhat extreme, but they were not untypical. What, then, has the nineteenth century – with its brilliant dawn illuminated by the genius of the young Gauss – in store?

14　Gauss

'Thou, Nature, art my goddess; to thy law
My services are bound.'
William Shakespeare, *King Lear* (I, 2)

1. Introduction

This sentence from *King Lear*, with the essential change from 'law' to 'laws', was adopted by Gauss as his motto. It epitomizes his lifelong devotion to mathematics and natural philosophy. While it is manifestly foolish to attempt to rank the greatest mathematicians in order of merit, we have already seen that three – Archimedes, Newton and Gauss – are generally held to be in a class of their own. The modest Gauss had no doubts as to the pre-eminence of the other two; his admiration for Archimedes and Newton comes through again and again.

Gauss himself was one of the last of the great all-rounders of mathematics. As with Euler, the range of his work is so extensive (much of it beyond the scope of this book) that we can but give a brief survey of his achievements and select a few topics for rather more detailed treatment.

2. The Prince of Mathematicians

Carl Friedrich Gauss (see Illustration 15) was born in 1777 in a Brunswick cottage exactly 50 years after Newton's death. His father was condemned to a life of hard physical labour as gardener, canal tender and bricklayer. This harsh but scrupulously honest man had

Illustration 15. Carl Gauss, 1777–1855 (Mary Evans Picture Library)

little understanding of or sympathy with the intellectual needs of his
unusual offspring, but fortunately his wife Dorothea was cast in a very
different mould. Her brilliant son was the centre of her life, her pride
and joy, from the day of his birth until her own death at the age of
97. The last twenty years of her life were spent in her son's house. It
was a source of great pleasure to Gauss that he was able to repay her
in this way for the support and protection she had given him during
his difficult early years. During her last illness, when Dorothea lost
her sight, Gauss would allow no one but himself to attend to her.

Gauss' precocity is legendary. At the age of 3 he was correcting his father's weekly wage calculations. When he was 7 he entered his first school, a squalid prison run by one Büttner, a brutal taskmaster. Two years later Gauss was admitted to the arithmetic class. Büttner had the endearing habit of giving out long problems of the kind, such as summing progressions, where the answer could readily be obtained from a formula – a formula known of course to the teacher, but not to the pupils. Each boy, on completing his task, had to place his slate on the master's desk. On one occasion, no sooner had Büttner dictated the last number than his youngest pupil flung his slate on the desk and waited for an hour while the other boys toiled. When Büttner looked at Gauss' slate, he found there a single number – no calculation at all. Gauss liked to recall this incident in his later years, and to point out that his was the only correct answer.

It is pleasant to be able to report that, faced with such evidence, a more humane aspect of Büttner's character soon asserted itself. He bought the best available textbooks for his young pupil, but was soon compelled to admit: 'I can teach him nothing more.' At this point Gauss had a remarkable stroke of luck. Büttner had a 17-year-old assistant, J. M. Bartels, who had a passion for mathematics and later became a professor of the subject. The two young people studied together and the schoolboy soon began to break new ground. At the age of 12 he was criticizing the axioms of Euclid; at 13 he had his first glimpse of non-Euclidean geometry; at 15 he mastered the concept of the convergence of infinite series and gave the first rigorous proof of the general binomial theorem (p. 181).

Gauss' good fortune held. Bartels, who had some influential acquaintances, was instrumental in bringing his young friend to the attention of Ferdinand, Duke of Brunswick. The result was that Gauss entered the local *Gymnasium*, the Caroline College, at the age of 15 with the Duke paying all the bills. The valiant Dorothea had overcome the opposition of her 'practical' husband to such useless book-learning. Gauss remained at the college for three years, during which time he studied the main works of Euler and Lagrange and, above all, Newton's *Principia*.

His abilities in classical languages matched those in mathematics. Indeed, when he left the college in 1795 to enter the University of Göttingen, he was undecided whether to devote himself to mathematics or to philology. Then, on 30 March 1796, something happened that tipped the balance in favour of mathematics (although the study of languages was to remain a lifelong hobby). On that day Gauss discovered how to construct a regular polygon of 17 sides by Euclidean methods (i.e. using only a straight-edge and compasses). The Greeks had constructed polygons of three and five sides, and it was generally believed that no other Euclidean construction of a regular polygon with a prime number of sides was possible. Now, after more than two thousand years, a young man of 18 had stormed the citadel! Gauss was so proud of his discovery that he expressed a wish – no doubt with Archimedes in mind – that a 17-sided polygon be carved on his tombstone. Although his exact wish was not granted, such a polygon was carved on a monument erected in his honour at Brunswick. The date 30 March 1796 marked a turning point in Gauss' life for another reason. On that day he began to keep his scientific diary (*Notizenjournal*) with the 17-sided polygon as the first entry. This slim volume of no more than 19 small octavo pages is one of the most remarkable mathematical documents of all time. It contains 146 very brief statements of Gauss' discoveries and results, the last being dated 9 July 1814. The diary came to light 43 years after its author's death, when in the possession of one of his grandsons. Felix Klein published its contents in 1901. Reference 2 contains a translation of the 49 entries for the year 1796.

The cryptic nature of the entries (a few remain undeciphered) may be illustrated by that for 10 July 1796. It reads:

EYPHKA! num $= \Delta + \Delta + \Delta$

This, with acknowledgements to Archimedes, records Gauss' discovery that every integer is the sum of at most three triangular numbers, i.e. numbers of the form $\frac{1}{2}n(n + 1)$.

The entries in the diary enable us not only to trace the development of his thought, but also to verify dates in several cases of disputed priority. It was not until after his death that the world learned how much of nineteenth-century mathematics had been anticipated before 1801 by this astonishing young man. There are enough unpublished ideas in the diary to have made half-a-dozen reputations.

Some explanation of the existence and style of the diary may be found in a remark Gauss made to a friend. He related how, in his youth, such a horde of new ideas surged uncontrollably in his mind that he could record but a small fraction. Gauss was in no hurry to publish his work. One reason for this was that, like Newton, he did not wish to be drawn into controversies with people who could not understand what he was doing. Another was his inflexible resolve, following the example of both Newton and Archimedes, to leave after him nothing but finished, perfect works of art in which all traces of the steps towards the final goal had been obliterated.

Gauss was a student at the University of Göttingen from October 1795 to September 1798 – the most prolific three years of his life. The Duke's generosity relieved Gauss of financial worries, and he was able to devote himself entirely to his studies. During these years he spent much of his time on original researches in the 'higher arithmetic' (what we now call the theory of numbers), a subject that remained a favourite with him throughout his life. By 1798 the *Disquisitiones arithmeticae* – the first and probably the greatest of his works – was nearly complete; it was published, after some printing delays, in 1801 and dedicated to his patron.

After leaving Göttingen, Gauss endeavoured to maintain himself by taking pupils but met with little success. Once again the good Duke came to the rescue: he granted Gauss a modest allowance. He even increased it in 1805 to enable Gauss to marry. Alas, his young wife died in 1809 after the birth of their third child. Gauss was devastated, but decided to marry again for the sake of his young family. There were three more children of his second and equally happy marriage.

Before this, in 1806, Gauss had suffered a severe blow in the death of Duke Ferdinand. At the age of 70 the Duke, who had been a distinguished soldier in his younger days, was put in charge of the Prussian forces who were attempting to stem Napoleon's advances. In the face of overwhelming odds, the old soldier was defeated and mortally wounded. Gauss, who was then living in Brunswick, saw his beloved patron pass through the town in a hospital wagon on his way to die in hiding. With the death of the Duke it became essential for Gauss to obtain a reliable livelihood. His fame had now spread throughout Europe and in 1807 he received a tempting offer from St Petersburg. This spurred some influential Germans to speedy action and Gauss was appointed Director of the Göttingen University Observatory, and later Professor of Astronomy. His duties included a limited amount of lecturing to university students, a job that seemed to give him little satisfaction. Here he is writing in 1810 to his close friend Friedrich Bessel:

This winter I am giving two courses of lectures to three students, of whom one is only moderately prepared, the other less than moderately and the third lacks both preparation and ability. Such are the burdens of a mathematical calling.

Gauss remained at Göttingen for the rest of his life. Indeed, he only once slept away from home – on the occasion of a scientific meeting in Berlin. His modest salary was sufficient for the simple needs of his family and himself; luxury and public position never attracted him – in contrast to the elderly Newton.

One facet of Gauss' character is illustrated by his reaction to the demands of the French occupying power. He was assessed for the sum of 2000 francs as a contribution to the Napoleonic war-chest, a sum far beyond his ability to pay. After he had refused offers of assistance from several German friends, Laplace wrote to say that he had paid the sum in Paris. He considered it 'an honour to assist the greatest mathematician in the world'. Later, when Gauss received an unexpected windfall, he repaid Laplace with full interest.

We have already mentioned Gauss' lifelong interest in languages. He had two other hobbies: modern and classical literature, and world politics. He was particularly attracted to some of the English writers, Gibbon and Macaulay being two of his favourites. When he read in Scott's *Kenilworth* that 'the moon rises broad in the north-west' he delighted in taking direct action on all the copies he could find. Gauss spent an hour or so every day in the Göttingen literary museum keeping abreast of events by reading all the newspapers. His political opinions were generally conservative; he took a sardonic view of the intellectual and moral qualities both of the ordinary people and of the men of power and position. He always refused to wear any of his many decorations.

From 1821 to 1848 Gauss was scientific adviser to the Hanoverian and Danish Governments on the comprehensive geodetic surveys that were being carried out at the time. The work, into which Gauss threw himself with enthusiasm, followed on naturally from his astronomical activities. It gave further scope for him to exercise his remarkable computational powers and his skill in handling large quantities of data, but even more important was the fact that his geodetic interests led directly to his fundamental researches in differential geometry.

During the last twenty years of his life Gauss devoted most of his energies to physical matters, making important contributions to analytical dynamics and optics. By the 1830s, however, the centre of interest in applied mathematics was moving from astronomy and mechanics to the rapidly growing subject of electricity and magnetism. The spectrum of Gauss' activities showed a corresponding shift. In 1839 he published an important paper on potential theory, in which he established several basic theorems about 'inverse square force-fields', some of which still bear his name. While Gauss was first and foremost a mathematician, he was also a gifted experimenter and a skilled observer – a rare combination. He made a number of improvements to astronomical and physical instruments; he invented the heliotrope and the bifilar magnetometer. In 1833, with his colleague Wilhelm Weber, he devised an improved electric telegraph and used it between the observatory and Weber's physics laboratory. His experimental and theoretical researches on magnetism extended over many years and were highly acclaimed: James Clerk Maxwell in his *Electricity and Magnetism* (1873) says that Gauss' studies of magnetism reconstructed the whole science – the instruments used, the methods of observation and the calculation of results. It is fitting that his name is associated with both the unit of magnetic intensity and the protection of ships in wartime against magnetic mines.

Gauss enjoyed good health throughout his life and preserved his mental vigour and originality to the end. After a great struggle for life, he died peacefully on 23 February 1855, in his seventy-eighth year.

3. The fundamental theorem of algebra

In 1799 Gauss obtained his doctor's degree at the University of Helmstädt for a dissertation in which he proved that every algebraic equation in one variable has a least one root. In later life he referred to this result as the 'fundamental theorem of algebra'. (In France, as we have seen, it was known as d'Alembert's theorem.) The title of Gauss' dissertation, in English translation, is 'A new proof that every rational integral function of one variable can be resolved into real factors of the first or second degree'. What Gauss provided was not merely a new proof, but the *first* proof. He showed why all earlier attempts – by Euler, d'Alembert and Lagrange among others – were, as he put it, 'unsatisfactory and illusory' and then gave a 'newly constructed rigorous proof'. The theorem, in its most general form, is equivalent to the statement that every polynomial equation $f(z) = 0$, where z is a complex number, has at least one root, either real or complex. The coefficients of the polynomial may themselves be either real or complex.

Before discussing Gauss' proof something more must be said about complex numbers. At the end of the eighteenth century three men – Caspar Wessel (1745–1818), a self-taught Norwegian surveyor, Jean Robert Argand (1768–1822), an equally self-taught Swiss book-keeper, and Gauss himself – working independently, put forward geometrical interpretations of complex numbers. Figure 14.1 represents what is called the *complex plane*, in which distances along the vertical axis are measured in units of i $(= \sqrt{-1})$, and those along the horizontal axis in the usual units of 1. Any complex number $a + ib$ can then be represented by a point P with Cartesian coordinates (a, b). Operations on complex numbers, such as addition and multiplication, can thus be given simple geometrical interpretations. An alternative approach is to regard a complex number as a directed line OP (nowadays called a *vector*) of length r, known as the *modulus*, which has rotated from the initial line OA through an angle θ known as the *amplitude*. Rotation through a right angle (anticlockwise) converts a real number, R say, into the (unhappily so-called) imaginary number iR; rotation through a further right angle yields another real number, $-R$,

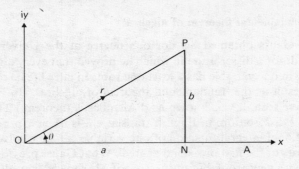

Figure 14.1.

and provides a geometrical representation of the fact that $i^2 = -1$.

We are now in a position to discuss Gauss' line of reasoning in his demonstration of the 'fundamental theorem'. His starting point is that a complex root $a + ib$ of the equation $f(z) = f(x + iy) = 0$ corresponds to a point (a, b) in the complex plane. If we separate $f(z)$ into its real and imaginary parts by writing

$$f(z) = f(x + iy) = u(x, y) + iv(x, y) \qquad (14.1)$$

then the point (a, b) must be a point of intersection of the two curves $u(x, y) = 0$ and $v(x, y) = 0$.

Consider, for example, the equation

$$f(z) = z^2 - 2z - (3 + 4i) = 0 \qquad (14.2)$$

Separating the real and imaginary parts yields

$$u(x, y) = x^2 - y^2 - 2x - 3 \quad \text{and} \quad v(x, y) = 2xy - 2y - 4$$
$$(14.3)$$

Setting $u = 0$ and $v = 0$, we see that the point (a, b) is given as the intersection of two hyperbolas, as shown in Figure 14.2, whose equations are, from equations (14.3),

$$(x - 1)^2 - y^2 = 4 \quad \text{and} \quad (x - 1)y = 2 \qquad (14.4)$$

The two curves clearly intersect at the point in the first quadrant marked P which corresponds to a root of equation (14.2); there is a

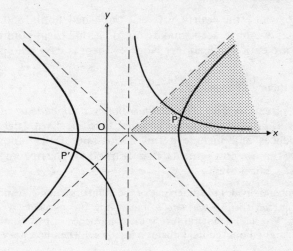

Figure 14.2.

second intersection at P′. Since each branch of each curve is continuous, it is apparent from the position of the asymptotes, shown as dashed lines in Figure 14.2, that there must be a point of intersection somewhere in the shaded region. Gauss was able to generalize and refine this type of argument to provide a rigorous proof that a branch of the '$u = 0$' curve must cross a branch of the '$v = 0$' curve. His proof is long and difficult; the technical details are beyond the scope of this book. Gauss puts the matter thus:

Both curves will, properly speaking, consist of several branches, which may be entirely separated from each other, but each by itself forms a continuous curve . . . Our problem has been reduced to the task of proving that there exists in the plane at least one point at which one of the branches of the first curve is intersected by one of the branches of the second curve.

He establishes this result by investigating the 'infinite branches' and the corresponding asymptotes of these curves, and uses the notion of continuity to demonstrate that the required intersection does exist.

Gauss returned to this theorem at various times during the rest of his life. He gave three other proofs: the second and third appeared in 1816, while the fourth was not published until 1850. His first proof, as we have seen, makes use of geometrical continuity; the others are entirely algebraic. None of them is easy reading.

4. The theory of numbers

Gauss' great treatise, the *Disquisitiones arithmeticae* of 1801, is organized in seven sections. The first three deal with the theory of congruences and include a full discussion of the binomial congruence that we now write as $x^n \equiv A \pmod{p}$. The opening words of the first section are:

> If a number a divides the difference of the numbers b and c, b and c are said to be *congruent with respect to a*: but if not, incongruent. We call a the *modulus*. We shall denote in future the congruence of two numbers by the sign $\equiv$, and adjoin the modulus in parentheses when necessary.

Thus $-16 \equiv 9 \equiv 4 \pmod 5$, and $-9 \equiv 13 \equiv 2 \pmod{11}$. To solve the congruence $x^3 \equiv 2 \pmod 3$, for example, we must look for values of x such that $x^3 = 3p + 2$, where p is an integer. Two solutions are 2 and -4, since $2^3 = 8 = (3 \times 2) + 2$ and $(-4)^3 = -64 = -(3 \times 22) + 2$.

The fourth section deals with the theory of quadratic residues and contains his first proof, obtained when he was 19, of the famous law (or theorem) of *quadratic reciprocity*. Once again Gauss had to point out that earlier alleged proofs, notably that by Legendre, were unsatisfactory. Gauss called this theorem, which had been conjectured but not proved by Euler, the 'gem of arithmetic'. It is of fundamental importance in number theory and in algebra generally, and Gauss obtained no fewer than seven other proofs (not all published) during the next 17 years. So let us explain it.

If there is at least one value of x for which $x^2 \equiv r \pmod m$, then the congruence is said to be solvable and r is called a *quadratic residue* of m. Thus, if $m = 5$, the congruence is solvable if $r = 4$ (having solutions 2, 3, 7, etc.), but if $r = 3$ it is not. The law of quadratic reciprocity expresses an elegant relationship between the pair of congruences

$$x^2 \equiv q \pmod p \quad \text{and} \quad x^2 \equiv p \pmod q$$

where both p and q are odd primes. At this point let us introduce the integer f, defined as $f = [\frac{1}{2}(p - 1)][\frac{1}{2}(q - 1)]$. The theorem states that if f is even, then either both or neither of the congruences are solvable; while if f is odd, then one and only one is solvable. Thus, if $p = 5$ and $q = 13$, giving $f = 12$, the theorem tells us that either both or neither of $x^2 \equiv 13 \pmod 5$ and $x^2 \equiv 5 \pmod{13}$ are solvable. In fact neither is. If, however, $p = 7$ and $q = 13$, giving $f = 18$, then both congruences are solvable. Examples of the other possibilities are easily constructed (try, for example, $p = 3, q = 11$; $p = 7, q = 5$; and $p = 3, q = 7$).

The fifth and sixth sections of the *Disquisitiones* deal with the theory of binary and ternary quadratic forms. Here Gauss is concerned with such questions as for what integral values of a, b, c and m the indeterminate equation

$$ax^2 + 2bxy + cy^2 = m$$

is solvable for integral values of x and y. It is interesting to note that Fermat's elegant result that every prime p of the form $p = 4n + 1$ is a unique sum of two squares appears as a simple consequence of Gauss' theory.

In the seventh section, usually considered the greatest part of the work, Gauss considers the equation $x^n = 1$, which is the algebraic equivalent of the geometrical problem of constructing a regular polygon of n sides. Here is his great discovery about the 17-sided regular polygon which confirmed him in choosing a mathematical career. Gauss was able to demonstrate exactly which regular polygons can be constructed by Euclidean methods and which cannot. It turns out that a polygon of N sides is so constructible if and only if N is of the form $2^m p_1 p_2 \ldots p_r$, where m is an integer greater than or equal to zero, and the p_i are distinct Fermat primes, i.e. prime numbers of the form $2^{2^n} + 1$. Only five such primes are known: 3, 5, 17, 257 and 65 537, corresponding to $r = 0, 1, 2, 3$ and 4 (see Chapter 12).

Gauss' approach was to show that the roots of the 'cyclotomic equation' $x^p - 1 = 0$, where p may be taken to be prime, may be rationally expressed in terms of the roots of a sequence of equations $S_1 = 0, S_2 = 0, S_3 = 0, \ldots$ whose coefficients are rational functions of the roots of preceding equations of the sequence. The degrees of these equations are the prime factors of $p - 1$. Thus if p is of the form $1 + 2^r$, all the equations in the sequence will be quadratics, and so solvable by a construction in which only straight-edge and compasses are used. In the *Disquisitiones arithmeticae* Gauss gives a full discussion of the equation $p = 17 = 1 + 2^4$. An account of his widely acclaimed treatment of this famous problem is given in Appendix 5.

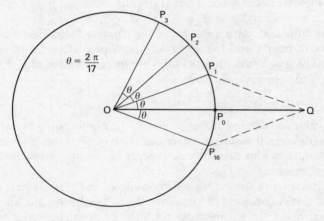

Figure 14.3.

The problem is illustrated geometrically in Figure 14.3. We wish to determine the length of the straight line $P_0 P_1$ (the side of the polygon), where the length of the corresponding arc $P_0 P_1$ is 1/17th of the circumference of the unit circle centred at O. Suffice it to say here that the expression for the length OQ in Figure 14.3, from which the side $P_0 P_1$ can readily be evaluated, turns out to be

$$\tfrac{1}{8}\left\{-1 + \sqrt{17} + \sqrt{34 - 2\sqrt{17}}\right.$$

$$\left. + \sqrt{68 + 12\sqrt{17} - 16\sqrt{34 + 2\sqrt{17}} + 2(-1 + \sqrt{17})\sqrt{34 - 2\sqrt{17}}}\right\}$$

A fearsome expression indeed! However, since it contains square roots but no other irrational quantities, it follows that the 17-sided regular polygon can be constructed using a straight-edge and compasses only.

The work presented in the *Disquisitiones* did not exhaust Gauss' discoveries in his favourite field. In a series of papers from 1808 to 1832 he continued his detailed studies of the properties of the higher congruences $x^n \equiv A \pmod{p}$ for $n < 2$. His search for a corresponding law of biquadratic reciprocity ($n = 4$) led him, after many false starts, to a major advance. This was the extension of the concept of an integer to include what are now called *Gaussian integers*. These are numbers of the form $a + \mathrm{i}b$, where a and b are integers. In this extended domain the number 5, for example, is no longer prime but is the product of two factors, $1 + 2\mathrm{i}$ and $1 - 2\mathrm{i}$ (or, alternatively, of $2 + \mathrm{i}$ and $2 - \mathrm{i}$). Gauss proved, among much else, that prime numbers of the form $4n - 1$ remain prime in the new domain, while those of the form $4n + 1$ do not. Here again, Gauss inaugurated a new branch of mathematics – the theory of *algebraic numbers*.

As a final example of Gauss' creativity in the number theory field, there is the following cryptic note written on the back page of one of his boyhood tables of logarithms:

$$\textit{Primzahlen unter } a \; (=\infty) \; \frac{a}{\ln a}$$

This is a shorthand statement of the celebrated prime number theorem: that the number of primes less than a given integer a approaches the quotient $a/\ln a$ asymptotically as a increases indefinitely. (The meaning of '$\ln a$' has been explained on p. 282.)

5. The theory of functions and infinite series

Gauss, in the words of G. H. Hardy, 'was the first mathematician to use complex numbers in a really confident and scientific way'. He also took the next step, and began to investigate the theory of *complex functions*. In 1811, in a letter to Bessel, he communicated his discovery – 'a very beautiful theorem' – which forms the foundation of the branch of mathematics now known as the 'theory of functions of a complex variable', and was to prove so important in the development of nineteenth-century mathematical physics. Gauss' discovery was not made public at the time, and was rediscovered soon afterwards by Cauchy, by whose name it is usually known.

What the theorem does is to specify the conditions that must be satisfied by the function $f(z)$ to ensure that its *line integral* from a point A to another point B in the complex plane is independent of the path along which the integral is evaluated: i.e. in Figure 14.4, that

$$\int_{ACB} f(z)\,dz = \int_{ADB} f(z)\,dz$$

An alternative formulation is that the line integral taken around a closed curve such as ACBDA is zero.

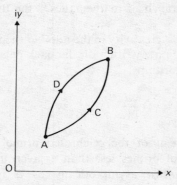

Figure 14.4.

A year later, however, Gauss did publish another major piece of work – his memoir on the *hypergeometric series*

$$F(a,b,c;\,x) = 1 + \frac{ab}{c}\,x + \frac{a(a+1)b(b+1)}{c(c+1)}\frac{x^2}{2!}$$
$$+ \frac{a(a+1)(a+2)b(b+1)(b+2)}{c(c+1)(c+2)}\frac{x^3}{3!} + \cdots$$

and its associated differential equation. He gives a complete discussion of the conditions for convergence (when $x = 1$ or -1, they are quite tricky!) and establishes that for particular values of a, b, c and x the series includes most of the elementary functions of analysis, and many higher transcendental functions as well. Here are some simple examples:

(1)   The general binomial expression: $(1 + x)^n = F(-n, 1, 1; -x)$.
(2)   The logarithmic series: $\log(1 - x) = xF(1, 1, 2; -x)$.
(3)   The 'arc tangent' series given by $x - \frac{1}{3}x^3 + \frac{1}{5}x^5 - \frac{1}{7}x^7 + \ldots$: $\arctan x = xF(\frac{1}{2}, 1, \frac{3}{2}; -x^2)$; 'arctan x' means the angle in radians whose tangent is x. Setting $x = 1$, so that $\arctan x = \frac{1}{4}\pi$, gives the Leibniz series for π discussed on p. 268.
(4)   The exponential function e^x. Here we must allow one parameter, say a, to become infinitely large; then e^x is the limit of $F(a, 1, 1; x/a)$ as a tends to infinity.

This memoir not only provides one of the first rigorous treatments of convergence, but also foreshadows much of the later nineteenth-century work on the differential equations of mathematical physics.

6. Error theory and numerical analysis

We must now go back a few years to the first day of the new century, when an event occurred that was destined to deflect Gauss for the rest of his life from his early concentration on 'pure' mathematics. On 1 January 1801 Giuseppe Piazzi discovered a small planet, later named Ceres, the first of the asteroids to be found. Ceres was badly placed for observation and the computation of its orbit from a very small number of observations was a challenge that Gauss could not resist. He later explained how:

An opinion had universally prevailed that a complete determination from observations embracing a short interval of time was impossible – an ill-founded opinion for it is now clearly shown that the orbit of a heavenly body may be determined quite nearly from good observations embracing only a few days.

In fact the observations of Ceres covered a period of 41 days, to 14 February 1801, after which the asteroid was lost to view for some nine months. During the time it had been observed Ceres had moved through a geocentric arc of only 3°. Any hope of locating it again after a lapse of nearly a year depended on the accuracy of Gauss' orbital calculations, based as they were on very meagre observations. In the event, however, his methods were triumphantly vindicated. On the first clear night of October 1801 astronomers looked for the asteroid in the position Gauss had predicted and, as he put it, 'the fugitive was restored to observation'.

Three more asteroids were discovered in the next six years, and 'furnished new opportunities for examining and verifying the efficiency and generality of the method'. Gauss goes on to say that he was urged to publish his methods 'immediately after the second discovery of Ceres', but he delayed because he wished to develop the subject more fully and to improve his computational techniques. This he did to great effect, and presented his results in his second great work, *Theoria motus corporum coelestium*, which was published in 1809.

Now it so happened that, with perfect ironic timing, the discovery of Ceres coincided with a pronouncement by the heavyweight philosopher Hegel that there must be precisely seven planets, neither more nor less. If astronomers would pay some attention to philosophy, they would not be so presumptuous as to waste their time searching for an eighth planet. It is not surprising that Gauss in his letters had some astringent comments to make on non-mathematical philosophers. 'Don't they make your hair stand on end with their definitions!'

The publication of the *Theoria motus* was a landmark in the application of mathematics to astronomy. It deals in great detail with the dynamical theory and practical computation of planetary and cometary orbits, including the difficult subject of orbital perturbations. Gauss' work on the effects of errors of observation laid the foundations of much of what we now know as mathematical statistics. Indeed, outside mathematical circles, he is probably best known for his 'normal law of errors' and its familiar bell-shaped distribution curve.

The status of this famous law has given rise to much controversy. Mathematicians, it has been said, regard it as an induction from experiment; experimentalists believe it can be proved by mathematics. Gauss, who was equally at home in both fields, approached the matter in several different ways. In the *Theoria motus* he first derived the law from the postulate of the arithmetic mean, which states that given a number of equally good observations of an unknown quantity, its most probable value is the arithmetic mean of the observations. Gauss then considered the problem of indirect observations where we wish to estimate the values of a set of quantities from measurements of some other, related quantities. Problems of this kind are usually solved by the 'method of least squares', which was put forward as a working principle by Legendre in 1805. A number of attempts were made to place it on a logical foundation, but Gauss, who had used the method since 1795, was the first to connect it with the theory of probability. His treatment in the *Theoria motus* assumes that the observations are subject to a large number of small, independent, accidental errors, and so would be 'normally' distributed. In a later work (*Theoria combinationis observationum erroribus minimis obnoxiae*, 1823) he derives the equations of the method of least squares in an entirely different – and, as he puts it, 'a less metaphysical' – manner. He introduces the concept of the 'weight' of an observation, and does not have to resort to the idea of the arithmetic mean.

Gauss also made a number of important contributions to what we now call 'numerical analysis'. He derived several central difference interpolation formulae (p. 200), some of which bear his name. The well-known numerical method of 'Gaussian integration' was published in 1814. The idea is to replace a definite integral, say $\int_0^1 f(x)\, dx$, by a linear function of the form

$$H_0 f(x_0) + H_1 f(x_1) + \ldots + H_n f(x_n)$$

and to choose the values of the $2n + 2$ arbitrary constants (i.e. the H_i and the x_i, with $i = 0, 1, 2, \ldots, n$) so as to give the best possible approximation to the integral for any specified choice of n. He also devised what are still the two most widely used procedures for solving sets of linear equations: the direct successive elimination and back-substitution method, and the indirect iterative method which was refined by Seidel in 1874.

In 1812 Gauss published a set of tables of what came to be known as Gaussian logarithms. The objective is to obtain $\log(a \pm b)$ by a single reference to the tables when $\log a$ and $\log b$ are known. A convenient arrangement is to tabulate

$$A = \log(n + 1)/n \quad \text{and} \quad S = \log n/(n - 1)$$

against $d = \log n$ (with $n > 1$). The procedure is as follows. The user subtracts $\log b$ from $\log a$, to obtain the required value of d. He can do this because

$$\log a - \log b = \log a/b = \log n = d$$

where we set $n = a/b$. For addition he then enters the table to get A, which he adds to $\log a$ to give $\log(a + b)$. ($A = \log\{(n - x)/n\} = \log\{(a + b)/a\}$, and so $\log(a + b) = \log a + A$.) For subtraction he enters the table to get S, which he subtracts from $\log a$ to give $\log(a - b)$. ($S = \log\{n/(n - 1)\} = \log\{a/(a - b)\}$, and so $\log(a - b) = \log a - S$.) Such tables were widely used by mariners to solve navigational problems. For Gauss, like Newton, theory and practice were two facets of a single endeavour.

7. Differential geometry

Gauss' third major treatise is the *Disquisitiones generales circa super-ficies curvas* of 1827. In this work he not only provided the definitive treatment of the differential geometry of surfaces lying in three-dimensional space, but also advanced the totally new concept that a surface is a space in itself – an idea that was soon to take on a new dimension in the work of Riemann and become generalized still further in the context of non-Euclidean geometry (to be discussed in the next section). Following the pioneering work of Euler (p. 289), Gauss' first task was to clarify the concept of the *total* (or *integral*) *curvature* of a part of a surface bounded by a closed curve. By letting the curve shrink to a point, he is able to define the *measure of curvature* (sometimes called the *Gaussian curvature*) at a point P on the surface, and to prove that it is proportional to the reciprocal of the product of the two principal radii of curvature (p. 290) at the point.

Gauss then introduces Euler's parametric representation, by which the coordinates (x, y, z) of any point on the surface may be expressed in terms of two new quantities, p and q. Thus $x = x(p,q)$, $y = y(p,q)$ and $z = z(p,q)$. He explains how, to quote his own words, this representation

facilitates the transition to another expression, which must be classed with the most remarkable theorems of this study. If the nature of the curved surface be expressed by this method, the general expression for any linear element upon it, or for $\sqrt{dx^2 + dy^2 + dz^2}$, has the form $\sqrt{E\,dp^2 + 2F\,dp\,dq + G\,dq^2}$, where E, F, G are again functions of p and q . . . In order to determine the measure of curvature, it is necessary to know only the general expression for a linear element; the expressions for the coordinates x, y, z are not required. A direct result from this is the remarkable theorem: if a curved surface can be developed upon another surface, the measure of curvature at every point remains unchanged after the development.

He goes on to explain how:

These theorems lead to the consideration of curved surfaces from a new point of view . . . where the very nature of the curved surface is given by means of the expression of any linear element in the form $\sqrt{E\,dp^2 + 2F\,dp\,dq + G\,dq^2}$.

Gauss proceeds to develop the mathematics in great detail, while not overlooking the practical applications to the study of the surface of the Earth. These he illustrates by reference to triangulation surveys that he had carried out himself. One of his more famous results is:

The excess of the sum of the angles of a triangle formed by shortest lines (on the surface) over two right angles is equal to the total (or integral) curvature of the triangle.

Gauss' map-making interests led him to investigate the subject of *conformal mapping*, the mapping from a curved surface to a plane so that angles are preserved. His memoir of 1822 forms the basis of the modern treatment of the subject.

8. Non-Euclidean geometry

We have seen in earlier chapters how Euclid's fifth postulate (the 'parallel postulate') had been a cause for concern ever since Greek times. Many attempts were made, over the centuries, to deduce this 'rogue' postulate from the other, more self-evident, basic notions, but without success. So strong, indeed, was the belief in the necessary truth of Euclid's system as the geometry of space itself, that it was not until the beginning of the nineteenth century that the possibility of other geometries, just as valid as Euclid's, was clearly recognized. The persistence of this cultural blind spot is all the more remarkable when it is realized that the surface of the Earth provides a familiar and well-studied example of such an alternative geometry.

The creation of non-Euclidean geometry is usually attributed to two men, János Bolyai (1802–60) and Nikolai Ivanovich Lobachevski (1793–1856), who published their ideas independently in the mid-1820s. Once again, Gauss had got there before, although he never published his work in this field – largely, it seems, for fear of ridicule – and we have to rely on letters and reviews. Gauss started thinking about the problem while he was still at school. In 1792 (when aged 15) he told a friend that he fully accepted the idea that there could be a logically consistent geometry in which the parallel postulate does not hold. A few years later he postulated an alternative: namely that there are at least two lines through a point parallel to a given line. He explored the consequences of this assumption and was able to deduce a self-consistent system containing many strange theorems.

Unlike some of his predecessors in the field, Gauss had the confidence to pursue this revolutionary approach, but not the courage – or perhaps, merely not the desire – to publish it. In a letter written in 1817 he says:

I am becoming more and more convinced that the necessity of our geometry cannot be proved, at least not by human reason. Perhaps in another life we will be able to obtain insight into the nature of space which is now unattainable. Until then we must place geometry not in the same class with arithmetic, which is purely *a priori*, but with mechanics.

We can go no further into this fascinating piece of mathematical history, but cannot leave the subject without making some mention of the network of personal relationships that existed between its three joint founders.

János Bolyai's father, Wolfgang, was a lifelong friend of Gauss and had himself spent years worrying about the parallel postulate, but with little to show for it. Ever since the 1790s, Gauss had kept him informed of the progress of his own thoughts on this tantalizing subject. János Bolyai was a Hungarian army officer who seemed to have inherited his father's obsession with the problem in spite of repeated warnings as to its dangers. Thus, Wolfang to János:

For God's sake, I beseech you, give it up. Fear it no less than sensual passions because it, too, may take all your time, and deprive you of your health, peace of mind, and happiness in life.

János was not to be dissuaded, however, and in time he reaped his reward. In a letter to his father dated 3 November 1823, he writes:

I have now resolved to publish a work on the theory of parallels as soon as I shall have put the material in order . . . I have made such wonderful discoveries that I have been almost overwhelmed by them . . . When you will see them, you too will recognize it. In the meantime I can only say this: *I have created a new universe from nothing.*

Lobachevski was a student, then a member of staff, and eventually Rector of the remote University of Kazan. One of the professors there, attracted from Germany, was J. M. Bartels, whom we have already met as the assistant master in Gauss' first school. The two friends kept in touch throughout their lives, and Bartels was almost certainly familiar with Gauss' ideas and conclusions on the theory of parallels. Opinions differ as to how much of this information Bartels passed on to his younger colleague at Kazan. It is perhaps worth remarking that when the prestigious Moscow Institute for Physical Problems was evacuated to Kazan – the University of Tolstoy and Lenin – in 1941 because of the German invasion, the Director and his family occupied a small house which had once been the home of Rector Lobachevski. The Director in question was Peter Kapitza, the distinguished and colourful physicist who spent 13 years at Cambridge with Lord Rutherford before being held in the USSR in 1934 to develop Soviet physics.

The works of Bolyai and Lobachevski were largely ignored until after Gauss' death in 1855. The posthumous publication of the great man's correspondence finally drew attention to this revolutionary departure and made it respectable. When Gauss read the work of the two younger men, he commended it and encouraged them privately – indeed, he was instrumental in getting Lobachevski elected to the Göttingen Scientific Society – but would not support them in print because he had held the same views for many years himself. Thus, in a letter to Wolfgang Bolyai after the proud father had sent a copy of his son's paper, Gauss wrote:

If I commenced by saying that I am unable to praise this work, you would certainly be surprised for a moment. But I cannot say otherwise. To praise it, would be to praise myself. Indeed the whole contents of the work, the path taken by your son, the results to which he is led, coincide almost entirely with my meditations, which have occupied my mind partly for the last thirty or thirty-five years.

Gauss goes on to explain why he did not intend to publish his own views on the subject during his lifetime:

On the other hand, it was my idea to write down all this later so that at least it should not perish with me. It is therefore a pleasant surprise for me that I am spared this trouble, and I am very glad that it is just the son of my old friend, who takes precedence of me in such a remarkable manner.

It is interesting to note that Gauss adopted the same attitude, and for the same reason, when in 1852 (three years before his death) he heard of Hamilton's work on quaternions, discussed in Chapter 15.

Enough, it is hoped, has been said to give some impression of the range and penetration of Gauss' contributions, not only to mathematics, but also to astronomy, geodesy and physics. Although his pre-eminence was recognized early in his life and has never been in question since, Gauss was in some ways a transitional figure. His universality and use of Latin were in the traditions of the eighteenth century; his discoveries and insights pointed the way to much of the best mathematics of the later nineteenth century. He was indeed, as his contemporaries styled him, the 'Prince of Mathematicians'.

15 Hamilton and Boole

'Pure mathematics was discovered by Boole, in a work which he called *The Laws of Thought*.' Bertrand Russell

1. Introduction

In this chapter we shall be mainly concerned with the development of algebra during the middle years of the nineteenth century, and especially with the contributions of two men in Ireland. By the 1820s mathematicians were using both real and complex numbers quite freely, but there was as yet no sound logical basis for the operations being performed on these numbers. What, in fact, was the justification for manipulating the symbolic expressions of algebra in accordance with the rules of arithmetic? One of the first to tackle this problem was George Peacock (1791–1858), a Cambridge mathematician who became Dean of Ely Cathedral. In the 1830s he formulated his 'principle of the permanence of equivalent forms', which states that the rules of what he called 'arithmetical algebra' may be carried over to the more general 'symbolical algebra', where the restriction to natural numbers is removed. Thus, for example, since the 'rule of indices', namely $a^m a^n = a^{m+n}$, is clearly valid when m and n are positive integers, it may be deemed to hold for all real m and n.

The main such laws, which were self-evident axioms to the classical algebraists, can be expressed as:

(1) $a + b = b + a$, the commutative law of addition;
(2) $(a + b) + c = a + (b + c)$, the associative law of addition;

(3) $ab = ba$, the commutative law of multiplication;
(4) $(ab)c = a(bc)$, the associative law of multiplication; and
(5) $a(b + c) = ab + ac$, the distributive law of multiplication.
In fact, Peacock's principle could not provide a broad enough foundation for nineteenth-century algebra and was soon undermined, as we shall see.

2. William Hamilton

William Rowan Hamilton (see Illustration 16), unquestionably Ireland's greatest man of science, was born at midnight on 3–4 August 1805 in Dublin, the son of a solicitor and successful businessman. His intellectually gifted mother died when William was 12; his less gifted father died two years later. However, William had exhibited early signs of such exceptional talent that at the tender age of 3 he was sent to live with his uncle, the Reverend James Hamilton, at the village of Trim, some twenty miles from Dublin. Uncle James was an accomplished linguist and polymath, and by the age of 8 the boy was at home in half a dozen languages, ancient and modern. His uncle reports that at 10 years old William's thirst for the Oriental languages was unabated: 'He is now master of most, indeed of all except the minor and comparatively provincial ones.' It was, we are told, difficult and expensive to obtain the necessary books 'in the Chinese'. At the age of 13, William claimed to have mastered one language for each year he had lived, and astonished the visiting Persian ambassador by composing a florid welcome in the ambassador's mother-tongue.

Hamilton's early interest in mathematics was attributed to his encounter with the young calculating prodigy, Zerah Colburn (1804–39); his lifelong obsession with the composition of poetry to his friendship with Wordsworth and Coleridge; his physical prowess and love of nature and animals to his abounding vitality and acute sensitivity. By the age of 17 he had acquired a passion for astronomy and was reading Newton and Lagrange, but classical literature remained his favourite study. He entered Trinity College, Dublin, in 1823, easily passing first in his year. Such indeed was his

Illustration 16. William Hamilton, 1805–65 (Mary Evans Picture Library)

fame, both as a mathematician and a classicist, that at the age of 22 and while still an undergraduate he was appointed Astronomer Royal for Ireland, Director of the Dunsink Observatory and Professor of Astronomy at Trinity College, Dublin.

In 1833, after two unhappy love affairs, William married Helen Bayley. Three years later, at the age of 30, he was knighted; in 1837 he became President of the Royal Irish Academy; in 1843 he was awarded a Civil List life pension by the British Government. Notwithstanding his public success, Hamilton's domestic life was causing increasing concern. A year before her marriage, Helen had suffered a serious illness which left her a semi-invalid for the rest of her life. She was unable to give her husband the domestic support he needed; instead, a pack of slovenly servants ran the house as they chose, and Hamilton's study, in particular, came to resemble a pigsty. As time went on, the regime of irregular meals, chronic overwork and marital unhappiness began to take its toll, and the poor man increasingly sought solace from the bottle. Eventually he went too far and got drunk at a scientific dinner, whereupon he resolved never to touch alcohol again. He kept his resolution for two years, but thereafter it was downhill all the way. In 1865, on his deathbed, he was delighted to learn that he had been elected the first foreign member of the National Academy of Sciences of the United States of America.

3. Researches in optics and dynamics

Most of Hamilton's early researches were in mathematical physics. His first published paper, on optics, appeared in 1822, and during the next ten years he produced a series of papers which established geometrical optics as a mature mathematical science. He became famous almost overnight when one of his theoretical predictions – that, in certain circumstances, a ray of light passing through a biaxial crystal would give rise to a cone of refracted rays – was confirmed experimentally. He then applied his optical ideas to the field of mechanics. The subject of analytical dynamics had been strongly developed during the eighteenth century, notably by Euler and Lagrange. One of its central concepts, ever since the time of Fermat, had been the 'principle of least action' (pp. 146 and 289). Hamilton generalized the principle in several directions. He introduced an 'action integral' S, which is the time integral of the difference between the kinetic energy T and the potential energy V of the dynamical system. (We can write S as $\int_{t_1}^{t_2} (T - V)\, dt$.) The Hamiltonian principle of least action asserts that the actual motion of the system is such as to make the value of S either a maximum or a minimum. In a later paper (1835) he gave an elegant and succinct formulation of the equations of motion of a general dynamical system in terms of what is now known as the 'Hamiltonian function',

H, a function of time and the positions and velocities of the constituent masses of the system. In many cases H is simply the total energy, $T + V$, of the system. Hamilton's detailed treatment of the subject, which uses sets of differential equations and the techniques of the calculus of variations, is beyond the scope of this book. Suffice it to say that many mathematicians regard Hamilton's contributions to analytical dynamics as his greatest achievement, but he himself gave pride of place to his work on quaternions, to which we now turn.

4. The creation of quaternions

The recognition that the real numbers can be interpreted as points along a line and the complex numbers as points in a plane led naturally to the search for 'hypercomplex' numbers that could be represented by points in three-dimensional space. In 1833 Hamilton read a paper to the Royal Irish Academy in which he pointed out that the plus sign in $a + ib$ was a misnomer, as a and ib cannot be added arithmetically. Following Gauss, he proposed that a complex number should be regarded as an *ordered pair* of real numbers (a, b) which obey certain operational rules, in particular

$$(a, b) + (c, d) = (a + c, b + d) \qquad \text{(addition)} \qquad (15.1)$$

$$(a, b) . (c, d) = (ac - bd, ad + bc) \qquad \text{(multiplication)} \qquad (15.2)$$

(Remember that $i^2 = -1$.)

Hamilton then sought to extend this idea to ordered 'number triples', (a, b, c), to be written as $a + ib + jc$, where i and j are two distinct and independent square roots of -1. The problem defeated him for many years: he could add and subtract his triples, but he could not multiply them.

With hindsight, the source of the difficulty is not hard to pin down. The geometrical effect of multiplying one complex number by another is to change the length (or modulus) of the corresponding directed line (or vector) and to rotate it in the complex plane. In the polar coordinate representation of complex numbers, the 'product' of $(r_1 \cos \theta, r_1 \sin \theta)$ and $(r_2 \cos \psi, r_2 \sin \psi)$ is

$$(r_1 r_2 \cos (\theta + \psi), r_1 r_2 \sin (\theta + \psi))$$

Since the direction of the axis of rotation is determined (it is normal to the complex plane), only one 'length change' and one rotational parameter are needed, i.e. two altogether. In three dimensions, however, we need two parameters to specify the direction of the axis of rotation, a third to determine the amount of rotation and a fourth to specify the change of length. The need to specify *four* parameters means that it is not possible to 'multiply' one ordered number triple by another.

Hamilton knew, of course, that when a complex number $a + ib$ is multiplied by its conjugate, $a - ib$, the product is the positive real number $a^2 + b^2$: the square of the modulus of either number. Let us consider the analogous operation for number triples. If we multiply $a + ib + jc$ by $a - ib - jc$, we find that most of the product terms cancel out and we are left with

$$a^2 + b^2 + c^2 - 2ijbc \tag{15.3}$$

Hamilton's difficulty lay in the existence of the product term. Setting $ij = 0$ will not do, because $ij.ij = i^2.j^2 = (-1).(-1) = 1$, so we have a contradiction. After pondering the matter for many years, he noticed that the product term actually consists of two terms, namely $-ijbc$ and $-jibc$. If we assume that $ij = -ji$, the unwanted term disappears. The crucial insight, which came to Hamilton in a sudden flash, as we shall see shortly, was the realization that he could break the commutative law of multiplication and still be left with a consistent mathematical structure.

The next question is: what is ij itself? Now,

$$ij.ij = i(ji)j = -i(ij)j = -(i^2)(j^2) = -(-1)(-1) = -1$$

so it appears that ij is yet another independent square root of -1; let us call it k. As the product of two number triples will, in general, involve this new 'imaginary' number k, we are led at once to the concept of 'number quadruples' (a, b, c, d) of the form $a + ib + jc + kd$, where $i^2 = j^2 = k^2 = -1$. These three square roots of -1 have a symmetrical relationship:

$$ij = k, \quad jk = i, \quad ki = j \quad \text{and} \quad ji = -k, \quad kj = -i, \quad ik = -j$$

Since $ij = k$, the result $ijk = -1$ follows at once.

Hamilton called these number quadruples *quaternions*. They obey all the fundamental laws of arithmetic with the single exception of the commutative law of multiplication. The operation of division can be defined by using the fact that

$$(a + ib + jc + kd).(a - ib - jc - kd) = a^2 + b^2 + c^2 + d^2$$

A quaternion can be used as an operator to change any directed line from the origin in three-dimensional space into any other such line. The four numbers a, b, c and d are sufficient for the purpose, whereas number triples are not.

Hamilton regarded his discovery of quaternions, after a gestation period of some fifteen years, as his greatest achievement. He describes his 'flash of inspiration' in a letter to one of his sons:

On the 6th day of October, which happened to be a Monday, and Council day of the Royal Irish Academy, I was walking to attend and preside, and your mother was walking with me along the Royal Canal; and although she talked with me now and then, yet an undercurrent of thought was going on in my mind, which gave at last a result, whereof it is not too much to say that I felt at once the importance. An electric circuit seemed to close; and a spark flashed forth, the herald (as I foresaw immediately) of many long years to come of definitely directed thought and work . . . Nor could I resist the impulse – unphilosophical as it may have been – to cut with a knife on a stone of Brougham Bridge, as we passed it, the fundamental formula with the symbols i, j, k: namely $i^2 = j^2 = k^2 = ijk = -1$, which contains the solution of the problem.

He goes on to say that he immediately obtained permission to read a paper at the Academy on quaternions, which he did at its next meeting, on 13 November 1843.

A quaternion has two parts: the single real number a, called the *scalar* part, and the *vector* part, $ib + jc + kd$, which can be interpreted as a directed line from the origin in three-dimensional space. In the course of time the two parts became conceptually separated, and the algebra of vectors became independent of the algebra of quaternions. The leaders in the development of *vector analysis* were the mathematical physicists, notably James Clerk Maxwell (1831–79) in Britain and Willard Gibbs (1839–1903) in America. There was fierce controversy between the 'quaternians' and the 'vectorians', with the latter finally carrying the day. The historical importance of quaternions is, however, secure. Once Hamilton had breached the principle of the permanence of equivalent forms, the way was open for the creation of all kinds of new abstract algebras, as we shall see.

5. George Boole

George Boole (see Illustration 17) – Russell's discoverer of pure mathematics – was born in 1815 in Lincoln, the son of an unsuccessful small shopkeeper. He attended the local National School and soon decided that the only way to get out of his miserable situation was to acquire some knowledge of Latin and Greek. Apart from a little help from the local bookseller, he taught himself the hard way. At the age of 16, faced with the urgent necessity of supporting his impoverished parents, Boole managed to get his foot on the bottom rung of the middle-class ladder: he became a teacher, and spent three years teaching in elementary schools. Meanwhile, he continued his grinding routine of private study in the hope of becoming a clergyman. His parents' increasing poverty forced him to abandon this ambition and, at the age of 19, he opened a school of his own. The need to teach his pupils mathematics aroused his interest in the subject; his innate ability enabled him to read and to understand, entirely on his own, some of the most difficult works of such masters as Laplace and Lagrange. It was not long before this remarkable young man was making discoveries of his own, mainly in the field of what we now call 'abstract algebra'.

Boole's greatest achievement was to create an algebra of logic, which he expounded in two books. The first was a slim volume entitled *The Mathematical Analysis of Logic, Being an Essay Towards a Calculus of Deductive Reasoning*, published in 1848, followed in 1854 by his masterpiece, *An Investigation of the Laws of Thought, on Which are Founded the Mathematical Theories of Logic and Probabilities* (see Illustration 18). A bold title indeed! It was probably the good reception accorded to the earlier work, supported by a number of miscellaneous mathematical papers, that eventually enabled Boole to get out of schoolteaching. In 1849 he was appointed Professor of Mathematics at the recently opened Queen's College in

Illustration 17. George Boole, 1815–64 (Mary Evans Picture Library)

the city of Cork. Although his academic load was a heavy one, he was able to use his new freedom from acute financial anxiety and from unremitting grind to excellent effect in polishing his *magnum opus*.

In 1855 Boole married Mary Everest, the niece of a colleague, the Professor of Greek. She became his devoted helpmate and disciple. After her husband's death she published their joint ideas on the more humane education of young children. Boole died in 1864 in his fiftieth year, honoured and famous at last. His premature death was caused by pneumonia, contracted after keeping a lecture engagement in severe weather.

6. The algebra of logic and sets

Boole begins his treatise *The Laws of Thought* with these words:

The design of the following treatise is to investigate the fundamental laws of those operations of the mind by which reasoning is performed; to give expression to them in the language of a Calculus, and upon this foundation to establish the science of Logic and construct its method; to make that method itself the basis of a general method for the application of the mathematical doctrine of probabilities; and, finally, to collect from the various elements of truth brought to view in the course of these enquiries some probable intimations concerning the nature or constitution of the human mind . . . There exist, indeed, certain general principles founded in the very nature of language, by which the use of symbols, which are but the elements of scientific language, are determined.

He goes on to say that the choice of symbols is, to a large extent, arbitrary, and their interpretation purely conventional, provided that in our reasoning we adhere strictly to the agreed operational rules – or 'laws', as Boole calls them. He continues:

In accordance with these principles, any agreement which may be established between the laws of the symbols of Logic and those of Algebra can but issue in an agreement of processes. The two provinces of interpretation remain apart and independent, each subject to its own laws and conditions.

Boole then proceeds to establish a new algebra, now known as *Boolean algebra* – the algebra of sets and of symbolic logic. Boole uses the letters x, y, z to represent the various subsets that can be selected from a complete or 'universal' set, which he calls the *universe of discourse*, denoted by U. He uses the symbol '1' to represent the set U, and the symbol '0' to represent the empty set or *null set*. The

AN INVESTIGATION

OF

THE LAWS OF THOUGHT,

ON WHICH ARE FOUNDED

THE MATHEMATICAL THEORIES OF LOGIC AND PROBABILITIES.

BY

GEORGE BOOLE, LL.D.

PROFESSOR OF MATHEMATICS IN QUEEN'S COLLEGE, CORK.

LONDON:

WALTON AND MABERLEY,

UPPER GOWER-STREET, AND IVY-LANE, PATERNOSTER-ROW.

CAMBRIDGE: MACMILLAN AND CO.

1854.

Illustration 18. Title page of Boole's *The Laws of Thought* (1854)

symbol '$=$' denotes the identity relationship. Thus, for example, x might represent the set (or class, or collection) of all Frenchmen over 60, y the set of all left-handed Frenchmen and z the set of all Frenchmen living in Paris. An appropriate universal set in this case might be the collection of all (living) Frenchmen. The *union* of x and y, i.e. the subset made up of those elements that are in either x or y, or both, is denoted by $(x + y)$; the *intersection* of x and y, i.e. the subset containing those elements that are in both x and y, is denoted by $x \times y$, or simply by $x . y$. The complement of x, denoted by $\bar{x}$, is the subset consisting of all the elements of U that are *not* in x. Clearly, then, $x + \bar{x} = 1$. Thus, in our example $(x + y)$ denotes the set of all Frenchmen who are either over 60 or left-handed, or both; $x . \bar{z}$ denotes the set of all elderly Frenchmen who live outside the capital city; while $(x . \bar{y} + \bar{x} . y)$ denotes the set of those Frenchmen who are either over 60, or are left-handed, but excludes those who have both these characteristics. Relations between subsets may be exhibited geometrically by means of *Venn diagrams*, named after the Cambridge logician John Venn (1834–1923). Figure 15.1 shows six such diagrams: they should be self-explanatory.

Although the five fundamental laws of ordinary algebra listed on pp. 335–36 remain valid in Boolean algebra, some other rules do not. Here are a few examples:

(1) $1 + 1 = 1$ (the universal set);

(2) $x + x = x . x = x$ if x is any subset of U;

(3)   the Boolean equation $x . y = 0$ does not imply that either $x = 0$ or $y = 0$, only that the two subsets have no elements in common;

(4)   in Boolean algebra, addition is distributive over multiplication, i.e. $x + y . z = (x + y).(x + z)$, as should be clear from Figure 15.2.

The rules of classical logic, as embodied in the syllogisms of Aristotle, are in fact statements about relationships between classes or sets; they can all be neatly expressed in Boolean terms. Thus the Boolean equation $x . y = x$ tells us that 'all x's are y's'. If also 'all y's are z's', we can write $y . z = y$. These two equations yield $x . (y . z) = x$, or $(x . y) . z = x$. We can replace $x . y$ by x to give $x . z = x$, which

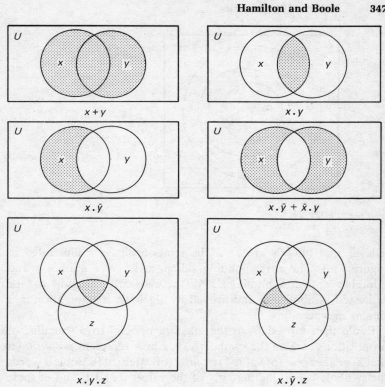

Figure 15.1.

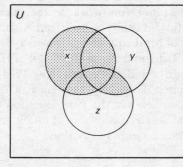

$x + y.z = (x + y).(x + z)$

Figure 15.2.

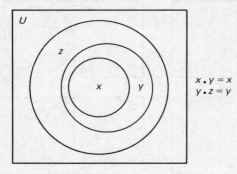

$$x \cdot y = x$$
$$y \cdot z = y$$

Figure 15.3.

tells us that 'all x's are z's'. The relationships are illustrated in Figure 15.3. The corresponding syllogism is: 'if x implies y and y implies z, then x implies z'. All the classical syllogisms can be expressed symbolically and proved by algebraic manipulation in a similar manner.

Boole then moved from the 'algebra of sets' to a 'calculus of propositions', where the symbols x, y, z now represent *propositions* that are either *true* (of value 1) or *false* (of value 0). In Boole's words the symbols 'admit indifferently of the values 0 and 1, and of these values alone'. With this interpretation, $x \cdot y = 1$ asserts that both x and y are true, while $x + y = 1$ asserts that either x or y is true, or that both are true. Boole's pioneering work on the calculus of propositions was continued after his death by his friend Augustus de Morgan (1806–71), Professor of Mathematics at University College, London, and in the U.S.A. by Benjamin Peirce (1809–80) of Harvard and his son, Charles (Santiago) Sanders Peirce (1839–1914). Both de Morgan and the elder Peirce discovered the 'laws of duality', whereby the intersection of two sets (or the joint assertion of two propositions) can be converted into a relationship of union, and vice versa. The laws may be expressed symbolically thus:

$$\overline{x \cdot y} \; = \; \overline{x} + \overline{y} \quad \text{and} \quad \overline{x + y} \; = \; \overline{x} \cdot \overline{y}$$

For every proposition involving logical addition and multiplication, there is a dual proposition in which the words addition and multiplication are interchanged. The validity of the duality laws may easily be demonstrated by Venn diagrams. Benjamin Peirce worked out multiplication tables for no fewer than 162 different algebras – a far cry from Peacock's principle of the permanence of equivalent forms. Charles Peirce showed that in only three of them – the algebras of real numbers, of complex numbers and of quaternions – is division uniquely defined.

The subject of mathematical logic was intensively cultivated during the early years of the twentieth century. Bertrand Russell and Alfred North Whitehead's monumental three-volume *Principia mathematica* appeared in 1910–13. In the middle of the century the study of Boolean algebra was given a new lease of life by the advent of the electronic computer. The basic building blocks of such computers are two-state devices known as switching circuits, each of which is, at any instant, either 'on' or 'off'. Boolean algebra is therefore the obvious basis for the 'logical design' (as opposed to the physical design) of computers, and is used to analyse the behaviour of complex assemblages of such two-state circuits.

Another application of Boolean algebra is to probability theory. The probability of an event occurring may, in general, assume any value betwen 0 and 1 inclusive. If p_1 and p_2 are the probabilities of occurrence of two independent events, the probability of both events occurring is given by $p_1 . p_2$ (intersection), and the probability of at least one occurring by $\{1 - (1 - p_1)(1 - p_2)\} = p_1 + p_2 - p_1 . p_2$. This is the counterpart of Boole's operation of 'union', but is not quite the same. The third term is clearly needed since $(p_1 + p_2)$ can exceed 1. Many probability problems are combinatorial in nature, being concerned with ways of combining events and counting possible outcomes. Such concepts can be elegantly expressed in terms of sets. Thus the possible outcomes of throwing a die form a set of six elements which may be denoted by $X = \{1, 2, 3, 4, 5, 6\}$. Any particular outcome of one or more throws may be represented by a subset of one or more of the X-sets, and such subsets may be combined and dissected on Boolean lines to give the probabilities of specified outcomes.

7. Matrices

Probably the best-known algebraic structure which breaches Peacock's principle is *matrix algebra*, first studied by Arthur Cayley (1821–95) at Cambridge. His first memoir on the theory of linear transformations was published in 1855. If, for example, the transformation

$$T_1 \begin{cases} x' = ax + by \\ y' = cx + dy \end{cases} \tag{15.4}$$

is followed by another transformation

$$T_2 \begin{cases} x'' = px' + qy' \\ y'' = rx' + sy' \end{cases} \tag{15.5}$$

the combined effect is equivalent to a single transformation

$$T_2 T_1 \begin{cases} x'' = (pa + qc)x + (pb + qd)y \\ y'' = (ra + sc)x + (rb + sd)y \end{cases} \tag{15.6}$$

If, however, we reverse the order in which the transformations are applied, we get a different result, namely the transformation

$$T_1 T_2 \begin{cases} x'' = (ap + br)x + (aq + bs)y \\ y'' = (cp + dr)x + (cq - ds)y \end{cases} \tag{15.7}$$

In the shorthand language of matrix algebra, equations (15.4) and (15.5) are written as

$$\begin{pmatrix} x' \\ y' \end{pmatrix} = M_1 \begin{pmatrix} x \\ y \end{pmatrix} \quad \text{and} \quad \begin{pmatrix} x'' \\ y'' \end{pmatrix} = M_2 \begin{pmatrix} x' \\ y' \end{pmatrix}$$

where M_1 and M_2, the matrices of the transformations T_1 and T_2, are given by

$$M_1 = \begin{pmatrix} a & b \\ c & d \end{pmatrix} \quad \text{and} \quad M_2 = \begin{pmatrix} p & q \\ r & s \end{pmatrix}$$

The product $M_2 M_1$, another matrix, is defined so that equation (15.6) may be written as

$$\begin{pmatrix} x'' \\ y'' \end{pmatrix} = M_2 M_1 \begin{pmatrix} x \\ y \end{pmatrix}$$

while the reverse product, $M_1 M_2$, is defined similarly to accord with equation (15.7). This yields, in matrix notation,

$$M_1 M_2 = \begin{pmatrix} a & b \\ c & d \end{pmatrix} \begin{pmatrix} p & q \\ r & s \end{pmatrix} = \begin{pmatrix} ap + br & aq + bs \\ cp + dr & cq + ds \end{pmatrix}$$

and

$$M_2 M_1 = \begin{pmatrix} p & q \\ r & s \end{pmatrix} \begin{pmatrix} a & b \\ c & d \end{pmatrix} = \begin{pmatrix} pa + qc & pb + qd \\ ra + sc & rb + sd \end{pmatrix}$$

Thus, in general, $M_1 M_2$ is not equal to $M_2 M_1$: matrix multiplication is a non-commutative operation.

8. The liberation of algebra

The main theme of this chapter has been the liberation of algebra from the straitjacket of arithmetic. The branch of mathematics known as *abstract algebra* was born and reared, as we have seen, in the nineteenth century; it came to maturity in the twentieth. Its subject matter is collections of discrete objects, such as sets, vectors, quaternions or matrices, as well as real or complex numbers, or some prescribed subset of them. These objects, or elements, are usually represented by symbols which can be manipulated or combined in accordance with operational rules which are specific to a particular collection. The rules themselves are to be taken as axioms – as something prescribed and accepted at the outset. They can be formulated, largely at will, by the creator of the algebraic structure – subject, of course, to the requirements of consistency and adequacy.

With the proliferation of algebraic structures during the nineteenth century, the need for some unifying concepts became increasingly apparent. One of the most powerful of these is that of the *group*. To define a group we need a set of elements and an operation (e.g. multiplication) for combining any two elements so as always to yield an element which is also a member of the set. In mathematical parlance, the set is said to be *closed* with respect to the operation. Various other conditions must also be met. The branch of mathematics known as *group theory*, with its specialized component, *Galois theory*, named after the brilliant young mathematician (p. 298) who initiated the subject, has provided a most fertile field of research and application ever since the early nineteenth century. We cannot pursue this large and technically demanding subject here; the interested reader is referred to Reference 41.

16 Dedekind and Cantor

'No one shall ever expel us from the paradise which Cantor has created for us.' David Hilbert

1. The arithmetization of analysis

By the beginning of the nineteenth century a clear distinction had been established between *analysis*, the study of infinite processes, and *algebra*, which deals with operations on discrete entities such as the natural numbers. A major objective of much nineteenth-century mathematical effort was to unify – or, at any rate, to build bridges between – these two branches of mathematics. This endeavour was termed 'the arithmetization of analysis'. It was realized that the prime task was to construct a sound logical foundation for the real number system. Although the basic concepts of analysis – function, continuity, limit, convergence, infinity and so on – were progressively clarified and refined during the first half of the nineteenth century, notably by Cauchy (p. 298), much remained to be done.

We have seen (p. 17) that real numbers can be divided into algebraic and transcendental numbers. An algebraic number is defined as a number which is a root of a polynomial equation of the form

$$a_0 x^n + a_1 x^{n-1} + \ldots + a_{n-1} x + a_n = 0$$

where n and the a_i are integers. This poses the question of whether there are indeed any transcendental numbers – that is, any irrational numbers that are not algebraic. In 1844 Joseph Liouville (1809–82) answered the question decisively in the affirmative by actually constructing such numbers. He proved, for example, that all numbers of the form

$$\frac{a_1}{10^{1!}} + \frac{a_2}{10^{2!}} + \frac{a_3}{10^{3!}} + \cdots$$

where the a_i are arbitrary integers in the range 0–9, are non-algebraic, and therefore transcendental. In 1873 Charles Hermite (1822–1901) proved that the exponential constant, e, is transcendental; in 1882 Ferdinand von Lindemann (1852–1939) did the same for π. To this day, few other transcendental numbers are known. However, we do have one general result, discovered as recently as 1934: if a is an algebraic number (not equal to 0 or 1) and b is an algebraic irrational number, then a^b is transcendental ($3^{\sqrt{2}}$ is a simple example).

A pioneering advocate of rigour in analysis was a Czech priest, Bernhard Bolzano (1781–1848) of Prague, but his theory of real numbers had little influence. Karl Weierstrass (1815–97) of Berlin took the matter a stage further by defining irrational numbers as 'aggregates' of real numbers. His treatment is not easy to understand and was superseded, like that of Bolzano, by the simpler approach taken by Dedekind – the subject of Section 3. Weierstrass did, however, make one lasting contribution to analysis: he was the first to formulate the definition of a limit that is used today. He defined the limit of the function $f(x)$ at the point x_0 as follows:

If, given any ε, however small, there is a number δ_0 such that for $0 < \delta < \delta_0$, the absolute value of the difference $\{f(x_0 \pm \delta) - L\}$ is less than ε, then L is the limit of $f(x)$ for $x = x_0$.

In this precise statement there is no suggestion of fluxions, ultimate ratios, differentials or infinitely small quantities – only real numbers, the operations of addition and subtraction, and the concept of 'less than'. Weierstrass had finally banished infinitesimals from mathematics.

2. Richard Dedekind

Richard Dedekind (1831–1916) – Gauss' most talented pupil – was born (like Gauss) in Brunswick, where his father was Professor of

Law. After attending the local *Gymnasium*, Richard moved, like Gauss before him, to the Caroline College before entering the University of Göttingen in 1850 to study mathematics and physics. At the age of 21 he received his doctor's degree – from Gauss, of course. In 1854 Dedekind obtained an unpaid lectureship at Göttingen; four years later he was appointed an 'ordinary professor' at the Zurich Polytechnic. In 1862 he returned to Brunswick as professor at the town's Technical High School and remained in this relatively modest position for 50 years. He died in 1916 at the age of 84, active in mind and body to the last. He never married, but lived with his sister Julie, a now forgotten novelist, until her death in 1914.

Dedekind lived long enough to become a shadowy, even ghostly figure. Indeed, he was reported in a mathematical calendar as having died on 4 September 1899. Dedekind, much amused, wrote to the editor:

According to my own memorandum I passed this day in perfect health and enjoyed a very stimulating conversation on 'system and theory' with my luncheon guest and honoured friend Georg Cantor of Halle.

3. Dedekind's theory of irrational numbers

Most of Dedekind's researches were concerned with the nature of number in its general aspect. His most important work, *Continuity and Irrational Numbers*, was published in 1872. He had realized as early as 1858 that the system of real numbers lacked a firm logical foundation: the validity of such a simple statement as $\sqrt{2} \times \sqrt{3} = \sqrt{6}$, for example, had never been rigorously established. We have seen how the Greeks, with their geometrical predilections, identified real numbers – first rational and, later, also irrational numbers – with line segments. They tacitly assumed that any such number could be represented by a unique point on an infinitely extended straight line with a specified origin. The rational numbers presented no difficulty; the problem lay with the irrationals. We can best convey Dedekind's approach by quoting from his 1872 essay:

The straight line is infinitely richer in point-individuals than the domain of rational numbers is in number-individuals. The comparison of the domain of rational numbers with a straight line has led to the recognition of the existence of gaps, of a certain incompleteness or discontinuity in the former; while we ascribe to the straight line completeness, absence of gaps or continuity. Wherein then does this continuity consist? Everything must depend on the answer to this question.

He goes on to tell us how, after much pondering, he found the answer – what he calls the essence of continuity – in the following principle:

If all points of a straight line fall into two classes, so that every point of the first class lies to the left of every point of the second class, then there exists one and only one point which produces this division of all points into two classes, this severing of the straight line into two portions.

Such a severance, or partition, is now known as a *Dedekind cut*. He goes on to remark that this proposition may be taken to command universal assent: it is one of those truths that may be held to be self-evident. This is just as well, for he admits that he is

utterly unable to adduce any proof of its correctness, nor has anyone else the power. The assumption of this property of the line is nothing else than an axiom by which we attribute to the line its continuity, by which we define its continuity.

A more formal statement of this axiom, now known as the *Dedekind–Cantor axiom*, is that the points on a line can be put into one-to-one correspondence with the real numbers. That is, to any point on the line can be assigned a unique real number; and conversely, any real number can be uniquely represented by a point on the line.

Let us now apply the 'Dedekind cut' principle to the class of *rational* numbers. (In developing this theory, Dedekind naturally sought to proceed from the known to the unknown.) Such a 'cut' will divide the rational numbers into two classes, L and R say, such that every number in L is less than every number in R. What, then, can we say about these classes? There are three possibilities:

(1) L has a greatest member, say l;
(2) R has a least member, say r; or
(3) L has no greatest and R has no least member.

(The possibility that L has a greatest *and R* a least member is ruled out because the rational numbers are 'everywhere dense': between any two such numbers, however close together, a third number can always be found.) Dedekind's crucial assertion is that a cut of type (1) or type (2) defines a rational number (either l or r), while a cut of type (3) defines an irrational number. Suppose, for example, that the cut is made at the point which defines the number we write as $\sqrt{2}$. Then, for any number, x_0 say, in class L, we can write $x_0^2 = 2 - \delta$ where $\delta > 0$. However small δ may be, we can always find another rational number, x_1 say, in class L such that x_1^2 differs from 2 by less than δ (e.g. $x_1^2 = 2 - \frac{1}{2}\delta$), and so $x_1 > x_0$. It follows that there can be no largest member of L and, by a similar argument, no smallest member of R: the cut is of type (3).

If, on the other hand, the cut is made at a 'rational number' point, say 2, then the number 2 itself must be assigned either to class L or to class R. So either L will have a greatest or R a least member: the cut is of type (1) or (2).

By his method of partitioning the class of rational numbers, Dedekind achieved a new insight into the nature of irrational numbers – and so, by extension, of real numbers. The essence of his achievement was to define the real numbers in terms of the rational numbers.

Dedekind completed his theoretical edifice in a series of papers written in the 1870s and 80s. First, he extended the range of the basic concepts and operational rules of ordinary arithmetic and algebra, and proved that they retain their validity in the domain of real numbers. Secondly, he demonstrated that with real numbers we reach, as it were, the end of the road: that any Dedekind cut of the class of real numbers does no more than specify a unique real number – it does not yield anything new. In more technical terms, what he proved is that the 'arithmetic continuum' of real numbers is 'closed' for infinite processes. This important result is usually known as Dedekind's theorem. Although Dedekind retained the Greek geometrical model of the number system as an aid to thought and exposition, the aim of most nineteenth-century mathematicians was to exclude geometrical considerations altogether, to base virtually the whole of mathematics on the concept of number. Broadly speaking, this objective had been achieved by the mid-twentieth century, although some foundational difficulties – for example, with the logical basis of set theory – still remained.

Finally, we must note the direct line of descent from Eudoxus to Dedekind. Eudoxus, as we have seen (pp. 29–30), gave a procedure for deciding whether or not two ratios of magnitudes (which may be incommensurable) are equal, and if they are not, which is the greater. What he did, in effect, was to separate the class of rational magnitudes into two classes (L and R) by a 'cut' at the point defined by the rational number m/n. He asserted that the two ratios are equal if and only if they are always to be found on the same side of the cut, wherever the cut is made (i.e. for *any* integral values of m and n). Dedekind handsomely acknowledged his debt to Eudoxus, and, more generally, to Book V of Euclid's *Elements*.

4. Georg Cantor

While Dedekind was quietly engaged in clarifying our perceptions of the real numbers, his friend Cantor was arousing acute controversy by his studies of infinite sets of such numbers. Georg Cantor (1845–1918) was born at St Petersburg. Both his parents were of Jewish descent; his father, a merchant, had come from Copenhagen and converted to Protestantism, while his artistic and musical mother was born a Roman Catholic. In 1856 the family moved to Frankfurt and Georg entered the Wiesbaden *Gymnasium* at the age of 15. He was already determined to be a mathematician but his father tried to force him into engineering. At first Georg submitted to parental pressure, but eventually his father – confronted with his son's determination and impressed by his mathematical abilities – accepted the inevitable. Georg began his university career at Zurich in 1862. A year later, on the death of his father, he moved to the University of Berlin to study mathematics, physics and philosophy. His early researches were in number theory, where he developed some ideas that had been suggested by Gauss. He then moved on to the study of trigonometric series. This inspired him to go deeper into the foundations of analysis; his first, revolutionary paper on the theory of infinite sets appeared in 1874 and marked the beginning of some 25 years of mathematical research of the highest originality.

In 1869 Cantor obtained an unpaid lecturing post at the minor University of Halle. He moved steadily up the academic ladder to a full professorship, but was not able to escape from this provincial backwater. He never achieved his cherished ambition of a Chair at the University of Berlin – at that time the world's centre of mathematical activity. As time went on, his theories aroused considerable opposition, and even some malicious personal attacks, notably from Kronecker at Berlin. Even as late as 1908 we find the great Henri Poincaré dismissing Cantor's theory of infinite sets as 'a disease from which one has recovered'.

Cantor married in 1874 and had two sons and four daughters. He spent his honeymoon at Interlaken, where he met Dedekind. Their originality of thought, common mathematical interests and rejection by the academic establishment drew the two men together, and they became close and lifelong friends. Ten years later Cantor suffered the first of the mental breakdowns that were to plague him for the rest of his life. He died in 1918 in a mental hospital at Halle. By that time his revolutionary ideas were becoming accepted by some of the leading figures of the new century – as exemplified by the quotation at the head of this chapter. Indeed, Hilbert went so far as to describe Cantor's new arithmetic, to which we now turn, as 'the most astonishing product of mathematical thought, one of the most beautiful realizations of human activity in the domain of the purely intelligible.'

5. Cantor's theory of infinite sets and transfinite numbers

From the time of the Greeks to the late nineteenth century, mathematicians drew a sharp distinction between the 'potentially infinite' and the 'actually infinite'. The former was a central concept of analysis; the latter was to be avoided like the plague. Here, for example, is Gauss writing to a colleague in 1831:

As to your proof, I must protest most vehemently against your use of the infinite as something consummated, as this is never permitted in mathematics. The infinite is but a figure of speech: an abridged form for the statement that limits exist which certain ratios may approach as closely as we desire, while other magnitudes may be permitted to grow beyond all bounds.

Indeed, discussion of the 'paradoxes of the infinite' goes back at least as far as Zeno in the fifth century BC. Some of the curious properties of infinite collections were expounded by Galileo in his *Dialogues Concerning Two New Sciences* of 1636, where he explains how the squares of the positive integers can be put into one-to-one correspondence with the integers themselves. This implies that an infinite set is 'equal', in some sense, to a part of itself – a most disturbing state of affairs! Galileo sums up the discussion thus:

So far as I can see, we can only infer that the number of squares is infinite and the number of their roots is infinite; neither is the number of squares less than the totality of all numbers, nor the latter greater than the former; and finally the attributes 'equal', 'greater' and 'less' are not applicable to infinite, but only to finite quantities.

There the matter rested for over 200 years until a book by Bolzano (p. 354) entitled *Paradoxien des Unendlichen* (Paradoxes of the Infinite) appeared in 1850. Some forty years later Dedekind suggested that the property of a set being 'equal' in size to a part of itself be adopted as the definition of an infinite set. Cantor took the matter a stage further when he proved that not all infinite sets are the same size. He expressed the size – or 'power' as he called it – of an infinite set by means of a *transfinite* number. He started with the infinite set of the natural numbers (i.e. the positive integers), and denoted its 'size' by the transfinite number $\aleph_0$ ($\aleph$ is 'aleph', the first letter of the Hebrew alphabet).

We have seen how Galileo showed that the natural numbers may be matched with the 'square numbers'. They may also be matched with the set of both positive and negative integers, with the odd numbers, with the integers greater than a million or, indeed, with any infinite set put together by selecting some of the whole numbers. All such sets are said to be *countable*, or *enumerable*, and their size is expressed by the transfinite number $\aleph_0$.

The first surprise is that the set of rational fractions is also countable: the fractions can be put into one-to-one correspondence with the natural numbers. Figure 16.1 shows how this can be done. (We have, for simplicity, limited ourselves to positive fractions.) The

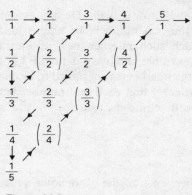

Figure 16.1.

rational fractions are arranged in order in a precise two-dimensional array as shown. To count them we merely follow the arrows, ignoring duplicates (shown in brackets). Cantor's next achievement was to prove that the set of all 'algebraic numbers' is countable. An algebraic number, as we have seen (p. 353), is a root of a polynomial equation

$$a_0 x^n + a_1 x^{n-1} + \ldots + a_{n-1} x + a_n = 0$$

where the a_i are integers. To any such equation Cantor assigned an integer, h, which he called the *height*, and defined it as

$$h = (n - 1) + |a_0| + |a_1| + \ldots + |a_n|$$

For example, the height of the equation $2x^3 - 4x + 7 = 0$ is $(3 - 1) + 2 + 4 + 7 = 15$. Cantor's argument is that to each value of h there corresponds only a finite number of algebraic equations, and hence only a finite number of algebraic numbers, say n_h. Starting with $h = 1$, we can label the corresponding algebraic numbers from 1 to n_1; the algebraic numbers of height 2 can then be labelled from $n_1 + 1$ to $n_1 + n_2$, and so on. In this way, each algebraic number will be reached in turn and will be paired with a unique integer. Thus the set of algebraic numbers is countable.

This unexpected result posed the question of whether the set of all real numbers is countable. In his 1874 paper Cantor was able to answer this question in the negative; some years later he gave an improved proof, which we shall now explain. He first assumes, as a working hypothesis, that the real numbers between 0 and 1 *are* countable. Let us write each such number as a non-terminating decimal (which means, for example, that we must write $\frac{1}{2}$ not as 0·5, but as 0·499 99 . . . . Since the numbers are assumed to be countable, we can arrange them in order so that each real number between 0 and 1 is matched with a positive integer, thus:

$$1 \quad \rightarrow 0 \cdot a_{11} a_{12} a_{13} \ldots$$
$$2 \quad \rightarrow 0 \cdot a_{21} a_{22} a_{23} \ldots$$
$$3 \quad \rightarrow 0 \cdot a_{31} a_{32} a_{33} \ldots$$

and so on, *ad infinitum*, where each of the a's denotes a digit in the range 0 to 9. Cantor's master-stroke was to exhibit a real number (between 0 and 1) that is different from all those listed. The procedure, known as the 'diagonal method', is delightfully simple. All we need do is to form a non-terminating number b, where $b = 0 \cdot b_1 b_2 b_3 \ldots$, and choose its digits such that $b_1 \neq a_{11}$, $b_2 \neq a_{22}$, $b_3 \neq a_{33}$ and so on. It follows that the real number b differs by at least one digit from every number in the list, which was assumed to contain *all* the real numbers between 0 and 1. So we have established a contradiction, and our initial assumption must be false. The conclusion is inescapable: the real numbers are *not* countable. Now, we know that the algebraic numbers are countable while the real numbers are not. It follows, therefore, that an uncountable collection of transcendental numbers must exist. Cantor's proof is what is known as a 'non-constructive existence proof' – and as such is not accepted, even today, by some mathematicians – in contrast to Liouville's constructive proof (p. 354) in which he presents some specific transcendental numbers. The transfinite number of the set of real numbers must therefore be greater than $\aleph_0$; let us, following Cantor, call it c, for the continuum.

Cantor had still not finished. He was able to prove some further surprising results, among them that the set of *all* real numbers (i.e. the set of points on an infinitely extended line) can be matched with the real numbers on any segment of the line, however short. We can show this geometrically. In Figure 16.2, let LOL′ be an unlimited line, and AB any segment of it, positioned obliquely through the

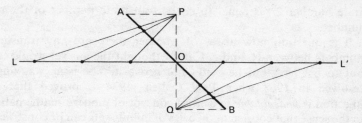

Figure 16.2.

origin O. By drawing pencils of lines from P and Q as shown, it is apparent that there is a one-to-one correspondence between points on AB and points on LL′. There are yet more surprises in store. It turns out that the 'size' (i.e. the transfinite number) of the set of points in a line segment of unit length, say, is the same as that of the set of points in a unit area, or a unit volume, or indeed of a unit 'quantity' of any number of dimensions. This result seemed so paradoxical that Cantor asked Dedekind in 1877 to check the proof. 'I see it, but I don't believe it,' he wrote.

So far we have established the existence of only two transfinite numbers, $\aleph_0$ and c. Cantor was able to prove that there are infinitely many such numbers. He did this by investigating the 'size' of the set of all possible subsets that can be selected from an infinite set. Let us first look at the situation for a finite set: for example, a set consisting of just three elements, a, b and c. From these we can select 8 (i.e. 2^3) subsets, namely abc; bc, ca and ab; a, b and c; and $\emptyset$ (the generally accepted symbol for the empty set). (It is a convention that the first and the last of these are to be included among the subsets.) For a set of n elements, the set of all possible subsets contains 2^n elements, where clearly $2^n > n$. Cantor was able to extend this result to infinite sets, in particular to establish the transfinite number of the set of all possible subsets of the natural numbers. Let us call it, by analogy with the situation for finite sets, $2^{\aleph_0}$. The argument concerning 'the set of all possible subsets' can be extended indefinitely to yield a sequence of increasingly 'large' transfinite numbers. Cantor was also able to prove that the transfinite number $2^{\aleph_0}$ is equal to c, the transfinite number of the real numbers.

The question now arises of whether there are any transfinite numbers between $\aleph_0$ and c. Cantor thought the answer would turn out to be 'No', but he could not prove it. The issue was finally resolved in 1963 when Paul Cohen (1934–) proved that the question is *undecidable* (a central concept of modern mathematics), in the sense that consistent theories of infinite sets can be constructed which either accept or deny the assumption – known as the *continuum hypothesis* – that there are no transfinite numbers between $\aleph_0$ and c. Its status is, therefore, that of an independent axiom of set theory, just as Euclid's parallel axiom was shown by Gauss and others to be an independent axiom of classical geometry.

The transfinite numbers we have been discussing so far are known as transfinite *cardinal* numbers. In a series of papers published between 1879 and 1897 Cantor introduced the concept of transfinite *ordinal* numbers. He developed a subtle theory of the relations between the two kinds of transfinite numbers, which is beyond the scope of this book. (Further information can be found in References 33 and 38.) Suffice it to say that Cantor was able to establish an 'aleph hierarchy' of transfinite cardinal numbers $\aleph_0, \aleph_1, \aleph_2 \ldots$, the smallest 'larger' such number being taken at each stage. The continuum hypothesis is that $\aleph_1 = c \equiv 2^{\aleph_0}$.

The study of the properties of infinite sets, with their paradoxes and apparent contradictions, has been a major mathematical activity of the twentieth century. We cannot go into the details of this highly technical field, but to give a glimpse of the kind of problems that arise, we end this section with a well-known story. A village has only one barber who claims that he shaves every man in the village who does not shave himself. The question is: who shaves the barber? Either answer leads to a contradiction. A recent suggestion is that the barber is a woman. This paradox is a simple example of a situation where the use of such words as 'all' or 'every' can set a trap for the unwary. Indeed, the argument as to how such difficulties are to be resolved has divided mathematical logicians into contending camps for most of this century.

6. The nineteenth century and after

During the whole of the nineteenth century mathematics advanced on a very broad front. Many new areas of research were opened up and the range of existing ones extended; the number of professional mathematicians and their research journals increased enormously. Although analysis, with its many specialized branches, continued to dominate the scene, algebra broke much new ground, as we saw in the last chapter. There was also a strong revival of geometry. The advent of non-Euclidean geometry (p. 332) provided a powerful stimulus to new thinking, as did the new approaches to projective geometry (neglected since the mid-seventeenth century) under the leadership first of Poncelet (p. 298) and others in France, and then of a strong German–Swiss school. Regrettably, the technical and conceptual complexity of most of these developments puts them beyond the scope of this book. In the last three chapters we have, so to speak, tasted a cupful of water from the vast reservoir of nineteenth-century mathematics; the adequately equipped reader who thirsts for more is referred to the list of references, particularly References 20, 21, 38 and 43.

The expansion and diversification of mathematical activity continued unabated into the twentieth century. Around the turn of the century many of the leading mathematicians, stimulated by the controversial insights of Cantor, sought to dig still deeper into the logical and philosophical foundations of their subject. There was much diversity of opinion and vigorous debate on the most fundamental questions. Three main schools of thought emerged: the Logicists, led by Bertrand Russell (1872–1970) and Alfred North Whitehead (1861–1947); the Intuitionists, founded by Leopold Kronecker (1832–92) (remembered for his aphorism quoted on p. 15), given qualified support by the great Henri Poincaré (1854–1912), and then vigorously led by Luitzen Brouwer (1881–1961) and Hermann Weyl (1885–1955); and the Formalists under the powerful leadership of David Hilbert (1862–1942).

Following Reference 38, we can describe these three mathematical philosophies very briefly as follows. The Logicists believed that mathematics can be derived from logic, and is indeed an extension of logic. The Intuitionists adopted a radically different approach: they conceived of a fundamental mathematical intuition based on our ability to perceive a sequence of events ordered in time (so giving us our notion of the natural numbers, as an obvious example). According to this view, mathematics is a constructive process that builds its own universe, independent of the world of everyday experience; mathematical ideas are embedded in the human mind, outside which they have no separate existence. The Formalists took something of a middle position. For them, mathematics consists of several branches, each a formal deductive system with its own axioms, concepts, rules and theorems. The objects of mathematical thought reduce to symbolic elements which no longer stand for idealized physical entities.

The 'optimists' among the contending parties were dealt a devastating blow in 1930 when a young Austrian logician, Kurt Gödel (1906–78), proved that no formal axiomatic system adequate to embrace arithmetic and number theory can be both consistent and complete. If such a system is consistent, then there must be some theorems which can be neither proved nor disproved; they are *undecidable* and so some problems are logically unsolvable. (Cohen's discovery, mentioned in the last section, established the continuum hypothesis as such an undecidable proposition.) In 1933 Gödel proved a second negative theorem: that there is no constructive procedure whereby an axiomatic system can establish its own consistency, i.e. freedom from internal contradictions. It was all very disturbing, especially to the Formalists. However, after the dust had settled most mathematicians took a deep breath and continued to pursue their specialized researches as before. Further information on Gödel's theorems can be found in References 5, 20 and 38.

Another striking feature of the twentieth century is the vast expansion in the applications of mathematics, not only to physical science (see the next chapter) and engineering, but increasingly to biology and medicine, and to social and economic problems. Since the 1950s the electronic computer has become an indispensable tool for many applied mathematicians. Even so, one would not expect the computer to contribute to the resolution of a 'classical' unsolved problem – Fermat's last theorem, for example – but we conclude this chapter with a note on a recent achievement which belies this expectation.

The four-colour theorem (or conjecture) asserts that four colours are sufficient to colour any map on a plane so that areas with common boundaries are coloured differently. The question was first raised in 1852 by one Francis Guthrie, who passed it on to his brother Frederick, a student of Professor Augustus de Morgan; Frederick brought it to his professor's attention, and de Morgan subsequently consulted a number of others, including William Hamilton. During the hundred years that followed, a variety of alleged proofs of the conjecture were put forward, but all were eventually found to be defective. (The full story is given in Reference 24.) Everyone believed the conjecture to be true, but no one could provide a conclusive proof. But in 1976 Kenneth Appel, Wolfgang Haken and John Koch of the University of Illinois presented their computer-assisted proof to a mathematical congress in Toronto, shortly after which they published a full account of their work in a mathematical journal.

The proof, which uses a process of successive reduction leading eventually to a contradiction (something like Euler's number theory proof given on p. 150), required more than 1000 hours of time on a large computer to check some 100 000 different cases (more details are given in Reference 2). The Toronto exposition was coolly received; many in the audience rejected a 'proof' that could not be checked in detail by any one person. Others took the position that traditional ideas as to what constitutes mathematical proof needed revision in the computer age. No error has been found in the Illinois proof, but the debate about its acceptability seems likely to continue for some time. Perhaps in the fullness of time the computer will help us to solve other 'real' mathematical problems.

17 Einstein

'The greatness of Einstein lies in his tremendous imagination, in the unbelievable obstinacy with which he pursues his problems.' L. Infeld

'To me he appears as out of comparison the greatest intellect of this century, and almost certainly the greatest personification of moral experience.' C. P. Snow

'Future historians . . . may find themselves wondering how it happened that an abstract scientist, whose work could be understood only by a handful of people, should nevertheless have become an idol of millions so that his name and face were known all over the globe.' H. le R. Finch

EINSTEIN ON THE WORLD

'The most incomprehensible thing about the world is that it is comprehensible.'

'God is subtle but he is not malicious.'

EINSTEIN ON HIMSELF

'I have no special gift, I am only passionately curious.'

'My passionate sense of social justice and social responsibility has always contrasted oddly with my pronounced lack of need for direct contact with other human beings and human communities. I am truly a "lone traveller" and have never belonged to my country, my friends, or even my immediate family with my whole heart.'

1. Introduction

This final chapter, which brings us to the fateful twentieth century, may be viewed as an epilogue. It differs from the previous chapters in that its subject is not primarily a mathematician, but a theoretical physicist. Nevertheless, his mathematical exposition of the General

Theory of Relativity is a veritable *tour de force*, but one far beyond the scope of this book; much of his finest work can be presented here only in general terms. As partial compensation, we shall give a fairly full account of his eventful and moving life.

Albert Einstein (see Illustration 19) is unquestionably the most widely known scientist since Darwin, and arguably the best loved of all time. During his lifetime he became a living legend, a folk hero, an oracle, and was hailed as a 'new messiah'. The first three quotations at the head of the chapter give some idea of the very special veneration and affection in which he was held by his contemporaries.

2. A simple genius

The old city of Ulm lies on the Danube, in what was formerly the state of Württemberg. It has three claims to fame: its cathedral has the tallest spire in Europe, it was the scene of a great Napoleonic victory in 1805, and it was the birthplace of Albert Einstein. This last event took place on 14 March 1879, in a house in the Bahnhofstrasse that was destroyed in an air raid in 1943. Einstein's ancestral roots, on both sides, were in the local Jewish community that had lived amicably with and prospered among its Swabian neighbours since the sixteenth century. A year after Albert's birth his father's business collapsed and the family moved to Munich – the future cradle of Nazism – where his father and his uncle Jacob set up a small electro-chemical works. His freethinking, easygoing parents sent Albert and his sister Maja to the nearby Catholic elementary school. Character-istically, Einstein reacted strongly and became, for a time, acutely conscious and intensely proud of his Jewish cultural and religious heritage.

From an early age Einstein was a 'loner', an outsider. He acquired a lifelong antipathy to externally imposed discipline and coercion, to military parades and uniforms, and to current teaching methods with their emphasis on rote learning. He was a late developer, with a poor memory for words (and for Greek texts in particular) but with a strong sense of wonder and awe when confronted with the natural world. He later recalled an incident when he was ill in bed at the age of 5. His father showed him a pocket compass, and the child was fascinated by the thought that since the needle aways pointed in the same direction it must be acted on by an invisible 'something' that could only exist in 'empty' space. 'I can still remember – at least I believe I can remember – that this experience made a deep and abiding impression on me.' At the age of 6, as a result of his mother's influence, he began to learn to play the violin. His delight in the instrument was a source of solace and comfort to him throughout his life. His favourite composer was Mozart.

In 1889 Einstein entered the Luitpold *Gymnasium*, a typical German high-school of the time. His six years there taught him the virtues of scepticism and confirmed his dislike of authority – especially of educational authority. He learnt little from his teachers but taught himself a lot of mathematics and physics. His love of these subjects was first encouraged by his uncle Jacob, and later by a medical student friend of the family, who lent Albert books. Like Bertrand Russell and many another, Einstein became entranced by a small textbook on Euclid – his 'holy geometry book' as he called it. He tells us that:

Here were assertions, as for example the intersection of the three altitudes of a triangle in one point, which – although by no means evident – could nevertheless be proved with such certainty that any doubt appeared to be out of the question. This lucidity and certainty made an indescribable impression on me.

He read popular science books and several mathematical texts 'with breathless attention'. One result was that he became, for a while, intensely antireligious.

In 1894 the Einstein brothers and their families moved across the Alps to Milan to try their luck with another factory. Albert was left in lodgings in Munich to finish the school year. The circumstances of his departure from the *Gymnasium* at the age of 15 are obscure, but he certainly left before acquiring the necessary diploma to ensure his entry to a university. He was undoubtedly a difficult pupil: argumentative, opinionated, not concealing his dislike – even contempt – for most of his teachers and their pedagogic methods. The sentiments were mutual. His Greek teacher told him: 'You will never

Illustration 19. Albert Einstein, 1879–1955 (Mary Evans Picture Library)

amount to anything.' Einstein describes in a letter written in 1940 how he was summoned by his form master and asked to leave the school at once: 'Your mere presence spoils the respect of the class for me.'

The year Einstein spent in Italy was an extremely happy one. Forty years later he was to write:

I was so surprised, when I crossed the Alps to Italy, to see how the ordinary Italian, the ordinary man and woman, uses words and expressions of a high level of thought and cultural content, so different from the ordinary Germans . . . This is due to their long cultural history. The people of northern Italy are the most civilized people I have ever met.

The state of the family finances made it urgent for Einstein to prepare for a career, but without a *Gymnasium* certificate university entry was barred. Fortunately there was a way out. There existed in Zurich the famous Swiss Federal Polytechnic, the Eidgenössische Technische Hochschule (or ETH, for short) to which entry was by a special examination. In 1895, although he was two years younger than the normal age, Einstein sat the entrance examination. Notwithstanding his exceptional knowledge of mathematics and physics, his performance in languages and biology let him down and he did not pass. Fortunately, however, the Principal of the ETH was so impressed by the candidate's mathematical ability that he arranged for him to attend the Swiss Cantonal School at nearby Aarau. After a year's preparation here, Einstein duly passed the ETH entrance examination in 1896.

The young man was delighted by the contrast between the Swiss School and the Munich *Gymnasium*, as he was generally between Swiss democracy and German militarism. Such, indeed, was his antagonism to all things German that he decided to give up his German nationality and become a Swiss citizen. He persuaded his father to write to the Württemberg authorities, and Albert's German nationality was formally ended on 28 January 1896. After five years as a stateless person he became, in February 1901, a citizen of the City and Canton of Zurich and so of Switzerland.

At the ETH Einstein took a four-year specialist course designed for intending teachers of mathematics and physics. The lectures aroused in him little enthusiasm, and he attended them fitfully. He preferred to explore for himself, either by performing experiments or by studying the works of the masters at first hand – Maxwell on electromagnetic theory, for instance. Some of this reading was done with a fellow student from Serbia named Mileva Maric. Einstein's cavalier attitude to the prescribed course might well have led to disaster in the examination room had he not had the good fortune to form a close friendship with a mathematical fellow student, Marcel Grossmann, who attended all the lectures and made excellent notes which he allowed his friend to study. Einstein duly graduated in 1900, but the four years of hard slog had taken their toll:

One had to cram all this stuff into one's mind for the examinations, whether one liked it or not. This coercion had such a deterring effect on me that, after I had passed the final examination, I found the consideration of any scientific problems distasteful to me for an entire year.

No academic opening could be found for him, and he was obliged to take a succession of temporary jobs in order to keep himself.

Fortunately his love of theoretical physics gradually returned and in 1901 he wrote his first research paper, on capillarity, and got it published in the prestigious *Annalen der Physik*. He sent copies to several leading academics, but to no avail. In the end it was Marcel Grossmann who once again came to the rescue. He spoke to his father about his friend's troubles, and Herr Grossmann spoke to his friend Friedrich Haller, the Director of the Swiss Patent Office in Berne. Einstein was duly called to an interview and in spite of his lack of technical training he seemed to have impressed the Director – particularly, according to one story, by his mastery of Maxwell's theory. There was then a considerable delay until a suitable vacancy occurred and was advertised. Einstein eventually started work at the Patent Office on 23 June 1902 as a probationary technical expert, third class. He now had a steady job and quickly mastered the undemanding work; he had time to spare for his own researches. And so, in the unlikely surroundings of a minor government office, his genius was able to mature. His seven years at the Patent Office were among the most creative of his life.

In October 1902 Einstein was shattered by the death of his father. In January 1903 he married Mileva Maric; a son Hans was born the following year, and a second son, Eduard, in 1910. The marriage was not a happy one and eventually ended in 1919 with an amicable divorce.

By 1904 Einstein had five scientific papers to his name, all of which were published in the *Annalen der Physik*. The last three deal with thermodynamical topics and develop Boltzmann's probabilistic interpretation of entropy in novel directions. However, these early papers were but a foretaste of the great things to come. It was in the year 1905, his *annus mirabilis*, that his genius burst into full flower. In the history of theoretical physics, 1905 ranks only with the plague year of 1665/66 that Newton spent at Woolsthorpe. Einstein, now aged 26, started the year by finishing his PhD dissertation for Zurich University (the Cantonal University, not to be confused with the ETH). The thesis was rejected as being too short. Einstein promptly resubmitted it with a single sentence added; it was accepted and eventually published in 1906. His five great papers of 1905 were all published in the *Annalen der Physik*, three of them in the famous Volume 17, now a valuable bibliographical rarity. It is not only their boldness and originality, but also the wide range of these papers that make them so impressive. The first paper – which Einstein described to a friend as 'very revolutionary' – deals with the energy of light and explains the photoelectric effect, while the second gives a molecular explanation of Brownian motion. (In December 1905 he sent off another paper on this topic.)

The third paper, 'On the electrodynamics of moving bodies', gives the first presentation of what we now know as the Special Theory of Relativity. In 9000 words the young iconoclast overthrew the accepted ideas of space and time in a manner which, as a reviewer in *The Times* put it, 'was an affront to common sense'. This famous paper reached the *Annalen der Physik* office on 30 June 1905. Within three months Einstein had produced three seminal papers and had completed his doctoral thesis.

In September 1905 he sent yet another paper to the *Annalen der Physik* in which he showed that, if a body gives off an amount of energy E in the form of radiation, its mass is reduced by an amount E/c^2, where c is the velocity of light. It took him another 18 months to realize that mass and energy are wholly equivalent. In 1907 he published a long paper in which he formulated his famous equation, $E = Mc^2$ (to be discussed in Section 6). Although this equation was not verified experimentally until some 25 years later, already by 1907 Einstein was speaking of it as the most important consequence of his Theory of Relativity. As a fitting recognition of his achievements during the extraordinary year of 1905 Einstein was promoted in April 1906 to the grade of technical expert, second class – the grade for which he had applied in the first place!

Einstein's first relativity paper appeared in print on 28 September 1905. By November that year no less a person than Max Planck had reported favourably on it. Most of the other leading physicists, Poincaré and Lorentz in particular, took rather longer to accept the revolution; indeed, some never did. Meanwhile even Einstein was beginning to feel the strain of combining such intense research activity with eight hours' work a day at the Patent Office. Once again he sought an academic post and submitted his relativity paper to Berne University as an inaugural thesis. It was rejected as being incomprehensible! So he carried on at the Patent Office. However, at long last his achievements were becoming known and in 1908 he became a *Privat Dozent* (a part-time unpaid assistant lecturer) at Berne University. He was an indifferent lecturer; no doubt he had more important things on his mind.

By this time Einstein had acquired a valuable new ally in Hermann Minkowski, a professor at Göttingen and one of Einstein's former teachers at Zurich. In 1907 Minkowski had provided a formal mathematical framework for the Special Theory with his concept of the four-dimensional space-time continuum (see Section 5). He opened a famous lecture in 1908 with the words:

From now on space by itself, and time by itself, are destined to sink completely into shadows, and only a kind of union of both to retain an independent existence.

Alas, by January 1909 Minkowski was dead, a victim of peritonitis.

The academic dam had to burst eventually, and in May 1909 Einstein was appointed Associate Professor of Theoretical Physics at Zurich University. The post was a new creation but, even so, the course of events did not run entirely smoothly. The job was in fact offered to Friedrich Adler, a friend of Einstein's and the son of Viktor Adler, the founder of the Austrian Social Democratic Party. (Viktor had sent his son to study in Switzerland in what proved to be a forlorn hope that the young man could be kept away from politics.) However, Friedrich insisted on withdrawing in favour of Einstein and his advocacy eventually convinced the members of the Zurich Board of Education, most of whom were Social Democrats. Adler told the Board that:

If it is possible to obtain a man like Einstein for our university, it would be absurd to appoint me. I must quite frankly say that my ability as a research physicist does not bear even the slightest comparison to Einstein's. Such an opportunity to obtain a man who can benefit us so much by raising the general level of the university should not be lost because of political sympathies.

In July 1909, at the age of 30, Einstein formally resigned from the Patent Office; he took up residence at Zurich in October.

His move up the academic ladder was now rapid: in 1911 he was offered a full professorship at the German University of Prague, despite some difficulties over his being both a Swiss citizen and a Jew. He remained in Prague for only 18 months, pondering and developing his General Theory of Relativity. During this period he also propounded a quantum law of photochemical processes that was soon confirmed experimentally. In January 1911 he was invited to attend the 1911 Solvay Congress in Brussels – a gathering limited to 21 highly selected physicists.

A year later Einstein was offered a ten-year appointment at the ETH. He moved there in October, having turned down offers from Utrecht, Leiden and Vienna. Once again his stay was brief. Planck and others were hatching plans to bring Einstein to Berlin, then the world centre of theoretical physics. They were able to offer a package deal consisting of election to the Royal Prussian Academy of Science with a special stipend, the rank of professor in the University, the directorship of the recently formed physical research branch of the Kaiser Wilhelm Institute, and, of course, complete freedom to do research or to teach as he wished. Einstein was very doubtful whether to accept; he hated German militarism and insisted that he should not be required to become a German citizen. Such, however, was his stature that the state and university authorities were prepared to accept all his terms and Einstein finally decided, with considerable reluctance, that the offer was irresistible. He moved to Berlin with his family in April 1914. He had now reached the summit of his profession. He was well known to (and highly esteemed by) the world's scientific community – but not, for another five years, to the general public.

All Einstein's hopes for a sheltered, carefree academic life were shattered by Germany's declaration of war on Russia on 1 August 1914. He took no part in the war. As a Swiss citizen he was not asked to join in the more patriotic activities of his German colleagues. He did what he could to further the cause of peace and was able to travel fairly freely to neutral Holland and Switzerland; he showed great courage when he signed a manifesto calling for the establishment of the League of Europeans and for cooperation between scholars of all the warring nations. But there was really very little he could do. Although he tried to immerse himself in his researches, he remained obsessed by feelings of guilt at the plight of stricken Europe.

The formulation of the General Theory of Relativity, after ten years of the most intense effort with many disappointments along the way, was Einstein's supreme achievement. The key paper, 'The foundations of the General Theory of Relativity', was published in *Annalen der Physik* early in 1916, following a short announcement to the Prussian Academy in 1915. The physicist Max Born describes the theory as:

the greatest feat of human thinking about Nature, the most amazing combination of philosophical penetration, physical intuition and mathematical skill. But its connections with experience were slender. It appealed to me like a great work of art, to be enjoyed and admired from a distance.

The Special Theory had unified Newtonian mechanics and Maxwellian electrodynamics in so far as they applied to bodies in uniform relative motion. The General Theory removed this restriction; it is concerned with accelerated relative motion, of which gravitational acceleration is the most conspicuous example. The full theory is, as the above quotation from Max Born suggests, extremely difficult to comprehend in all its richness. A brief discussion of some of its main features will be attempted in Section 7.

Einstein had begun his attack on the problem of gravitation in his 1907 paper on matter and energy. He returned to it in 1911 in the most important publication of his Prague period. In these two papers he propounded what was later to be called the 'principle of equivalence'. This states that a gravitational field at any point in space is in every way equivalent to an artificial field of force resulting from acceleration, so that no experiment can distinguish between the two. One of the consequences of this principle is that gravitation will bend light rays. In his 1911 paper Einstein calculated that a ray of starlight which grazes the Sun ought to be deflected by 0·87 seconds of arc, and suggested that it should be possible to detect this during a total eclipse of the Sun. The astronomer Finlay-Freundlich arranged to make such a test in Russia, where a suitable eclipse would occur in 1914. However, the expedition was abandoned on the outbreak of war, but this, as we shall see, was to prove a blessing in disguise.

In his younger days Einstein had rather neglected mathematics because, as he tells us, he could not decide in which branches of the subject to specialize. It was while he was in Prague that he came to the conclusion that what needed to be done was to express the equations of physics in such a way that would place all space-time coordinate systems on an equal footing. He realized that this would pose formidable mathematical difficulties, and that he would need to enlist some expert help. On his return to Zurich in 1912 he wrote to a friend:

I occupy myself exclusively with the problem of gravitation and now believe that I will overcome all difficulties with the help of a friendly mathematician here. But this one thing is certain: that all my life I have never before laboured at all as hard, and that I have become imbued with a great respect for mathematics, the subtle parts of which, in my innocence, I had till now regarded as pure luxury.

The helper referred to was none other than his old friend Marcel Grossmann. By a happy chance Grossmann's speciality was the *tensor calculus*, which had been developed by two Italian mathematicians during the previous 30 years. It proved to be just the tool that Einstein needed. Tensor calculus may be thought of, in Einstein's own words, as the study of 'the form of the equations which are invariant with respect to arbitrary point transformations'. In 1913 and 1914 Einstein and Grossmann published two joint papers in which, as we can now see with hindsight, they came very close to achieving their goal. Einstein's move to Berlin in 1914 effectively brought this collaboration to an end, but it had proved its value in providing him with the specialized mathematical equipment he needed. In 1915 Einstein found the field equations that he sought, and the great edifice of the General Theory was essentially complete. Gravitation was revealed not as a force but as intrinsic curvature of a four-dimensional space-time continuum. Such a continuum is no longer merely a background for events, but possesses an autonomous structure of its own which is governed by the distribution of material bodies in it.

While its mathematical complexities put the full understanding of the General Theory beyond the reach of most of Einstein's scientific colleagues, it made three predictions about the physical world that were readily understandable and were, in principle, amenable to experimental testing. The motion of the planet Mercury had long been a puzzle to astronomers. The perihelion of its orbit (that is, the position of its closest approach to the Sun) was observed to advance by about 5600 seconds of arc per century. Most of this could be accounted for by Newtonian perturbations of other members of the Solar System, but a residuum of 45 ± 5 seconds of arc remained stubbornly unexplained. Einstein's General Theory accounts for an additional 43 seconds of arc per century, a fact which he triumphantly announced in his 1915 note to the Prussian Academy. He wrote to his close friend Paul Ehrenfest in Holland in January 1916: 'Imagine my joy . . . at the result that the equations yield the correct perihelion motion of Mercury. I was beside myself with ecstasy for days.' The theory also made two more testable predictions: the bending of light by a gravitational field and the gravitational shift of the spectral lines. We have already mentioned the first of these in connection with the abortive 1914 expedition to Russia. In fact, Einstein's 1911 prediction of a bending of 0·87 second of arc had been deduced when he was working with an incomplete formulation of the General

Theory. The full theory predicted a deflection of twice this value, namely 1·7 seconds.

In 1916 Willem de Sitter in neutral Holland was able to send a copy of Einstein's paper to Arthur Eddington, Professor of Astronomy at Cambridge. He was captivated, and called it 'one of the most beautiful examples of the power of general mathematical reasoning.' Eddington enlisted government support for two expeditions, one to Brazil and the other (led by himself) to the Island of Príncipe on the west coast of Africa; a favourable eclipse of the Sun would occur at both places on 29 May 1919. The photographic results triumphantly confirmed the prediction of the theory. Eddington said that these findings gave him the greatest moment of his life. Communications with Germany were still bad, and it was 22 September before Lorentz sent Einstein a telegram containing the good news. The official announcement was made on 6 November in London at a joint meeting of the Royal Society and the Royal Astronomical Society. J. J. Thomson, the President of the Royal Society, hailed Einstein's work as 'one of the greatest – perhaps the greatest – of achievements in the history of human thought.'

It is interesting to remind ourselves that if Finlay-Freundlich had been able to observe the 1914 eclipse he would presumably have found a deflection of about 1·7 seconds of arc, and not the then predicted value of 0·87. Einstein's 1915 prediction of 1·7 seconds of arc would have been construed as a belated change of mind after having been shown to be wrong, and the dramatic impact of the 1919 verification would have been lost. As it was, Einstein awoke in Berlin on the morning of 7 November to find himself world-famous. The public had been dazzled by a deflection of starlight that amounted to the size of a pea as seen from at a distance of about 1000 metres. Writing to Einstein in December, Eddington says that:

all England has been talking about your theory. It has made a tremendous sensation . . . It is the best possible thing that could have happened for scientific relations between England and Germany.

The other prediction – of the red shift of lines in the spectra of distant objects – was not confirmed experimentally until 1926. In Section 8 we shall look further into how these three predictions emerge from the General Theory.

It is worth remarking that as early as 1916 Einstein found time to write *Relativity: The Special and the General Theory. A Popular Exposition* (Reference 1). Eleven editions of the authorized English translation, from which we shall quote extensively in later sections, were issued between 1920 and 1936.

During the war years in Berlin Einstein often stayed with his father's cousin Rudolf, whose household included Rudolf's widowed daughter Else and her two children. In 1917 Else nursed Einstein through a serious illness; in June 1919 they were married. In Einstein's notes for his weekly lectures on relativity one finds for 9 November 1918 the entry: 'Cancelled because of revolution.' Einstein rejoiced at the downfall of Prussian militarism and as a gesture of encouragement to the new German republic he became a German citizen. He remained in starving Berlin, declining highly attractive offers from Zurich and Leiden. He had become, whether he liked it or not, a symbol of the new democratic Germany and he knew he could not leave it during its birth-pangs.

It was, however, inevitable that Einstein should come under attack from those who wished to attribute the blame for the defeat of Imperial Germany to Jews, pacifists and communists. With the resurgence of antisemitism Einstein realized that he had a special responsibility towards the Jews. In 1921 he reluctantly decided to support the cause of Zionism and agreed to join Chaim Weizmann, later to be the first President of the State of Israel, in a fund-raising visit to the U.S.A. Einstein's own efforts were primarily directed to enlisting support for the proposed Hebrew University of Jerusalem. Needless to say, he was fêted all over America – equally by the national and academic leaders and by the general public. On his way back to Europe he lectured in England and during the next two years visited France, Japan, Palestine and Spain. He probably did more than any other single person to heal the scars left by the war.

In 1921 Einstein was awarded the Nobel Prize for Physics 'for his services to theoretical physics and in particular for his discovery of the law of the photoelectric effect'. Relativity was not mentioned – it was still too controversial. In 1923 he returned to Berlin, then at the centre of a short-lived flowering of art and science. He had a serious illness in 1928 from which he took some time to recover.

His main relaxations were music and sailing, and the city fathers of Berlin planned to honour Einstein on the occasion of his fiftieth birthday by presenting him with some land and a house among the Berlin lakes. The scheme ran into political opposition, and in the end Einstein used his savings to build the house himself.

Einstein and Else spent the winters of 1930/31 and 1931/32 in the U.S.A.; he also visited England several times. Meanwhile the world scene was darkening ominously and in January 1933 Hitler became the German Chancellor. The Einsteins, who were then in Pasadena, realized that they could not return to Germany. He immediately resigned from the Prussian Academy and renounced his German citizenship, narrowly forestalling official action in each case. In May 1933, Max Planck made a courageous statement in support of his former colleague. This and other such activities so enraged Hitler that he told Planck that it was only his age that saved him from being sent to a concentration camp.

On leaving the U.S.A. the Einsteins went first to Belgium, where his friend the King insisted on providing a personal bodyguard. Up to this time Albert Einstein had been an outspoken pacifist; in July 1933 he, like so many others, reluctantly changed his position. 'Were I a Belgian, I would not, in the present circumstances, refuse military service; I would enter such service cheerfully in the belief that I should thereby be helping to save European civilization.' Einstein and his wife spent the summer of 1933 in England; in October they left Europe for good. He accepted a post at the newly created Institute for Advanced Study at Princeton, and it was in this quiet academic town that he spent the rest of his days. His wife died there in December 1936.

After the publication of the General Theory, Einstein's scientific work inevitably gives an impression of anticlimax. During the years between the World Wars, his work centred on three areas of research. The first was the application of the General Theory to the Universe as a whole – the development of a relativistic cosmology. The second stemmed from his intense involvement in contemporary developments in theoretical physics, especially in the field of quantum theory. The third was the search for a unified field theory: for a set of equations that would link Einsteinian gravitation with

Maxwellian electrodynamics. (We shall look further into Einstein's contributions in each of these areas in Section 9.) As a cosmologist, Einstein was but one among a number of workers, and in the two other areas he achieved no major successes. Indeed, as the years went by, the ageing master became increasingly isolated from the majority of his colleagues and from the mainstream of thought in theoretical physics.

During most of the 1920s and 30s the theoreticians undoubtedly dominated the scene, but the experimental physicists also did great things. The year 1932 at Cambridge saw the discovery of the neutron and the first clear verification of Einstein's 1907 result that $E = Mc^2$. In 1934 Fermi split the uranium atom. After that, work on nuclear fission intensified in many parts of the world, and the production of an atomic bomb within a few years began to emerge as a feasible proposition. Could such a bomb be made in Germany? The refugee physicists in the U.S.A. became increasingly alarmed. During the summer of 1939 Einstein was visited by two Hungarian-born physicists, Leo Szilard and Edward Teller. The outcome was that Einstein helped to draft and then signed a letter to President Roosevelt that has become famous. (Many years later Einstein said that it was the worst thing he had ever done.) The letter is dated 2 August 1939. Here is an extract from it:

Some recent work by E. Fermi and L. Szilard, which has been communicated to me in manuscript, leads me to expect that the element uranium may be turned into a new and important source of energy in the immediate future. Certain aspects of the situation seem to call for watchfulness and, if necessary, quick action on the part of the Administration. I believe therefore that it is my duty to bring to your attention the following . . . it is conceivable that extremely powerful bombs of a new type may . . . be constructed. A single bomb of this type, carried by boat or exploded in a port, might very well destroy the whole port together with some of the surrounding territory.

Einstein then pointed out that the Germans had already stopped the sale of uranium from the Czechoslovakian mines. Unfortunately, for reasons that are not clear, the letter was not delivered to the President until 11 October. Roosevelt at once set up an Advisory Committee on Uranium, but matters moved so slowly that Einstein was persuaded to write another letter. As a result he was invited in

April 1940 to attend a meeting of the Committee. He declined the invitation, but stressed the urgency of some governmental action. In fact, an official decision to go all out to manufacture an atomic bomb was not made until 6 December 1941, the day before the Japanese attack on Pearl Harbor.

During the war Einstein acted as a consultant to the U.S. Navy and donated some of his early manuscripts to raise funds for the war effort. He had not kept the manuscript of his 1905 paper on Special Relativity, so he rewrote the whole paper in his own hand as his secretary read from the printed version. When it was auctioned, it fetched $6 million. In 1940 Einstein, his stepdaughter Margot and his devoted secretary Helen Dukas were sworn in as U.S. citizens. During the years of the Third Reich Einstein was very active in helping Jewish and other anti-Nazi European scientists to escape to the U.S.A.

In March 1945 Einstein wrote a third letter to Roosevelt urging him to prevent the new atomic bomb from being dropped on Japan. The letter was found unopened on Roosevelt's desk on the day he died. The dropping of the two atomic bombs in August 1945 filled Einstein with despair and grief. He felt a profound moral obligation to use his unique influence to save mankind from the new horror to which it had become exposed, and threw himself into a variety of activities in support of world government and peace. This time, however, he imposed some limits on his advocacy of reconciliation. His attitude to the Germany of 1945 was very different from his attitude to the young republic of 1919. He never forgave the Nazi atrocities against the Jews, and would not join or be associated with any German organization. In 1946 he refused to rejoin the Bavarian Academy: 'The Germans slaughtered my Jewish brethren; I will have nothing more to do with the Germans.' In 1949 he refused in even stronger terms when asked to renew official ties with the Kaiser Wilhelm Institute, now renamed the Max Planck Institute:

The crime of the Germans is truly the most abominable ever to be recorded in the history of the so-called civilized nations. The conduct of the German intellectuals – seen as a group – was no better than that of the mob. And even now there is no indication of any regret or any real desire to repair whatever little may be left to restore after the gigantic murders. In view of these circumstances I feel an irrepressible aversion to participating in anything that represents any aspect of public life in Germany.

He maintained this attitude to the end of his life.

In 1948 Einstein underwent an abdominal operation, and thereafter the years took an increasing toll. In June 1952 he wrote to a cousin:

As to my work, it no longer amounts to much: I don't get many results any more and have to be satisfied with playing the Elder Statesman and the Jewish Saint, mainly the latter.

Later that year he was asked to succeed Weizmann as President of the State of Israel. Einstein was deeply moved, but declined on the grounds that he lacked the necessary experience and aptitude. In 1953 he delivered a broadside against the activities of Senator McCarthy's notorious House Committee on Un-American Activities. By 1955 Einstein was able to write: 'I have come to look upon death as an old debt, at long last to be discharged.' His last public act was to assist Bertrand Russell in preparing a statement to warn the world of the perils ahead: 'Shall we put an end to the human race; or shall we renounce war?' The Russell–Einstein Statement, signed by eleven intellectual leaders, was published after Einstein's death, which occurred in the early hours of 19 April 1955. He faced death calmly, even cheerfully; he had asked that there be no funeral service, no grave and no monument. He was cremated near Trenton and the manner of disposal of his ashes was kept secret at his own wish. During the last few days of his life he was in fact working on a statement to be broadcast on the seventh anniversary of the founding of the State of Israel.

Albert Einstein was at once a towering genius and a profoundly simple man. He is revered throughout the world for a combination of qualities that must surely be unique: his intellectual and intuitive powers, his beauty and sweetness of character, and his unique moral authority based on a passionate involvement in the follies and hopes of a suffering world.

3. The pre-relativity papers of 1905

Einstein submitted two classic papers for publication during the first half of 1905. The first, 'On a heuristic viewpoint concerning the production and transformation of light', explained the photoelectric effect which had been puzzling scientists for some time. The paper's influence, with its revolutionary idea of the particle-like nature of light, increased steadily over the years and, as we have seen, eventually won for its author the Nobel Prize for Physics. Louis de Broglie, writing in 1955, said that the paper 'fell like a bolt from the blue, so much so that the crisis which it ushered in some fifty years ago is not yet passed today.'

During the nineteenth century, Newton's corpuscular theory of light (see Chapter 8) had been superseded, first by the wave theory of Young and Fresnel, and subsequently by the more comprehensive electromagnetic theory of Maxwell. Then, at the turn of the century, came the discovery of the photoelectric effect – that light falling on a metal can knock electrons out of it. When the intensity of the light is increased, one would expect the speeds of the ejected electrons to increase in proportion. This does not happen: instead, more electrons are ejected, but at the same speed. Furthermore, this speed is characteristic of the wavelength of the incident light – the higher the frequency, the greater the speed of the electrons.

At the same time, in 1900, Planck had put forward his quantum hypothesis as 'an act of desperation' to explain some curious results on the distribution of radiant energy with wavelength. Planck postulated that such energy is emitted in discrete bursts, or *quanta*, the size of the quantum being proportional to the frequency of the wave-like radiation. Einstein's point of departure was the contrast presented by the different ways in which physicists regarded matter and radiation. Matter was treated as being made up of discrete particles; radiation as electromagnetic waves, smooth and continuous. He was able to show mathematically that the two traditional theories could not be meshed together in a satisfactory manner. His solution was to propose a daring hypothesis – that light be thought of as consisting – like matter – of particles, of discrete packets of energy (we now call them 'photons'). The quantum concept had been almost completely neglected since 1900. Einstein not only revived it, but was able to show that the ratio of the energy of his light 'particles' to the frequency of their wave-like manifestation was exactly equal to Planck's constant, h.

This hypothesis enabled Einstein to explain not only the observed features of the photoelectric effect, but also several other hitherto unexplained phenomena concerning the interaction between matter and radiation. He calculated that the maximum kinetic energy E_e of the emitted electron is given by the very simple formula

$$E_e = h\nu - w \tag{17.1}$$

where ν is the frequency of the incident light, considered as wave-like, and w, the 'work function', is the work required to knock an electron out of the metal, and differs from one metal to another. The theory explained much that had previously been obscure, but Einstein still had to face an embarrassing contradiction: for some purposes, light was to be thought of as consisting of particles; for other purposes, as a wave motion. He believed that a satisfactory reconciliation would emerge in due time. His hopes were partially realized, but never to his complete satisfaction.

His second 1905 paper is entitled 'On the motion of small particles suspended in a stationary liquid according to the molecular kinetic theory of heat'. The continuous random movement of small pollen grains suspended in water had been reported by the botanist Robert Brown some 75 years earlier. This never-diminishing agitation, known as Brownian motion, seemed contrary to all experience, but Einstein was able to explain how the random motion of the individual particles, as observed in a microscope, derived from the kinetic energy of the invisible molecules with which the particles are constantly colliding. Using statistical methods, he was able to predict the mass and the number of molecules involved. Another prediction of the theory was that the mean kinetic energy of agitation of the particles would be the same as that of a gas molecule, which could be computed by the classical dynamic theory of gases. This prediction was confirmed experimentally in 1908 by Jean Perrin in Paris.

To appreciate the significance of Einstein's paper we must remember that at this time many of the leading physicists did not believe in the objective reality of atoms and molecules. (Do they now?) As Einstein put it many years later:

My main aim . . . was to find facts that would guarantee as far as possible the existence of atoms of definite finite size . . . The experimental verification of the statistical laws . . . of the Brownian motion . . . coupled with Planck's determination of the true molecular size from the law of radiation . . . convinced the sceptics.

The name of Einstein has become so closely identified with relativity that we are apt to forget his other contributions. It was Max Born, a close colleague, who said that 'Einstein would be one of the greatest theoretical physicists of all time if he had not written a single line on relativity.'

4. The Special Theory of Relativity

The third and most famous of Einstein's 1905 papers has the innocuous title 'On the electrodynamics of moving bodies'. In it he overturned the basic ideas about the nature of the physical world that had held sway since the publication of the *Principia* in 1687. From our twentieth-century vantage-point we can see that the Newtonian structure was built on the shakiest of foundations – that of absolute space and absolute time. At the beginning of the *Principia* we are told that:

Absolute, true and mathematical time, of itself, and from its own nature, flows equably without relation to anything external . . . Absolute space, in its own nature, without relation to anything external, remains always similar and immovable . . . Absolute motion is the translation of a body from one absolute place into another.

Newton was not entirely happy with this approach, and the rest of the long first scholium does little to clarify the matter (p. 208). However, since his laws of motion make no distinction between a state of absolute rest and a state of uniform rectilinear motion, he was able to sidestep the problem. In his fifth corollary to the laws of motion he states explicitly that:

The motions of bodies included in a given space are the same among themselves, whether that space is at rest, or moves uniformly forward in a right line without any circular motion.

We are, however, faced with a difficulty from another direction. When Newton's corpuscular theory of light was superseded by a wave theory, space had to be thought of as being filled with a 'something' – called the *aether* – to carry the waves. It should therefore be possible to devise an optical experiment that would detect motion (of the Earth, for example) through the aether. Maxwell proposed such an experiment in the final year of his life, although he did not believe it could ever be performed. However, improvements in measuring techniques enabled Michelson and Morley to carry out their celebrated experiment in 1887. The null result (i.e. the failure to detect any motion through the aether) posed a major problem for the physicists of the time.

The first attempt to retrieve the situation was made by George Fitzgerald in Dublin in 1893. He suggested that all moving objects were shortened along the axis of their movement, but this explanation could not be tested and was dismissed as a mere trick. Then in 1895 Hendrik Lorentz at Leiden gave the hypothesis a measure of respectability by deriving the required contraction as a consequence of his electric theory of matter. Poincaré also considered the problem, and came very near to discovering Special Relativity himself. How, then, did Einstein approach the problem? He begins, characteristically, by drawing attention to a basic contradiction in Maxwell's theory: to the fact that it makes unwarranted distinctions between rest and motion. Einstein illustrates his point by discussing Maxwell's treatment of a magnet and a coil of wire in relative motion. He then remarks that it is impossible to detect any motion of the Earth relative to the aether, although he makes no explicit reference to the Michelson–Morley result.

Einstein is now ready to put forward the two basic propositions on which the Special Theory rests. The first is that no experiment of any sort can detect absolute rest or absolute uniform motion in a straight line. This he called the 'principle of relativity' (in the restricted sense). What it does is to extend Newton's fifth corollary from the limited field of mechanics to the whole of physics. The second proposition is that in empty space light travels in straight lines with a definite speed c that does not depend on the relative motion of source and observer. This formulation enabled Einstein to reject the concept of an aether as superfluous. Here, then, were two simple propositions, each logically convincing and well established by experiment, but irreconcilable within the framework of contemporary physical theory. To reconcile them Einstein had to jettison the concepts of absolute duration and absolute distance: to allow measuring rods to contract and clocks to slow down when measured by a receding observer; to allow that two events which occur simultaneously for one observer may be separated in time for another. The equations which specify the extent of such curious behaviour are in fact exactly those that had already been put forward by Lorentz. (Here we shall merely state the transformation equations; we shall prove them in the next section.)

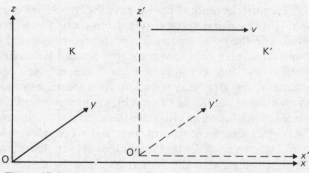

Figure 17.1.

Let us consider two observers moving apart with relative velocity v along their common x-axis. (This last assumption is made purely for simplicity.) The frames of reference of the two observers (K and K′) are shown in Figure 17.1. The position and time of occurrence of an event P will be measured by one observer in terms of the coordinates x, y, z and time t, and by the other in terms of x', y', z' and t', each with respect to their own frame of reference. How must these two tetrads of numbers be related so as to satisfy our two propositions? Einstein's answer is given by the famous equations

$$x' = \frac{x - vt}{\sqrt{1 - (v^2/c^2)}}$$

$$y' = y$$

$$z' = z \qquad (17.2)$$

$$t' = \frac{t - (vx/c^2)}{\sqrt{1 - (v^2/c^2)}}$$

Setting $1/c = 0$ gives the classical transformation. Although these equations are identical to those obtained previously, the interpretation that Einstein places on them is fundamentally different. For him the strange behaviour of clocks and measuring rods is a completely mutual effect applying equally to the measurements of any two observers in uniform relative motion. For Lorentz one observer is regarded as being at rest in the aether, and the other as moving through it.

Equations (17.2) can be inverted to give

$$x = \frac{x' + vt'}{\sqrt{1 - (v^2/c^2)}}$$

$$y = y'$$

$$z = z' \qquad (17.3)$$

$$t = \frac{t' + (vx'/c^2)}{\sqrt{1 - (v^2/c^2)}}$$

so they too satisfy the first basic proposition. We also find that either group, (17.2) or (17.3), yields

$$x^2 + y^2 + z^2 - c^2t^2 = x'^2 + y'^2 + z'^2 - c^2t'^2 = s^2 \text{ (say)} \quad (17.4)$$

The 'track' of a light ray is given by $s = 0$, and the equality of the two expressions in (17.4) means that the transformation equations (17.2) also satisfy the second basic proposition.

A physical event can be specified by a tetrad of numbers (x, y, z, t). Following Minkowski, we can regard these numbers as the coordinates of a 'point-event' in a particular frame of reference in a four-dimensional space-time continuum. Each observer dissects, so to speak, space-time in their own way. However, these dissections are not arbitrary, but are such that certain combinations of the measurements remain unchanged; they are independent of the motion of a particular observer. An example of such an invariant is the quantity we have denoted by s – usually called the 'interval' – in equation (17.4).

If we set $x = x_1, y = x_2, z = x_3$ and $ict = x_4$ (where $i = \sqrt{-1}$), the interval is given by

$$s^2 = \sum_{i=1}^{4} x_i^2$$

and we can regard Minkowski's continuum as a four-dimensional Euclidean space, but with an imaginary time coordinate. The transformations (17.2) correspond to a rotation of the four-dimensional coordinate framework. The interval s between two events is analogous to the 'straight-line' distance between two points in two or three dimensions; it is the shortest 'distance' between the events.

Once Einstein has established the foundations of his theory, the 1905 paper becomes highly mathematical. He shows, among other things, how the original form of Maxwell's general electrodynamic equations is exactly preserved when they are transformed to new variables (x', y', x', t') in accordance with the relations (17.2). Indeed, Lorentz had reached much the same conclusion in 1904, but Einstein was almost certainly unaware of this work when he published his own paper in 1905. In any case, as we have seen, the two men placed very different interpretations on what was virtually the same piece of mathematical analysis.

5. Derivation of the transformation equations

The treatment here follows closely that given by Einstein himself (Reference 1). Let us return to Figure 17.1. It will be convenient to consider first events that are localized on the x-axis. Any such event can be specified by coordinates (x, t) with respect to frame K and (x', t') with respect to frame K'; the other coordinates can be ignored for the moment. A light signal travelling along the positive x-direction is transmitted relative to frame K in accordance with the equation $x - ct = 0$; it is also transmitted relative to frame K' in accordance with the similar equation $x' - ct' = 0$. Both will be satisfied, provided

$$x' - ct' = \lambda(x - ct) \tag{17.5}$$

where λ is an arbitrary constant.

By applying similar considerations to a light signal transmitted along the negative x-direction, we get

$$x' + ct' = \mu(x + ct) \tag{17.6}$$

Adding and subtracting equations (17.5) and (17.6), and writing a for $\frac{1}{2}(\lambda + \mu)$ and b for $\frac{1}{2}(\lambda - \mu)$, gives

$$x' = ax - bct \tag{17.7}$$

and

$$ct' = act - bx \tag{17.8}$$

We now have to determine a and b.

At the origin of frame K', $x' = 0$, and so, from equation (17.7), $x = (bc/a)t$. We can thus write $v = bc/a$, giving

$$x' = a(x - vt) \tag{17.9}$$

and

$$t' = a\{t - (vx/c^2)\} \tag{17.10}$$

Now, the principle of relativity tells us that, as viewed from frame K, the length of a unit rod which is at rest with respect to frame K' must be exactly the same as the length, as viewed from frame K', of another unit rod that is at rest with respect to frame K. To see how a point on the x'-axis appears as viewed from frame K, we can take a snapshot of K' from K at any convenient time, say $t = 0$. From equation (17.9) we get $x' = ax$. Two points on the x'-axis which are separated by unit distance when measured in the K' frame are therefore separated in our snapshot by a distance d, given by

$$d = 1/a \qquad (17.11)$$

If we take the snapshot from frame K′ at $t' = 0$, we find, on eliminating t from equations (17.9) and (17.10), that

$$x' = a\{1 - (v^2/c^2)\}x$$

Hence two points on the x-axis a unit distance apart in the K frame will be separated in the snapshot by a distance d', given by

$$d' = a\{1 - (v^2/c^2)\} \qquad (17.12)$$

Now, the two snapshots must be identical and so d in equation (17.11) must be equal to d' in equation (17.12). This yields $a^2 = 1/\{1 - (v^2/c^2)\}$, and inserting this value of a in equations (17.9) and (17.10) gives the first and fourth of the transformation equations (17.2) in Section 4. The argument can clearly be extended to deal with events which do not occur on the x-axis by adding the equations $y' = y$ and $z' = z$ to give the full set of transformation equations.

There is no difficulty in generalizing these results so that they may be applied when the axes of frames K and K′ are not parallel, and the relative velocity is not in the x-direction. Einstein sums up the matter in these words:

Mathematically, we can characterize the generalized Lorentz transformation thus: it expresses x', y', z', t' in terms of linear homogeneous functions of x, y, z, t of such a kind that the relation

$$x'^2 + y'^2 + z'^2 - c^2 t'^2 = x^2 + y^2 + z^2 - c^2 t^2$$

is satisfied identically.

6. Relativistic mass and energy

We have seen how Einstein drove a coach and horses through the classical concepts of space and time. He then turned his attention to the third basic concept of Newtonian mechanics – mass – and tells us that:

In accordance with the theory of relativity the kinetic energy of a material point of mass m is no longer given by the well-known expression $\frac{1}{2}mv^2$ but by the expression

$$mc^2 / \sqrt{1 - (v^2/c^2)} \tag{17.13}$$

Furthermore, the premises of the Special Theory, in conjunction with Maxwell's electromagnetic equations, lead to the conclusion that:

A body moving with the velocity v, which absorbs an amount of energy E_0 (as judged by an observer moving with the body) in the form of radiation without suffering an alteration in velocity in the process, has, as a consequence, its energy increased by an amount

$$E_0 / \sqrt{1 - (v^2/c^2)} \tag{17.14}$$

Expressions (17.13) and (17.14) may be combined to give the total kinetic energy of the body as

$$\frac{mc^2 + E_0}{\sqrt{1 - (v^2/c^2)}} \quad \text{or} \quad \frac{\{m + (E_0/c^2)\}c^2}{\sqrt{1 - (v^2/c^2)}} \tag{17.15}$$

Thus the body has the same energy as a body of mass $m + (E_0/c^2)$ moving with velocity v. Hence we can say: If a body takes up an amount of energy E_0, then its inertial mass increases by an amount E_0/c^2; the inertial mass of a body is not a constant, but varies according to the change in the energy of the body. The inertial mass of a system can even be regarded as a measure of its energy.

If such a system neither gives out nor takes up energy, then the law of conservation of mass becomes identical with the law of conservation of energy.

We have seen in Chapter 8 how Newton was already speculating on these lines in his *Opticks*, specifically in the Query, quoted on p. 190, which asks: 'Are not gross Bodies and Light convertible into one another . . . ?'

If we expand the expression (17.13) for the kinetic energy we get

$$mc^2 + \frac{1}{2} mv^2 + \frac{3}{8} m \frac{v^4}{c^2} + \dots$$

When v^2/c^2 is very small, terms after the second may be neglected. The second term is the classical approximation. What is the significance of the first term? Einstein's answer derives from the expression (17.15), which shows that: 'The term mc^2 is nothing else than the energy possessed by the body (as judged by an observer moving with the body) before it has absorbed the energy E_0.'

The conclusions we have summarized in this section so far are, in essence, those that Einstein reached in his two 1905 papers. In 1907 he published a long paper in which he elaborated the arguments that are epitomized in the famous equation

$$E = Mc^2 \qquad (17.16)$$

Mass and energy are now wholly equivalent: all mass is congealed energy, all energy is liberated matter. (Here, as we explain below, M represents an extended concept of mass, and is to be distinguished from what we call the 'rest mass', denoted by m.)

To preserve the simplicity of equation (17.16) in the general case, we must allow the mass of a body to vary, not only when it absorbs or emits radiation, but also with its speed relative to the observer. Expressions (17.13) and (17.16) combine to yield

$$M = m/\sqrt{1 - (v^2/c^2)} \qquad (17.17)$$

which shows the dependence of mass on relative velocity.

The validity of equation (17.17) was confirmed experimentally when it became possible to project electrons and other atomic particles with velocities comparable to that of light. The result embodied in (17.16) explained the puzzling phenomenon of radioactivity whereby certain elements are able to emit particles at high speeds for long periods, so drawing on virtually unlimited sources of energy. Forty years later, alas, the mass–energy relation received its tragic confirmation at Hiroshima and Nagasaki.

7. The General Theory of Relativity

We have seen how the Special Theory enables laws of mechanics to be formulated that are equally applicable from the point of view of either of two observers who are moving uniformly relative to each other. This at once poses a question: why should uniform motion be something special – can we construct a system in which *all* motion is acknowledged to be relative? Now, the most familiar example of non-uniform (i.e. accelerated) motion is that caused by gravity. Clearly, the Newtonian theory of gravitation could not survive even the Special Theory intact. Gravitation was thought of as an instantaneous 'action-at-a-distance' force operating in a world of absolute masses and absolute distances. Indeed, Newton himself was fully aware of the philosophical objections. In a letter to a friend he wrote:

That gravity should be innate, inherent and essential to matter so that one body may act upon another at a distance through a vacuum, without the mediation of anything else, by and through which their action and force may be conveyed from one to another, is to me so great an absurdity, that I believe no man who has in philosophical matters a competent facility of thinking, can ever fall into it.

In attacking the gravitational problem, Einstein took as his starting point the remarkable experimental fact, enshrined in the story of Galileo dropping his weights from the Leaning Tower of Pisa, that bodies falling freely (in a vacuum and at a specific place) 'receive an acceleration, *which does not in the least depend either on the material or on the physical state of the body*' (Einstein's italics). He then develops his argument (Reference 1) as follows:

According to Newton's laws of motion we have

$$\text{force} = \text{inertial mass} \times \text{acceleration}$$

where the 'inertial mass' is a characteristic constant of the accelerated body. If now gravitation is the cause of the acceleration, we then have

$$\text{force} = \text{gravitational mass} \times \text{intensity of the gravitational field}$$

where the 'gravitational mass' is likewise a characteristic constant for the body. From these two expressions for 'force' we obtain:

$$\text{acceleration} = \frac{\text{gravitational mass}}{\text{inertial mass}} \times \text{intensity of the gravitational field}$$

If now, as we find from experience, the acceleration is independent of the nature and the condition of the body and is always the same for a given gravitational field, then the ratio of the gravitational to the inertial mass must be the same for all bodies. By a suitable choice of units we can make this ratio equal to unity. We then have the following law:

The *gravitational* mass of a body is equal to its *inertial* mass.

A satisfactory interpretation of this law can be obtained, Einstein remarks, 'only if we recognize the following fact: The same quality of a body manifests itself according to circumstances as *inertia* or as *weight*.'

Einstein now invites us to consider a simple conceptual model of a box in empty space that is pulled by an external force so as to give it a constant acceleration, as viewed by an external observer. An observer inside the box, on the other hand, will take the view that he and the box are at rest in a gravitational field of constant intensity. The general principle of relativity assures us that either interpretation of the facts is equally valid, so we are forced to conclude that, *so far as mechanical effects are concerned*, acceleration is relative and not absolute. Einstein now takes one of his bold, intuitive steps. He removes the italicized words and proposes a general 'equivalence principle' which states that:

A gravitational field of force is precisely equivalent to an artificial field of force, so that in any small region it is impossible by any conceivable experiment to distinguish between them.

We experience an artificial field of force when a lift starts from rest or an aircraft turns or dives.

In these days of spaceflight we are all familiar with the concept of 'weightlessness'. The sensation of heaviness is not felt when we are free to respond to the force of gravitation, but only when something interferes to prevent our falling freely. A little thought should convince us that an artificial field of force is always associated with some kind of bending of the path of a body in space and time. To be more precise, we can say, with Eddington (Reference 28), that 'whenever the observer's track through the four-dimensional world is curved he perceives an artificial field of force.'

Unfortunately, however, this statement is unsatisfactory on two counts. First, it follows from the equivalence principle that we need a single rule for determining how an observer will perceive a field of force of any kind: whether he thinks of it as artificially created or as the natural force of gravitation. Secondly, the statement is meaningless because a straight track is no longer an absolute concept, since straightness or otherwise depends on the observer's frame of reference. Einstein saw, however, that the underlying meaning of this loosely worded statement is that an observer believes himself to be in a field of force only when he is deflected from what he regards as his proper, or natural, track. So long as he is unconstrained, the field of force round him vanishes. Perhaps, indeed, the existence of a field of force is nothing more than the reflection of the fact that the observer is leaving his natural track.

What, then, do we mean by 'natural' tracks? Clearly they must be tracks in space-time that are marked out in some absolute way. In the restricted situations covered by the Special Theory, we have encountered one such invariant – a quantity that is independent of the motion of the observer and so has an absolute significance. This quantity is the interval s between two events as defined in equation (17.4). Two events, P_1 and P_2, in space-time can be joined by a variety of tracks, and the interval-length from P_1 and P_2 along any track can be measured using the obvious summation procedure, and hence expressed as

$$s = \int_{P_2}^{P_1} \mathrm{d}s$$

All observers will agree on the measurement of the interval for each 'infinitesimal' element $\mathrm{d}s$, and so the interval-length of the whole track is something that can be measured absolutely. It follows that all observers will agree on which track is the shortest (or longest, depending on the sign convention adopted for specifying the interval) track between the two events, judged in terms of interval-length.

This gives a means of defining certain tracks in space-time as having an absolute significance, and it is plausible to identify such tracks with the 'natural' tracks followed by freely moving particles. This is exactly what Einstein did when he proposed a new law of motion in the following terms:

Every particle moves so as to take the track of greatest (or least) interval-length between two events, except in so far as it is disturbed by impacts of other particles or by electric forces.

The two events, P_1 and P_2, might, for example, be the position of the Earth now and its position ten years ago. In those ten years the Earth has moved so as to take the shortest track from P_1 to P_2. We might choose to think of this track as an approximate spiral – a near-circle drawn out by continuous displacement in time – but the important point is that any other track would have a longer interval-length. The study of fields of force is thus reduced to a study of the geometrical properties of a four-dimensional space-time continuum in which the tracks of shortest (or longest) length (known as *geodesics*) are identified with the actual tracks of freely moving particles.

Einstein's problem, with which he struggled for nearly ten years, was to express the laws of this geometry in a concise form. Clearly such geometry must be non-Euclidean. In a Euclidean geometry the shortest track is always straight, but we know that planets, for example, move freely in curved tracks. This applies not only to freely moving bodies, but also to light. It turns out that in gravitational fields light rays are no longer propagated either in straight lines or at a constant speed. One of the basic propositions of the Special Theory is now seen to be an approximation – valid only when the influence of gravitational fields may be disregarded.

In order 'to investigate the laws satisfied by the gravitational field itself', as Einstein put it, he had to investigate the properties of a very general type of four-dimensional non-Euclidean geometry. The theory and the techniques for doing this had been worked out by Gauss for two-dimensional surfaces and generalized for n dimensions by Riemann and others. It is convenient to replace the Gaussian parameters (p, q) of p. 331 by (x_1, x_2), as in the mesh system illustrated in Figure 17.2, and to use the Leibnizian 'd' notation for small increments. Denoting by ds the distance between two neighbouring points $P(x_1, x_2)$ and $Q(x_1 + dx_1, x_2 + dx_2)$, we can write

$$ds^2 = g_{11}dx_1^2 + 2g_{12}dx_1 dx_2 + g_{22}dx_2^2 \qquad (17.18)$$

where the g's, which may be constants or functions of (x_1, x_2), determine both the character of the mesh system and the intrinsic nature of the surface, which is of course independent of the mesh system.

Here are some familiar examples:

$$ds^2 = dx_1^2 + dx_2^2 \qquad \text{(rectangular plane coordinates)}$$
$$ds^2 = dx_1^2 + 2kdx_1 dx_2 + dx_2^2 \quad \text{(oblique plane coordinates)}$$
$$ds^2 = dx_1^2 + x_1^2 dx_2^2 \qquad \text{(plane polar coordinates)}$$
$$ds^2 = dx_1^2 + \cos^2 x_1 dx_2^2 \qquad \text{(latitude and longitude)}$$

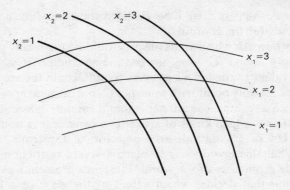

Figure 17.2.

Thus, in the first example, $g_{11} = g_{22} = 1$ and $g_{12} = 0$. The first three examples describe different mesh systems on a flat surface; the fourth relates to the surface of a sphere. To sort out the effect of different kinds of space from that of different mesh systems on the same kind of space involves quite complicated mathematical analysis. Thus the g's in our first three examples satisfy a certain partial differential equation which is the condition for a flat surface; the g's in the fourth example do not.

To generalize from two dimensions to four-dimensional space-time, we must replace distance by interval (an absolute quantity, as we have seen) and equation (17.18) by

$$ds^2 = \sum_{i=1}^{4} \sum_{j=1}^{4} g_{ij}\, dx_i\, dx_j \quad (g_{ij} = g_{ji}) \tag{17.19}$$

There are ten terms altogether. The values of the g's provide the only means of telling what kind of space-time is being described. What these values are must be found experimentally by measuring the intervals between events in space-time. For an inertial type of space-time of the kind envisaged in the Special Theory, the relation (17.19) could take the form

$$ds^2 = -dx^2 - dy^2 - dz^2 + c^2\, dt^2 \tag{17.20}$$

where, for convenience, we have changed the sign of ds^2 and used new symbols for the coordinates.

We may now ask which kinds of space-time are physically possible in an empty region, i.e. one not containing matter. If we assign arbitrary values to the ten g's at every 'point', could the space-time so specified actually occur with some possible arrangement of matter around such a region? The answer is that it cannot; indeed, the rule that determines which kinds of space-time can occur is none other than the law of gravitation – the objective of Einstein's ten-year search. What this law does is to place a severe restriction on the permissible geometries of the world. Einstein's problem was to find the criterion that decides which values of the g's give a kind of space-time that is possible in nature, and which do not. He had only two clues to guide him:

(1) The criterion must refer to those properties of the g's that distinguish different kinds of space-time; the formulae must not be altered if we merely change the mesh system.

(2) Flat (or Euclidean) space-time can occur at great distances from all matter, so any set of g-values that specifies a flat space-time must satisfy the criterion.

Such was Einstein's physical intuition that these two clues sufficed to suggest a particular line of approach which eventually led him to the solution of the problem.

To sort out the various possibilities Einstein had to use the techniques of tensor calculus. One important result in this field is that the condition for flat space-time is the vanishing of a certain tensor, known as the Riemann–Christoffel tensor. This condition is usually expressed as

$$B^{\rho}_{\mu\nu\sigma} = 0 \qquad\qquad (17.21)$$

Since each of the Greek letters can take any value from 1 to 4, we have here 256 separate equations. Many of them, however, are repetitions and only 20 are essentially different. Now, clearly the criterion given by (17.21) is far too severe: it applies only to the special case when all matter is infinitely remote. One important tensor operation is known as contraction. The contracted Riemann–Christoffel tensor is formed by setting $\rho = \sigma$ and adopting the convention that if a suffix occurs twice, the value of the tensor is obtained by summing over all values of that suffix. This new tensor is denoted by $G_{\mu\nu}$ and is equal to $\Sigma_{\sigma=1}^{4} B_{\mu\nu\sigma}^{\sigma}$. Now, it turns out that the condition

$$G_{\mu\nu} = 0 \qquad\qquad (17.22)$$

does in fact satisfy all the requirements for a general law of nature. It can be proved to be independent of the mesh system, and since (17.22) is satisfied whenever (17.21) is, flat space-time is admissible as a particular case. The law expressed by equation (17.22), which embodies six independent equations, was in fact the one Einstein chose for his law of gravitation; it is the essence of the General Theory.

It must be emphasized that the law expressed by equation (17.22) represents the state of an empty region of space-time: a region containing no matter, light or electromagnetic fields of any kind, but in the neighbourhood of any or all of these forms of energy. The truth or falsity of the proposed law can, of course, be settled only by experiment or observation. To begin with, it must yield the Newtonian law of gravitation as a first approximation. Where, exceptionally, there are observable differences between the consequences of the two laws, these differences must be investigated. We shall return to this matter in the next section.

To get some impression of the implications of the General Theory, let us consider a very simple case. Einstein's law in the form (17.22) expresses the conditions to be satisfied in a gravitational field produced by any arbitrary distribution of matter. Let us particularize, and ask what kind of space-time exists in the region round a single attracting particle, or point-mass, of mass m. To simplify matters still further, we consider a space of two dimensions only – sufficient for the plane orbit of a planet – together with a third dimension of time. It turns out that the geometry of the space-time round such a particle is given by

$$ds^2 = -\frac{1}{\gamma}dr^2 - r^2 d\theta^2 + \gamma c^2 dt^2 \qquad\qquad (17.23)$$

where $\gamma = 1 - (2Gm/c^2 r)$ and the coordinate notation is an obvious one suggested by polar coordinates in a Euclidean space;

G is the gravitational constant and c is the velocity of light in the absence of a gravitational field. To begin with, equation (17.23) should be thought of simply as a particular solution of the field equations (17.22). It is only after a successful appeal to experiment that we are entitled to identify this solution with the physical situation we have just specified. If $m = 0$, then $\gamma = 1$ and the space-time is perfectly flat. In most practical cases $2Gm/c^2r$ will be extremely small. The other extreme case, given by $\gamma = 0$, is, however, of some interest – it represents a simple example of what we have now come to know as a 'black hole'.

The coefficient γ appears twice in equation (17.23), but it is the second appearance, as the coefficient of $c^2\,dt^2$, that is responsible for the main features of Newtonian gravitation. A body in this kind of space-time will appear to be under the influence of an attractive force directed towards the origin and proportional to m/r^2. The appearance of $1/\gamma$ as the coefficient of dr^2 is responsible for the main deviations of the new law from the old. Setting $dt = 0$ in equation (17.23) and changing the sign of ds^2 so that space-like intervals are real, we have

$$ds^2 = \frac{1}{\gamma}dr^2 + r^2\,d\theta^2 \qquad (17.24)$$

The presence of $1/\gamma$ shows that the space is distorted (i.e. is non-Euclidean) near an attracting body of mass m. We can see this by considering the operation of surveying a circle of radius r. If we lay our scale transversely so as to measure the circumference, we see that with $dr = 0$, we have $ds = r\,d\theta$. The measured circumference will therefore be the 'Euclidean' $2\pi r$. If, on the other hand, we lay the scale radially to measure the diameter, we have $d\theta = 0$ and so $ds = dr/\sqrt{\gamma} > dr$. Thus the measured diameter will be slightly greater than $2r$ and the ratio of the measured circumference to the measured diameter will come out as slightly less than π.

Finally, let us consider the motion of a light pulse, defined by the condition $ds = 0$. For radial motion $d\theta = 0$ and so $dr/dt = c\gamma$; for transverse motion $dr = 0$ and so $r\,d\theta/dt = c\sqrt{\gamma}$. Thus the velocity of light is reduced by the influence of an attracting particle (i.e. by a gravitational field) and the actual velocity depends on the direction of motion.

It is quite impossible to present the General Theory convincingly without a liberal dose of tensor calculus. The intention in this section has been to convey some impression of both the power and the beauty of Einstein's achievement. We must now look at the question of validation.

8. Experimental tests of the General Theory

Although the old and new theories of gravitation are so very different in form, in concept and in their basic assumptions, the deductions that may be made from them about the physical world agree extremely closely. Indeed, only three testable differences have been found, and in each case the effect is extremely minute. We shall discuss each of them in turn.

(a) *The motion of the perihelion of Mercury*

The General Theory leads to a set of differential equations for the orbit of a planet which are the same as that given by the Newtonian theory, except that one of the equations contains a small correction term. Einstein describes the effect of this term:

According to the General Theory of Relativity a small variation from the Kepler–Newton motion of a planet in its orbit should take place, and in such a way that the angle described by the radius Sun–planet between one perihelion and the next should exceed that corresponding to one complete revolution (i.e. to the angle 2π) by an amount [in radians] given by

$$+ \frac{24\pi^3 a^2}{T^2 c^2 (1 - e^2)} \tag{17.25}$$

In (17.25), a is the semi-major axis of the orbital ellipse, e its eccentricity, c the velocity of light and T the period of revolution of the planet. Mercury is the only planet for which this very small angle can be measured. It comes out to be 43 seconds of arc per century – exactly the amount left unaccounted for by the Newtonian theory, as we mentioned in Section 2 (p. 379).

(b) *The deflection of light by a gravitational field*

Another effect of the correction term introduced by the General Theory is that the path of a ray of light is bent when it passes through a gravitational field. In particular, a light ray passing close to a star of mass m will be deflected towards the star through an angle α (in radians) given by

$$\alpha = 4Gm/c^2 R \qquad (17.26)$$

where R is the perpendicular distance of the undeflected light ray from the centre of the star. The best way to test the result is, as we explained in Section 2, to photograph the stars in the neighbourhood of the Sun during a solar eclipse, and to compare their recorded positions with their normal positions in the night sky. Figure 17.3 should make the situation clear. In the absence of the Sun, S, a distant star would be seen from the Earth, E, to lie in the direction D_1. During a solar eclipse the light from the star will be deflected by the gravitational field of the Sun and will be seen to come from the direction D_2. For a ray of light which, if undeflected, would pass the Sun (of mass m and radius r) at a distance $R = kr$, the formula (17.26) reduces to $\alpha = 1 \cdot 7/k$ seconds of arc. (We had $k = 1$ on p. 380.) The experimental confirmation of this result in 1919 was, as we have seen, a second triumph for the General Theory.

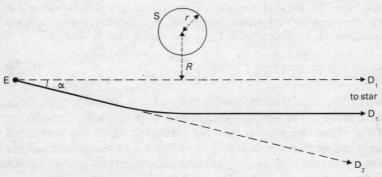

Figure 17.3.

(c) *Displacement of spectral lines towards the red*

Another consequence of the General Theory is that the frequency with which an atom absorbs or emits light should depend on the strength of the gravitational field in which it is situated. To see this, let us consider a number of similar atoms vibrating at different points in a region of space-time characterized, once again, by the interval equation (17.23). Now, a vibrating atom may be thought of as an ideal type of clock. The beginning and end of a single vibration constitute two events, and the interval ds between them is an absolute quantity independent of the mesh system. Let us suppose that the atoms are all momentarily at rest in the coordinate system of (17.23). We may therefore set dr and $d\theta$ equal to zero to give $ds^2 = \gamma c^2 dt^2$, or $c\,dt = ds/\sqrt{\gamma}$. This means that the vibration times of similar atoms situated in different parts of the region will be proportional to $1/\sqrt{\gamma}$, so the frequencies of vibration will be proportional to $\sqrt{\gamma}$. In particular, the frequency of vibration of an atom situated on the surface of a star (such as the Sun) will be somewhat less than the frequency of an atom of the same chemical element situated either in free space ('empty space') where $\gamma = 1$, or, for all practical purposes, on the surface of a much smaller celestial body such as the Earth. Since $\gamma = 1 - 2Gm/c^2r$, we can set $\sqrt{\gamma} = 1 - Gm/c^2r$, to a close approximation. Our result can then be expressed as

$$(v_0 - v)/v_0 = Gm/c^2r \qquad (17.27)$$

where v is the frequency of vibration of an atom on the surface of a star of mass m and radius r, and v_0 is the frequency of a similar atom in free space, or, effectively, on the surface of the Earth.

The spectrum of a beam of sunlight or starlight is found to be crossed by many dark lines, known as Fraunhofer lines after their discoverer, Joseph von Fraunhofer. They are caused by light from the Sun or star itself being selectively absorbed by atoms in the solar or stellar atmosphere. The frequency of red light is lower than that of blue light, so the effect of a gravitational field (equation (17.27)) is to displace the Fraunhofer lines very slightly towards the red end of the spectrum. For the Sun, the predicted displacement amounts to about two-millionths of the wavelength. Such minute displacements are extremely difficult to measure, and Einstein's third prediction was not definitely confirmed until 1924, by W. S. Adams at Mount Wilson Observatory. Adams examined spectra of the dense companion star of Sirius, where the effect (which depends on the size of the ratio of m to r) is about 30 times greater than that on the Sun.

It is interesting now to read Einstein's laconic comment on the concept of red shift in his 1916 book:

At all events, a definite decision will be reached during the next few years. If the displacement of spectral lines towards the red by the gravitational potential does not exist, then the general theory of relativity will be untenable.

We may be sure that he had no doubts about the outcome!

9. Einstein's later researches

Einstein's scientific activities after the publication in 1916 and the general acceptance by 1920 of the General Theory fell mainly into three broad areas. Let us look briefly at each in turn.

(a) *Cosmology*

Einstein inaugurated the subject of relativistic cosmology in a paper that was published in 1917. In it he sought to apply the General Theory to the basic cosmological problem of the structure of the Universe as a whole. The concept of infinite space presented some serious difficulties which Einstein made a number of attempts to overcome. Eventually he decided to ban infinite space altogether. Instead, he postulated a static, finite but unbounded Universe, uniformly filled with matter that could be regarded as being in an overall state of rest (i.e. for every particle in motion, the motion is exactly 'cancelled' by another, identical particle with opposite motion). To do this he needed to add a small term, λ, to the basic gravitational equation (17.22) to give, in its simplest form, the equation

$$G_{\mu\nu} = \lambda g_{\mu\nu} \qquad (17.28)$$

His calculations led him to what he thought was a unique model of the Universe in which the total size and the total mass are determined by the value of λ.

During the next few years, however, other solutions of these equations were discovered. In 1924 Edwin Hubble's observations at Mount Wilson Observatory established that the Universe is expanding. Einstein's assumption of a static Universe was no longer acceptable, so he was able to get rid of his λ term; he had never liked it, regarding it as 'gravely detrimental to the formal beauty of the theory'.

A further consequence of setting $\lambda = 0$ was that Einstein, in a paper published in 1931, could restore the uniqueness of the closed finite model of the Universe. Unfortunately this model gave too low a figure for the age of the Universe. However, retaining the λ term enabled the age to be sufficiently extended to accord with the observational evidence. But Einstein would have none of it – for him, beauty and mathematical simplicity were paramount. He trusted his pristine field equation (17.22) more than the astronomical observations with which they were in conflict. In taking this line, Einstein set himself against the mainstream of opinion. The controversy remains unresolved.

(b) *The quantum theory controversies*

The years between the two World Wars saw the second great modern upheaval in physics, an upheaval which led to the creation of quantum mechanics. In 1911 Ernest Rutherford proposed a 'sun and planet' model of the atom; in 1913 Niels Bohr put forward the wildly implausible theory that the electrons forming the planetary system can remain indefinitely in certain permitted steady orbits, and that light is radiated or absorbed only when an electron makes a 'quantum jump' from one permitted orbit to another. The ratio of the energy change to the frequency of the light was assumed to be equal to Planck's constant, h. Bohr's theory, with its mixture of classical and quantum concepts, was a piece of inspired nonsense, a marvel of intuition. Einstein spoke of it as 'one of the greatest discoveries . . . the highest form of musicality in the sphere of thought'.

When Planck produced his formula for black-body radiation in 1900, he could not avoid using an inelegant mixture of conflicting ideas – quantum and Maxwellian. In 1916 Einstein found a new quantum approach that basically avoided Maxwellian concepts. Building on Bohr's energy levels, and using probabilistic arguments, Einstein derived Planck's formula 'in an amazingly simple' manner, and also established a direct link with some of Bohr's results. He wrote of this theory that 'it commends itself by its simplicity and generality'; he regarded it as among his best work.

By 1922 Bohr's 'old' quantum theory was facing serious difficulties, and the 1920s saw a succession of startling theoretical formulations that are now loosely grouped together as the 'new' quantum theory. In 1927–28 Paul Dirac harmonized the rival theories in a brilliant synthesis, now known as quantum mechanics.

Meanwhile, Werner Heisenberg was stating his famous principle of indeterminacy: the more accurately we observe the position of a particle, the less accurately can we observe its momentum, and vice versa. This principle was highly subversive – it was a mortal blow to the whole idea of causality. The issue was intensely debated for years, and many ingenious attempts were made to extract a modicum of sense from the general confusion. Einstein's position was quite clear: he did not like the new ideas at all. The dethronement of causality offended all his scientific instincts. His view is epitomized in one of his well-known remarks: 'God does not play at dice.' (Einstein's God was the God, not of Moses, St Paul or Calvin, but of Spinoza.)

Throughout his working life Einstein had struggled unsuccessfully with the problem of making physical sense of the concept of light quanta which he had put forward as a young man. As he put it in 1951:

All these fifty years of conscious brooding have brought me no nearer to the answer to the question, 'What are light quanta?' Nowadays every Tom, Dick and Harry thinks he knows it, but he is mistaken.

For many years Einstein was at the centre of all the arguments, with his great friend, Niels Bohr, his main antagonist. Eventually the two masters had to agree to differ.

(c) *The search for a unified field theory*

In the General Theory, Einstein reinterpreted gravitational forces in terms of the geometry of space-time, but he was not able to deal with electromagnetic forces in the same way. Efforts to repair this weakness by constructing a *unified field theory* occupied some of the ablest mathematical physicists during the inter-war years – and Einstein for the rest of his days. Several times he thought he had found the key to success, only to have his hopes dashed. He published what was to be his final theory in 1945. It was highly mathematical, impossible to explain in simple terms, and could not be tested.

By this time the interior of the atom had become, and has remained, the centre of interest. Few theoretical physicists now envisage a Universe that can be adequately modelled by a set of field equations of the kind Einstein searched for so devotedly for so many years. Even so, his position as the greatest mathematical physicist of the century – based on his unique achievements during the first half of his working life – is secure. Let his close friend and colleague, Leopold Infeld, have the last word on him:

The most amazing thing about Einstein was his tremendous vital force directed towards one and only one channel: that of original thinking, of doing research. Slowly I came to realize that in exactly this lies his greatness.

*

This is, perhaps, a fitting note on which to bring our story to an end. In fact, however, provided we can avoid a global catastrophe, the story need have no end. More mathematics is being done today than ever before: some 25 000 articles are published in mathematical research journals each year. Man's creativity – powered by his curiosity, concentration and commitment – is unbounded and inexhaustible.

Appendices

Appendix 1. A Eudoxan treatment of similarity

The basic similarity theorem that we wish to prove is that, in Figure A.1, if the transversals AA′, BB′ and CC′ are parallel, then

$$AC:CB = A'C':C'B'$$

If the line segments AC and CB are commensurable, their ratio being equal to the rational fraction, p/q say, the proof is straightforward. If we divide AB into $p + q$ equal parts, AC will occupy just p of them. If we now draw lines through the points of division parallel to AA′, it is easy to prove that they cut off equal intervals on A′B′. (We have a set of congruent triangles, two of which are shown shaded.) There will therefore be p equal intervals in A′C′ and q more in C′B′; the proof follows at once.

How are we to proceed, however, when AC and CB are incommensurable? Let us assume first that

$$AC:AB < A'C':A'B'$$

It is a simple deduction from the Eudoxus–Archimedes axiom (p. 29) that between any two real numbers we can always find a *rational* number. In particular, we can find a rational number r such that

$$AC/AB < r < A'C'/A'B' \tag{A 1}$$

Now, let $AD = r.AB$, and draw DD′ parallel to AA′. Then, by the result proved for the commensurable case, $A'D' = r.A'B'$.

412

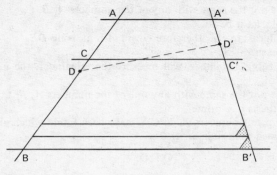

Figure A.1.

Furthermore, it follows from (A 1) that AD > AC and A'D' < A'C'. This implies that DD' cuts CC', a direct contradiction since both lines are parallel to AA'. Similarly, we can derive a contradiction from the assumption that AC:AB > A'C':A'B'. It follows, therefore, that the two ratios must be equal, as was to be proved.

Appendix 2. Euclid's proof that the number of primes is infinite

Here, without comment, is the proof of Proposition IX, 20 of the *Elements*, taken from the 1956 edition of Heath's English translation:

Prime numbers are more than any assigned multitude of prime numbers.
Let *A, B, C,* be the assigned prime numbers; I say that there are more prime numbers than *A, B, C*.
For let the least number measured by *A, B, C** be taken, and let it be *DE*; let the unit *DF* be added to *DE*.
Then *EF* is either prime or not.
First, let it be prime;
then the prime numbers *A, B, C, EF* have been found which are more than *A, B, C*.
Next, let *EF* not be prime; therefore it is measured by some prime number [VII, 31].
Let it be measured by the prime number *G*.

* Measured by *A, B, C* means the volume computed from *A, B, C*, i.e. the product *A.B.C*.

I say that G is not the same with any of the numbers A, B, C.
For, if possible, let it be so.
Now A, B, C measure DE; therefore G will also measure DE.
But it also measures EF.
Therefore G, being a number, will measure the remainder, the unit DF; which is absurd.
Therefore G is not the same with any one of the numbers A, B, C.
And by hypothesis it is prime.
Therefore the prime numbers A, B, C, G have been found which are more than the assigned multitude of A, B, C.

Q.E.D.

Appendix 3. Viète's solution of Van Roomen's problem

The 45th-degree equation presented to Viète (p. 119) was

$$x^{45} - 45x^{43} + 945x^{41} - 12300x^{39} + \ldots - 3795x^3 + 45x = A$$

It seems that Van Roomen proposed several specific values of A. How, then, did Viète solve this equation? We must assume that he recognized its form and realized that the problem amounted to: 'given $\sin 45\theta$, find $\sin\theta$'. Presumably he also knew that, since 45 has factors 5, 3 and 3, the problem could be broken down into three stages, involving the solution of one fifth-degree and two third-degree equations. Let us try to reconstruct the procedure. In Viète's time, as we have seen, trigonometrical formulae were based on the chord of a circular arc, not on the arc (or angle) itself. So let us define

$$c = 2\sin 45\theta, \quad c_1 = 2\sin 15\theta, \quad c_2 = 2\sin 5\theta, \quad x = 2\sin\theta$$

We are given c and wish to find x. The familiar formula for $\sin 3\alpha$ can be written

$$2\sin 3\alpha = 3(2\sin\alpha) - (2\sin\alpha)^3$$

Thus we get, in turn,

$$c = 3c_1 - c_1^3, \quad c_1 = 3c_2 - c_2^3$$

and, from the similar formula for $\sin 5\alpha$,

$$c_2 = 5x - 5x^3 + x^5$$

Hence

$$c = 3(3c_2 - c_2^3) - (3c_2 - c_2^3)^3$$
$$= 9(5x - 5x^3 + x^5) - 3(5x - 5x^3 + x^5)^3$$
$$-\{3(5x - 5x^3 + x^5) - (5x - 5x^3 + x^5)^3\}^3$$

When multiplied out, this equation agrees with that presented by Van Roomen. The complete set of solutions is given by the equation $x_n = 2\sin(\theta + 2\pi n/45)$ where $n = 0, 1, 2, \ldots, 44$. Of these 22 are negative, and so were rejected by Viète; he gave the 23 positive solutions.

Appendix 4. Viète's formula for π

We start with a Viète speciality – multiple-angle formulae. We have

$$\sin\theta = 2\sin\theta/2 \,.\, \cos\theta/2$$
$$= 2^2 \sin\theta/2^2 \,.\, \cos\theta/2^2 \,.\, \cos\theta/2$$
$$\vdots$$
$$= 2^n \sin\theta/2^n \,.\, \cos\theta/2 \,.\, \cos\theta/2^2 \,.\, \cos\theta/2^3 \ldots \cos\theta/2^n$$

giving

$$\frac{\sin\theta}{\theta} = \left\{\frac{\sin\theta/2^n}{\theta/2^n}\right\}(\cos\theta/2 \,.\, \cos\theta/2^2 \,.\, \cos\theta/2^3 \ldots \cos\theta/2^n)$$

As n tends to infinity, $\cos\theta/2^n$ approaches 1 from below, as also does the expression in the curly brackets. Putting $\theta = \pi/2$ and proceeding to the limit, we obtain the convergent infinite product

$$2/\pi = \cos\pi/4 \,.\, \cos\pi/8 \,.\, \cos\pi/16 \ldots$$

Since $\cos\alpha/2 = \sqrt{\tfrac{1}{2}(1 + \cos\alpha)}$, this formula can be expressed as

$$\frac{2}{\pi} = p_1 \,.\, p_2 \,.\, p_3 \ldots$$

where

$$p_1 = \sqrt{\tfrac{1}{2}} \quad \text{and} \quad p_{n+1} = \sqrt{\tfrac{1}{2}(1 + p_n)}$$

($2/\pi$ is expressed as a single formula on p. 122.)

Appendix 5. Gauss' Euclidean construction of the 17-sided regular polygon

We have seen in Chapter 14 that the young Gauss was able to prove that the roots of the equation

$$z^p - 1 = 0 \qquad (A\,2)$$

where p is prime, may be solved in terms of the roots of a sequence of equations whose degrees are the prime factors of $p - 1$. So if $p = 2^n + 1$, all such equations will be quadratics and the corresponding geometrical problem is solvable by Euclidean methods. In fact, the primality of $p = 2^n + 1$ ensures that n is of the form 2^r. Thus p is a Fermat prime, $F_r = 2^{2^r} + 1$, where $r = 0, 1, 2, 3, 4$. Gauss proved (p. 323) that a Euclidean solution of equation (A 2) is possible if and only if p is a Fermat prime.

Before discussing Gauss' treatment of the $F_2 = 17$ case, it will be helpful to look at the simpler case of $F_1 = 5$. The Euclidean construction of the regular pentagon was, as we saw in Chapter 3 (p. 43), well known to the Greeks. The problem is equivalent to that of dividing the unit circle into five equal parts, or to solving the equation

$$z^5 - 1 = 0 \qquad (A\,3)$$

which can be factorized to give

$$(z - 1)(z^4 + z^3 + z^2 + z + 1) = 0$$

The roots consist, therefore, of the trivial root $z = 1$ and four other numbers (real or complex) satisfying the equation

$$z^4 + z^3 + z^2 + z + 1 = 0$$

or, more symmetrically,

$$z^2 + z + 1 + z^{-1} + z^{-2} = 0 \qquad (A\,4)$$

The fact that $(z + z^{-1})^2 = z^2 + 2 + z^{-2}$ suggests the introduction of a new variable,

$$w = z + z^{-1} \qquad (A\,5)$$

which satisfies the quadratic equation

$$w^2 + w - 1 = 0 \tag{A 6}$$

Furthermore, z can be obtained from the roots of equation (A 6) by solving a second quadratic equation,

$$z^2 - wz + 1 = 0 \tag{A 7}$$

So equation (A 3) can be solved by extracting square roots only, and the regular pentagon can be constructed by Euclidean methods.

The case where $p = 17$ involves solving the equation

$$z^{17} - 1 = 0 \tag{A 8}$$

by solving a sequence of four quadratic equations. Now, since $\cos 2k\pi = 1$ and $\sin 2k\pi = 0$ for all integral values of k, it follows that

$$e^{2ik\pi} = \cos 2k\pi + i \sin 2k\pi = 1$$

for all such k. We may therefore write equation (A 8) as $z^{17} - e^{2ik\pi} = 0$, whose roots are given by

$$z_k = e^{2ik\pi/17} = \cos(2k\pi/17) + i \sin(2k\pi/17), \tag{A 9}$$

$$\text{for } k = 0, 1, 2, \ldots, 16$$

The factor $z - 1$ may be removed as before to give, from equation (A 8),

$$z^{16} + z^{15} + \ldots + z^2 + z + 1 = 0 \tag{A 10}$$

Note that $z_0 = 1$, $z_k = (z_1)^k$ and $z_{17n+k} = z_k$ with n integral, and $z_{17-k} = 1/z_k$ for $k = 1, 2, 3, \ldots, 16$.

We solve the problem by obtaining the value of

$$z_1 + z_{16} = z_1 + 1/z_1 = 2\cos(2\pi/17)$$

represented by the length OQ in Figure 14.3 (p. 324).

Gauss' approach was to arrange the 16 roots of equation (A 10) in a particular order, and then decompose the ordered sum into sums (called periods) containing 8, 4 and 2 roots respectively, and to do this in such a way that the values of the periods can be calculated successively as the roots of quadratic equations. How, then, do we find the particular order that enables this to be done? Gauss' researches in number theory had provided the clue: we must find a *primitive root* of the associated congruence $a^s - 1 \equiv 0 \pmod{17}$. The number a is said to be a primitive root if this congruence has a solution when $s = 17 - 1 = 16$, but for no smaller value of s. (We assume s is a positive integer.) Now, it turns out that the number 3 has this property, and is therefore a primitive root of our congruence. In fact we have, modulo 17,

$$3^1 \equiv 3, \qquad 3^5 \equiv 5, \qquad 3^9 \equiv 14, \qquad 3^{13} \equiv 12$$
$$3^2 \equiv 9, \qquad 3^6 \equiv 15, \qquad 3^{10} \equiv 8, \qquad 3^{14} \equiv 2$$
$$3^3 \equiv 10, \qquad 3^7 \equiv 11, \qquad 3^{11} \equiv 7, \qquad 3^{15} \equiv 6$$
$$3^4 \equiv 13, \qquad 3^8 \equiv 16, \qquad 3^{12} \equiv 4, \qquad 3^{16} \equiv 1$$

Let us now arrange the roots z_k so that their subscripts are these remainders in order, thus:

$$z_3, \ z_9, \ z_{10}, \ z_{13}, \ z_5, \ z_{15}, \ z_{11}, \ z_{16}, \ z_{14}, \ z_8, \ z_7, \ z_4, \ z_{12}, \ z_2, \ z_6, \ z_1$$

In this series each root is the cube of its predecessor.

We now proceed with the decomposition. We first form two periods, each of 8 roots, by taking first the roots that are in the even positions of our series, and then those that are in the odd positions. This gives

$$u_1 = z_9 + z_{13} + z_{15} + z_{16} + z_8 + z_4 + z_2 + z_1$$
$$u_2 = z_3 + z_{10} + z_5 + z_{11} + z_{14} + z_7 + z_{12} + z_6$$

Note that $u_1 + u_2 = -1$, from equation (A 10). Operating on u_1 and u_2 in the same way, we obtain four periods each of 4 roots:

$$v_1 = z_{13} + z_{16} + z_4 + z_1$$
$$v_2 = z_9 + z_{15} + z_8 + z_2$$
$$v_3 = z_{10} + z_{11} + z_7 + z_6$$
$$v_4 = z_3 + z_5 + z_{14} + z_{12}$$

Continuing similarly, we obtain eight periods of 2 roots each. The first two are

$$w_1 = z_{16} + z_1 \quad \text{and} \quad w_2 = z_{13} + z_4$$

the others are readily found. It is apparent that all the 2-root periods are real quantities of the form $2\cos(2k\pi/17)$, with $k = 1, 2, \ldots, 8$; our objective is to evaluate w_1.

We must now show how the periods can be calculated successively as roots of quadratic equations. By definition,

$$w_1 + w_2 = v_1$$

and

$$w_1 w_2 = (z_{16} + z_1)(z_{13} + z_4)$$
$$= z_{12} + z_3 + z_{14} + z_5 = v_4$$

Hence w_1 and w_2 are the roots of the quadratic equation

$$w^2 - v_1 w + v_4 = 0 \tag{A 11}$$

A simple geometric argument suffices to determine the relative magnitudes of the different periods. In particular, we find that

$$w_1 > w_2 \quad \text{and so} \quad w_1 = \tfrac{1}{2}(v_1 + \sqrt{v_1^2 - 4v_4})$$

We must now determine v_1 and v_4. Associating v_1 with v_2, since $v_1 + v_2 = u_1$, and expanding as before, we find that $v_1 v_2 = -1$; associating v_3 with v_4, we get

$$v_3 + v_4 = u_2 \quad \text{and} \quad v_3 v_4 = -1$$

We thus have two quadratic equations (although they are not independent) corresponding to equation (A 11):

$$v^2 - u_1 v - 1 = 0 \tag{A 12a}$$

with roots v_1 and v_2, and

$$v^2 - u_2 v - 1 = 0 \tag{A 12b}$$

with roots v_3 and v_4. Since $v_1 > v_2$ and $v_4 > v_3$, we have

$$v_1 = \tfrac{1}{2}(u_1 + \sqrt{u_1^2 + 4}) \quad \text{and} \quad v_4 = \tfrac{1}{2}(u_2 + \sqrt{u_2^2 + 4})$$

Finally, we have $u_1 + u_2 = -1$ and $u_1 u_2 = -4$, so that u_1 and u_2 are the roots of

$$u^2 + u - 4 = 0 \qquad\qquad (A\,13)$$

Since $u_1 > u_2$, we get $u_1 = \frac{1}{2}(-1 + \sqrt{17})$ and $u_2 = \frac{1}{2}(-1 - \sqrt{17})$. By working backwards and solving the quadratic equations (A 13), (A 12) and (A 11) in turn, we can express w_1 as an algebraic expression containing a series of square roots, but no other irrational quantities. It follows, therefore, that the corresponding polygon can be constructed using a straight-edge and compasses only. The final expression for w_1 is in fact

$$\tfrac{1}{8}\Big\{ -1 + \sqrt{17} + \sqrt{34 - 2\sqrt{17}}$$
$$+ \sqrt{68 + 12\sqrt{17} - 16\sqrt{34 + 2\sqrt{17}} + 2(-1 + \sqrt{17})\sqrt{34 - 2\sqrt{17}}}\, \Big\}$$

as noted on p. 324. Since $w_1 = z_1 + 1/z_1$, z_1 can be obtained as the solution of a final quadratic equation

$$z_1^2 - w_1 z_1 + 1 = 0$$

as in the case of the pentagon.

The geometrical construction is quite complicated; details are given in References 35 and 37. In the nineteenth century, the quadratic equations for the two remaining cases, the 257-gon (F_3) and the 65 537-gon (F_4), were calculated, and the geometrical construction for the former case was laid out.

Appendix 6. Leonardo's generalized 'squares' problem

In Chapter 6 (p. 104) we presented the problem that Leonardo of Pisa succeeded in solving at one of the mathematical contests that were a feature of the more civilized courts of the Middle Ages. The 'court' problem invites an obvious generalization, thus: given an integer, N, find a rational number, r, such that both $(r^2 + N)$ and $(r^2 - N)$ are rational square numbers (of the form p^2/q^2). Diophantus treated some particular cases and an anonymous tenth-century Arab manuscript contains a statement of the generalized problem, a simple method of treatment and a number of solutions. The method is based on the theorem that if (x, y, z) is a Pythagorean number triple such that $x^2 + y^2 = z^2$ with $x < y < z$, then $z^2 \pm 2xy$ are both squares. If (x, y, z) is a primitive number triple (see p. 9), then x and y are of opposite parity and so $2xy$ is divisible by 4. We can then write

$$(z/2)^2 \pm (xy/2) = \{(y \pm x)/2\}^2 \qquad (A\ 14)$$

where $(xy/2)$ is our positive integer, N.
Thus the triple (3, 4, 5) yields

$$(5/2)^2 \pm 6 = \{(4 \pm 3)/2\}^2$$

and (5, 12, 13) yields

$$(13/2)^2 \pm 30 = \{(12 \pm 5)/2\}^2$$

Any square factors of $(xy/2)$ can be removed by division to produce what we shall call a primitive solution. Thus, the triple (8, 15, 17) yields

$$(17/4)^2 \pm 15 = \{(15 \pm 8)/4\}^2$$
and (7, 24, 25) yields

$$(25/4)^2 \pm 21 = \{(24 \pm 7)/4\}^2$$

Leonardo's 'court' case ($N = 5$) can be derived from the triple (9, 40, 41) to give

$$(41/12)^2 \pm 5 = \{(40 \pm 9)/12\}^2$$

It can, indeed, be quite easily proved that *any* primitive solution of the 'squares' problem has a corresponding Pythagorean triple from which it can be derived by the Arabic method. The converse result is also true.

In 1225 – some years after the 'court' challenge – Leonardo completed a major work entitled *Liber quadratorum*. An excellent annotated English translation by L. E. Sigler under the title *The Book of Squares* was published in 1987 (Academic Press). Leonardo dedicated his book to the Holy Roman Emperor, Frederick the Second. It consists of 24 Propositions and establishes a number of important results. The generalized 'squares' problem is treated in Proposition 14; the particular case of $N = 5$ – the 'court' problem

– is the subject of Proposition 17. Leonardo's approach is quite different from that of the Arabs. It is based on the fact that any square number is equal to the sum of a sequence of consecutive odd numbers. (The general formula is given on p. 16.) For most of his demonstrations Leonardo uses the methods of geometrical algebra (see p. 39), with numbers being envisaged as line segments. Not surprisingly, the treatment is often verbose and difficult to follow – hence the value of Sigler's commentaries –and will not be further discussed here.

Instead, we will present one of Dr Thornhill's ingenious methods of solution (see p. xiv). It utilizes a special kind of Pythagorean triple in which one of the elements is itself a perfect square; let us call them *Thornhill triples*. There are two types which we may write as (l^2, m, N) where $l^4 + m^2 = N^2$, and (N, m, l^2), where $N^2 + m^2 = l^4$.

The first type yields the algebraic identity

$$\{(l^4 + N^2)/2lm\}^2 \pm N = \{(2Nl^2 \pm m^2)/2lm\}^2 \qquad \text{(A 15)}$$

(We will omit Thornhill's derivation; the result can easily be verified.)

Thus the familiar triple $(3, 2^2, 5)$ yields at once Leonardo's result, namely

$$(41/12)^2 \pm 5 = \{(40 \pm 9)/12\}^2$$

Another example is the triple $(3^2, 12, 15)$ which yields

$$(51/12)^2 \pm 15 = \{(45 \pm 24)/12\}^2,$$

while, as a Thornhill *tour de force*, we have $(5^2, 312, 313)$, which yields

$$(49297/1560)^2 \pm 313 = \{(48672 \pm 7825)/1560\}^2$$

The second type of Thornhill triple – where it is the hypotenuse of the triangle that is a perfect square – yields a similar identity to (A 15). Every triple of this type yields two solutions, since N and m are interchangeable. For example, the triple $(7, 24, 5^2)$ yields

$$(1201/140)^2 \pm 6 = \{(1200 \pm 49)/140\}^2$$

and

$$(337/120)^2 \pm 7 = \{(288 \pm 175)/120\}^2$$

It is perhaps worth remarking that any Pythagorean triple can be used to generate a Thornhill triple of the second type. Thus, the triple (x, y, z), where $x^2 + y^2 = z^2$, gives the Thornhill triple

$$\{(y^2 - x^2), 2xy, z^2\},$$

and the process can clearly be extended indefinitely. Thus, from the familiar (3, 4, 5) triple we derive the Thornhill triple just mentioned, i.e. $(7, 24, 5^2)$.

Another interesting feature is that there can be more than one solution of the 'squares' problem for a single value of N. Thus, for $N = 210$ we have a triplet of solutions:

$$(29/2)^2 \pm 210 = \{(21 \pm 20)/2\}^2$$
$$(37/2)^2 \pm 210 = \{(35 \pm 12)/2\}^2$$

and

$$(113/4)^2 \pm 210 = \{(112 \pm 15)/4\}^2$$

We can, indeed, go much further. There is a remarkable theorem which states that for any given value of N, either there is no solution or there are infinitely many. The result can be proved by using the method of infinite descent, described on p. 149, but applying it in reverse. Thus, if there is a known solution, the method will produce an unending sequence of ever 'higher' solutions, i.e. solutions involving larger and larger square numbers.

The 'squares' problem is in fact insoluble for many values of N; in particular for $N = 1$, 2, 3 and 4, so that the 'court' problem specifies the smallest value of N for which the general problem is soluble. Indeed, the only primitive values of N less than 20 for which the problem is soluble are 5, 6, 7, 13, 14 and 15.

Thornhill has devised several other methods of treatment which yield many further solutions. A number of mathematicians, from the eighteenth century to the present day, including Euler, have added their contributions and the subject remains a live one. There is, however, still no procedure which will decide whether *any* arbitrary value of N does or does not yield a solution of the 'squares' problem. This was described in 1988 by Professor J. H. Coates, the President of the London Mathematical Society, as the oldest unsolved problem in mathematics.

It will be apparent that this Appendix is no more than a very brief introduction to a large subject. It is to be hoped, however, that enough has been said to stimulate the interested reader to attempt some explorations of his own.

Appendix 7. A case of mistaken Newtonian identity

In December 1989 I received a letter from Mr A. R. Thatcher (see p. xiv), which I will quote in full.

> On p. 173 you mention that William and Caroline Lamb used to read *Newton on the Prophecies*. By an extraordinary coincidence, I actually have a copy of a book which on the spine has the title *Prophecies* and author 'Newton'. The inside title is *Dissertations on the Prophecies* but alas, the author is not Isaac Newton but 'Thomas Newton D. D., late Lord Bishop of Bristol'.
>
> This book came to me from the library of my uncle, who was a clergyman. In appearance the book looks to me like eighteenth or very early nineteenth century. It is a revised edition of sermons which were commissioned by the Hon. Robert Boyle in 1756. It was revised by the Revd W. S. Dobson, which is the same surname as my uncle's so it may have been in the family. Incidentally, the footnotes include some quotations from Sir Isaac Newton's book on Daniel.

The title of Sir Isaac Newton's book that I presumed the Lambs were reading was, as mentioned on p. 173, *Observations on the Prophecies*.

In a later letter, Mr Thatcher sums up thus: 'What an extraordinary thing (a) that William and Caroline read this book at all; (b) that you came across this fact; and (c) that you had a reader with a copy!' A fascinating story, indeed.

The worthy prelate was, it seems, a well-known and controversial figure in the late-eighteenth century. Dr Johnson was not one of his greatest admirers. Boswell records several of the Doctor's astringent comments. Here is one exchange of views with Dr Adams, the Master of Pembroke College, Oxford (Johnson's old college):

ADAMS: I believe that *Dissertations on the Prophecies* is his great work.
JOHNSON: Why, sir, it *is* Tom's great work; but how far it is great, or how

much of it is Tom's, are other questions. I fancy a considerable part of it was borrowed.

And again:

ADAMS: He was a very successful man.

JOHNSON: He was late in getting what he did get, and he did not get it by the best of means. I believe he was a gross flatterer.

Appendix 8. Fermat's Last Theorem proved?

In June 1993 about one hundred of the world's most distinguished mathematicians were enjoying their summer conference in Cambridge at the new Isaac Newton Institute. Professor Andrew Wiles of Princeton had been lecturing for some two and a half hours, spread over three days, when he calmly announced that he had proved Fermat's Last Theorem. (It will be recalled, p. 148, that the theorem states that the equation $x^n + y^n = z^n$ has no solution in integers when n is greater than 2.) After a few seconds of stunned silence, his statement, which he had delayed until the end of his long discourse, was greeted with tremendous applause. Within hours the University had issued a press release entitled 'Mathematical result of the century'.

Although Wiles' line of thought had its origins in the *Arithmetica* of Diophantus (p. 85), his approach was indirect and extremely complex, drawing on some of the latest ideas from several different fields of mathematics. In 1955 a Japanese mathematician, Yutaka Taniyama, posed some difficult questions relating to elliptic curves with equations of the form $y^2 = x^3 + ax + b$, where a and b are integers. More work by André Weil (brother of the philosopher, Simone Weil) and another Japanese mathematician led to an esoteric answer known as the Shimura–Taniyama–Weil conjecture. The next important step was taken in 1986 when Gerhard Frey from Saarbrucken (reinforced by two colleagues) established a most unexpected link between this very abstract conjecture and Fermat's Last Theorem. Frey proved that a counter-example to Fermat's Theorem would directly contradict the S.–T.–W. conjecture. Professor Wiles enters the story at this point. His crucial contribution was to *prove* the S.–T.–W. conjecture for the complete class of elliptic curves that are relevant to the proof of Fermat's Last Theorem. There can be no counter-example to the famous theorem, which is therefore proved for *all* values of *n*.

It is said that no more than half a dozen people in the world are capable of fully understanding all the details of Wiles's lengthy argument, and that all but one were present at the Cambridge conference. Be that as it may, a proof of such enormous complexity, requiring some 1,000 pages to present, will need to be checked and rechecked in every detail by the few mathematicians capable of doing so. However, as one of Wiles's senior Princeton colleagues put it: 'the whole structure of this beautiful proof is very tight and very solid . . . Wiles is an extremely careful mathematician who does not make rash statements.' We shall just have to wait and see.

A word, finally, about Professor Wiles himself. He is a slightly built forty-year-old who graduated from Oxford before moving to Cambridge as a Ph.D. student. The high quality of his early work led to a research fellowship at Clare College, Cambridge. He was soon lured to Harvard, where he stayed until he moved to Princeton as a full professor in the early 1980s. A few years later he briefly accepted a Royal Society research professorship, but decided to return to Princeton.

(The material in this Appendix is based on an article in the *Guardian* of 24 June 1993.)

References

Source books

1. Einstein, Albert, *Relativity: The Special and the General Theory. A Popular Exposition* (Translated by R. W. Lawson), Methuen, 1920.
2. Fauvel, J. and Gray, J. (Editors), *The History of Mathematics: A Reader*, Macmillan Education/Open University, 1987.
3. Galilei, Galileo, *Dialogues Concerning Two New Sciences* (Translated by H. Crew and A. deSalvio), Dover, 1944.
4. Midonick, H. (Editor), *The Treasury of Mathematics* (2 vols), Penguin, 1968.
5. Newman, J. R. (Editor), *The World of Mathematics* (4 vols), Allen & Unwin, 1956.
6. Newton, Sir Isaac, *Principia* (Translated by A. Motte in 1729, edited by F. Cajori) (2 vols), University of California Press, 1924.
7. Newton, Sir Isaac, *Opticks* (4th edition, London, 1730), Dover, 1952.
8. Smith, D. E. (Editor), *A Source Book in Mathematics* (2 vols), Dover, 1939.
9. Struik, D. K. (Editor), *A Source Book in Mathematics, 1200–1800*, Harvard University Press, 1969.
10. Whiteside, D. T. (Editor), *The Mathematical Papers of Isaac Newton* (8 vols), Cambridge University Press, 1967–81.

General histories and commentaries

11. Aaboe, A., *Episodes from the Early History of Mathematics*, Random House, 1964.
12. Ball, W. W. Rouse, *A Short Account of the History of Mathematics* (4th edition), Macmillan, 1927.

13. Bell, E. T., *Men of Mathematics* (2 vols), Penguin, 1953.
14. Boyer, C. B., *A History of Mathematics*, Wiley, 1968.
15. Dantzig, T., *Number: The Language of Science*, Doubleday, 1954.
16. Dedron, P. and Itard, J., *Mathematics and Mathematicians* (2 vols), Transworld, 1971.
17. Eves, H., *An Introduction to the History of Mathematics* (5th edition), Saunders, 1983.
18. Hardy, G. H., *A Mathematician's Apology*, Cambridge University Press, 1940.
19. Kline, M., *Mathematics in Western Culture*, Penguin, 1972.
20. Stewart, I., *Concepts of Modern Mathematics*, Penguin, 1975.
21. Stewart, I., *The Problems of Mathematics*, Oxford University Press, 1987.
22. Struik, D. K., *A Concise History of Mathematics*, Dover, 1948 (revised edition 1967).
23. Wilder, R. L., *The Evolution of Mathematical Concepts*, Transworld, 1974.

More specialist or advanced books

24. Biggs, N. L., Lloyd, E. K. and Wilson, R. J., *Graph Theory, 1736–1936*, Clarendon Press, 1976.
25. Cantor, Georg, *Contributions to the Founding of the Theory of Transfinite Numbers* (Translated by P.E.B. Jourdain), Dover, 1915.
26. Cohen, I. B., *Introduction to Newton's 'Principia'*, Cambridge University Press, 1971.
27. Cohen, I. B., *The Newtonian Revolution*, Cambridge University Press, 1980.
28. Eddington, A. S., *Space, Time and Gravitation*, Cambridge University Press, 1920.
29. Eddington, A. S., *The Mathematical Theory of Relativity*, Cambridge University Press, 1923.
30. Einstein, Albert, *The Meaning of Relativity* (Translated by E. P. Adams), Methuen, 1920.
31. Fowler, D. H., *The Mathematics of Plato's Academy: A New Reconstruction*, Clarendon Press, 1987.
32. Gowing, R., *Roger Cotes: Natural Philosopher*, Cambridge University Press, 1981.
33. Grattan-Guiness, I. (Editor), *From Calculus to Set Theory: 1630–1910*, Duckworth, 1980.

34. Hall, A. R., *Philosophers at War: The Quarrel Between Newton and Leibniz*, Cambridge University Press, 1980.
35. Hardy, G. H. and Wright, E. M., *An Introduction to the Theory of Numbers* (5th edition), Clarendon Press, 1979.
36. Hofman, J. E., *Leibniz in Paris: 1672–1676*, Cambridge University Press, 1974.
37. Klein, F., *Famous Problems of Elementary Geometry* (Translated by W. W. Beman and D. E. Smith), Dover, 1956.
38. Kline, M., *Mathematical Thought from Ancient to Modern Times*, Oxford University Press, 1972.
39. Needham, J., *Science and Civilisation in China* (vol. 3), Cambridge University Press, 1959.
40. Pais, A., *'Subtle is the Lord': The Science and the Life of Albert Einstein*, Oxford University Press, 1982.
41. Stewart, I., *Galois Theory*, Chapman & Hall, 1973.
42. Westfall, R. S., *Never at Rest: A Biography of Isaac Newton*, Cambridge University Press, 1980.
43. Woodcock, A. and Davis, M., *Catastrophe Theory*, Penguin, 1980.

Index

Discover more about our forthcoming books through Penguin's FREE newspaper...

READ MORE IN PENGUIN

In every corner of the world, on every subject under the sun, Penguin represents quality and variety – the very best in publishing today.

For complete information about books available from Penguin – including Puffins, Penguin Classics and Arkana – and how to order them, write to us at the appropriate address below. Please note that for copyright reasons the selection of books varies from country to country.

In the United Kingdom: Please write to *Dept. JC, Penguin Books Ltd, FREEPOST, West Drayton, Middlesex UB7 OBR.*

If you have any difficulty in obtaining a title, please send your order with the correct money, plus ten per cent for postage and packaging, to *PO Box No. 11, West Drayton, Middlesex UB7 OBR*

In the United States: Please write to *Consumer Sales, Penguin USA, P.O. Box 999, Dept. 17109, Bergenfield, New Jersey 07621-0120.* VISA and MasterCard holders call 1-800-253-6476 to order all Penguin titles

In Canada: Please write to *Penguin Books Canada Ltd, 10 Alcorn Avenue, Suite 300, Toronto, Ontario M4V 3B2*

In Australia: Please write to *Penguin Books Australia Ltd, P.O. Box 257, Ringwood, Victoria 3134*

In New Zealand: Please write to *Penguin Books (NZ) Ltd, Private Bag 102902, North Shore Mail Centre, Auckland 10*

In India: Please write to *Penguin Books India Pvt Ltd, 706 Eros Apartments, 56 Nehru Place, New Delhi 110 019*

In the Netherlands: Please write to *Penguin Books Netherlands bv, Postbus 3507, NL-1001 AH Amsterdam*

In Germany: Please write to *Penguin Books Deutschland GmbH, Metzlerstrasse 26, 60594 Frankfurt am Main*

In Spain: Please write to *Penguin Books S. A., Bravo Murillo 19, 1° B, 28015 Madrid*

In Italy: Please write to *Penguin Italia s.r.l., Via Felice Casati 20, I–20124 Milano*

In France: Please write to *Penguin France S. A., 17 rue Lejeune, F–31000 Toulouse*

In Japan: Please write to *Penguin Books Japan, Ishikiribashi Building, 2–5–4, Suido, Bunkyo-ku, Tokyo 112*

In Greece: Please write to *Penguin Hellas Ltd, Dimocritou 3, GR–106 71 Athens*

In South Africa: Please write to *Longman Penguin Southern Africa (Pty) Ltd, Private Bag X08, Bertsham 2013*

READ MORE IN PENGUIN

PHILOSOPHY

What Philosophy Is   Anthony O'Hear

'Argument after argument is represented, including most of the favourites
... its tidy and competent construction, as well as its straightforward style,
mean that it will serve well anyone with a serious interest in philosophy'
– *The Journal of Applied Philosophy*

Montaigne and Melancholy   M. A. Screech

'A sensitive probe into how Montaigne resolved for himself the age-old
ambiguities of melancholia and, in doing so, spoke of what he called the
"human condition"' – Roy Porter in the *London Review of Books*

Labyrinths of Reason   William Poundstone

'The world and what is in it, even what people say to you, will not seem
the same after plunging into *Labyrinths of Reason* ... Poundstone's book
merits the description of *tour de force*. He holds up the deepest
philosophical questions for scrutiny and examines their relation to reality
in a way that irresistibly sweeps readers on' – *New Scientist*

I: The Philosophy and Psychology of Personal Identity
Jonathan Glover

From cases of split brains and multiple personalities to the importance of
memory and recognition by others, the author of *Causing Death and
Saving Lives* tackles the vexed questions of personal identity.

Ethics Inventing Right and Wrong   J. L. Mackie

Widely used as a text, Mackie's complete and clear treatise on moral
theory deals with the status and content of ethics, sketches a practical
moral system, and examines the frontiers at which ethics touches psy-
chology, theology, law and politics.

The Central Questions of Philosophy   A. J. Ayer

'He writes lucidly and has a teacher's instinct for the helpful pause and
reiteration ... an admirable introduction to the ways in which philosophic
issues are experienced and analysed in current Anglo-American academic
milieux' – *Sunday Times*

READ MORE IN PENGUIN

HISTORY

The World Since 1945 T. E. Vadney
New edition

From the origins of the post-war world to the collapse of the Soviet Bloc in the late 1980s, this masterly book offers an authoritative yet highly readable one-volume account.

Ecstasies Carlo Ginzburg

This dazzling work of historical detection excavates the essential truth about the witches' Sabbath. 'Ginzburg's learning is prodigious and his journey through two thousand years of Eurasian folklore a *tour de force*' – *Observer*

The Nuremberg Raid Martin Middlebrook

'The best book, whether documentary or fictional, yet written about Bomber Command' – *Economist*. 'Martin Middlebrook's skill at description and reporting lift this book above the many memories that were written shortly after the war' – *The Times*

A History of Christianity Paul Johnson

'Masterly ... It is a huge and crowded canvas – a tremendous theme running through twenty centuries of history – a cosmic soap opera involving kings and beggars, philosophers and crackpots, scholars and illiterate exaltés, popes and pilgrims and wild anchorites in the wilderness'– Malcolm Muggeridge

The Penguin History of Greece A. R. Burn

Readable, erudite, enthusiastic and balanced, this one-volume history of Hellas sweeps the reader along from the days of Mycenae and the splendours of Athens to the conquests of Alexander and the final dark decades.

Modern Ireland 1600–1972 R. F. Foster

'Takes its place with the finest historical writing of the twentieth century, whether about Ireland or anywhere else' – Conor Cruise O'Brien in the *Sunday Times*

READ MORE IN PENGUIN

HISTORY

The Guillotine and the Terror   Daniel Arasse

'A brilliant and imaginative account of the punitive mentality of the revolution that restores to its cultural history its most forbidding and powerful symbol' – Simon Schama.

The Second World War   A J P Taylor

A brilliant and detailed illustrated history, enlivened by all Professor Taylor's customary iconoclasm and wit.

Daily Life in Ancient Rome   Jerome Carcopino

This classic study, which includes a bibliography and notes by Professor Rowell, describes the streets, houses and multi-storeyed apartments of the city of over a million inhabitants, the social classes from senators to slaves, and the Roman family and the position of women, causing *The Times Literary Supplement* to hail it as a 'thorough, lively and readable book'.

The Anglo-Saxons   Edited by James Campbell

'For anyone who wishes to understand the broad sweep of English history, Anglo-Saxon society is an important and fascinating subject. And Campbell's is an important and fascinating book. It is also a finely produced and, at times, a very beautiful book' – *London Review of Books*

The Making of the English Working Class   E. P. Thompson

Probably the most imaginative – and the most famous – post-war work of English social history. 'A magnificent, lucid, angry historian ... E. P. Thompson has performed a revolution of historical perspective' – *The Times*

The Habsburg Monarchy 1809 –1918   A J P Taylor

Dissolved in 1918, the Habsburg Empire 'had a unique character, out of time and out of place'. Scholarly and vividly accessible, this 'very good book indeed' (*Spectator*) elucidates the problems always inherent in the attempt to give peace, stability and a common loyalty to a heterogeneous population.

READ MORE IN PENGUIN

HISTORY

Citizens Simon Schama

The award-winning chronicle of the French Revolution. 'The most marvellous book I have read about the French Revolution in the last fifty years' – Richard Cobb in *The Times*

To the Finland Station Edmund Wilson

In this authoritative work Edmund Wilson, considered by many to be America's greatest twentieth-century critic, turns his attention to Europe's revolutionary traditions, tracing the roots of nationalism, socialism and Marxism as these movements spread across the Continent creating unrest, revolt and widespread social change.

Jasmin's Witch Emmanuel Le Roy Ladurie

An investigation into witchcraft and magic in south-west France during the seventeenth century – a masterpiece of historical detective work by the bestselling author of Montaillou.

Stalin Isaac Deutscher

'The Greatest Genius in History' and the 'Life-Giving Force of socialism'? Or a despot more ruthless than Ivan the Terrrible and a revolutionary whose policies facilitated the rise of Nazism? An outstanding biographical study of a revolutionary despot by a great historian.

Aspects of Antiquity M. I. Finley

Profesor M. I. Finley was one of the century's greatest ancient historians; he was also a master of the brief, provocative essay on classical themes. 'He writes with the unmistakable enthusiasm of a man who genuinely wants to communicate his own excitement' – Philip Toynbee in the *Observer*

British Society 1914–1945 John Stevenson

'A major contribution to the *Penguin Social History of Britain*, which will undoubtedly be the standard work for students of modern Britain for many years to come' – *The Times Educational Supplement*

READ MORE IN PENGUIN

SCIENCE AND MATHEMATICS

QED Richard Feynman
The Strange Theory of Light and Matter

'Physics Nobelist Feynman simply cannot help being original. In this quirky, fascinating book, he explains to laymen the quantum theory of light – a theory to which he made decisive contributions' – *New Yorker*

Does God Play Dice? Ian Stewart
The New Mathematics of Chaos

To cope with the truth of a chaotic world, pioneering mathematicians have developed chaos theory. *Does God Play Dice?* makes accessible the basic principles and many practical applications of one of the most extraordinary – and mind-bending – breakthroughs in recent years.

Bully for Brontosaurus Stephen Jay Gould

'He fossicks through history, here and there picking up a bone, an imprint, a fossil dropping and, from these, tries to reconstruct the past afresh in all its messy ambiguity. It's the droppings that provide the freshness: he's as likely to quote from Mark Twain or Joe DiMaggio as from Lamarck or Lavoisier' – *Guardian*

The Blind Watchmaker Richard Dawkins

'An enchantingly witty and persuasive neo-Darwinist attack on the anti-evolutionists, pleasurably intelligible to the scientifically illiterate' – Hermione Lee in the *Observer* Books of the Year

The Making of the Atomic Bomb Richard Rhodes

'Rhodes handles his rich trove of material with the skill of a master novelist ... his portraits of the leading figures are three-dimensional and penetrating ... the sheer momentum of the narrative is breathtaking ... a book to read and to read again' – Walter C. Patterson in the *Guardian*

Asimov's New Guide to Science Isaac Asimov

A classic work brought up to date – far and away the best one-volume survey of all the physical and biological sciences.